NATO ASI Series

Advanced Science Institutes Series

A series presenting the results of activities sponsored by the NATO Science Committee, which aims at the dissemination of advanced scientific and technological knowledge, with a view to strengthening links between scientific communities.

The Series is published by an international board of publishers in conjunction with the NATO Scientific Affairs Division

A Life Sciences B Physics	Plenum Publishing Corporation London and New York
C Mathematical and Physical Sciences D Behavioural and Social Sciences E Applied Sciences	Kluwer Academic Publishers Dordrecht, Boston and London
F Computer and Systems Sciences G Ecological Sciences H Cell Biology I Global Environmental Change	Springer-Verlag Berlin Heidelberg New York London Paris Tokyo Hong Kong Barcelona Budapest

PARTNERSHIP SUB-SERIES

1. Disarmament Technologies	Kluwer Academic Publishers
2. Environment	Springer-Verlag/Kluwer Academic Publishers
3. High Technology	Kluwer Academic Publishers
4. Science and Technology Policy	Kluwer Academic Publishers
5. Computer Networking	Kluwer Academic Publishers

The Partnership Sub-Series incorporates activities undertaken in collaboration with NATO's Cooperation Partners, the countries of the CIS and Central and Eastern Europe, in Priority Areas of concern to those countries.

NATO-PCO DATABASE

The electronic index to the NATO ASI Series provides full bibliographical references (with keywords and/or abstracts) to about 50000 contributions from international scientists published in all sections of the NATO ASI Series. Access to the NATO-PCO DATABASE is possible via a CD-ROM "NATO Science & Technology Disk" with user-friendly retrieval software in English, French and German (© WTV GmbH and DATAWARE Technologies Inc. 1992).

The CD-ROM can be ordered through any member of the Board of Publishers or through NATO-PCO, Overijse, Belgium.

Series H: Cell Biology, Vol. 103

Springer
Berlin
Heidelberg
New York
Barcelona
Budapest
Hong Kong
London
Milan
Paris
Santa Clara
Singapore
Tokyo

Molecular Microbiology

Edited by

Stephen J. W. Busby

The University of Birmingham, School of Biochemistry
P.O. Box 363, Birmingham, B15 2TT, U.K.

Christopher M. Thomas
Nigel L. Brown

The University of Birmingham, School of Biological Sciences
P.O. Box 363, Birmingham, B15 2TT, U.K.

With 78 Figures

Springer

Published in cooperation with NATO Scientific Affairs Division

Proceedings of the NATO Advanced Study Institute "Molecular Microbiology", held at Chamberlain Hall, The University of Birmingham, Birmingham, U.K., April 7–17, 1997

Library of Congress Cataloging-in-Publication Data

Molecular microbiology / edited by Stephen J. W. Busby, Christopher M. Thomas, Nigel L. Brown.
p. cm. – (NATO ASI series. Series H, Cell biology; vol. 103)
Includes bibliographical references and index.
ISBN-13:978-3-642-72073-4
1. Molecular microbiology. I. Busby, Stephen J. W., 1951– . II. Thomas, Christopher M.
III. Brown, Nigel., 1948– . IV. Series.
QR74.M653 1998 527.8'29–dc21 97-32739 CIP

ISSN 1010-8793
ISBN-13:978-3-642-72073-4 e-ISBN-13:978-3-642-72071-0
DOI: 10.1007/978-3-642-72071-0

Softcover reprint of the hardcover 1st edition 1998

Typesetting: Camera ready by authors/editors
Printed on acid-free paper
SPIN 10531516 31/3137 - 5 4 3 2 1 0

PREFACE

Molecular Microbiology is a rapidly expanding area of contemporary science: the application of molecular biology has opened up the microbial world in many remarkable ways. The attraction of microbes is that they are self contained and that they offer complete solutions to understanding the phenomenon of life. Perhaps the clearest index of this growth is the number of doctoral students and young postdoctoral workers currently engaged in this research area. The NATO Advanced Study Institute, reported in this volume, was organised in order to cater for the needs of these research workers. The aim was to provide a course that would provide essential background information for anyone recently recruited to (or about to enter) Molecular Microbiology. Thus the Advanced Study Institute united nearly 100 students and postdoctoral workers who enjoyed a gourmet guide to the subject. Contributors were urged to focus on the background and the principles of the subject rather than the detail and, for this reason, we believe that the Advanced Study Institute and the ensuing book are unique. Clearly the main defect is that neither the taught course nor the book is comprehensive and concentrates on four rather arbitrarily chosen research themes. However we believe that, despite this shortcoming, this book will have lasting value as an introductory text to this exciting research area.

Birmingham, England
Steve Busby
Chris Thomas
Nigel Brown

CONTENTS

Part 1

Bacterial Biochemistry

Life and Death in Stationary Phase

Steven E. Finkel, Erik Zinser, Srishti Gupta, and Roberto Kolter

Department of Microbiology and Molecular Genetics, Harvard Medical School, 200 Longwood Avenue, Boston, MA 02115, USA

Introduction

Bacteria are remarkable for their ability to occupy virtually all environmental niches. In addition, many bacterial organisms transit from one environment to another as part of their normal lifestyle. From the bacterial point of view most of these environments are nutrient poor; more often than not, bacteria find themselves growing under conditions where nutrients are scarce and competition for those nutrients is fierce. Therefore, most bacteria spend much of their lives in a state of starvation, only occasionally finding themselves in a nutrient-rich environment where balanced growth can be achieved. In order to survive within these disparate and changing environments bacterial populations have developed mechanisms to protect both their cellular and genetic integrity. These mechanisms involve changes in cellular properties and morphology, the production of protective enzymes and agents, and, in some cases, entry into developmental programs resulting in sporulation, dormancy and programmed cell death.

The well-characterized gram-negative bacterium *Escherichia coli* is an ideal model organism for the study of bacterial responses to starvation. It occupies many environmental niches, including fresh water, soil, and the gut of many higher organisms. It can transit between all of these environments and has developed mechanisms to survive and reproduce within each of them.

The *E. coli* Life Cycle

E. coli undergoes a well-characterized cycle of growth under standard laboratory conditions (Fig. 1). These conditions typically consist of growth at a constant temperature, with or without aeration, in complex or minimal media. Under these batch culture conditions cells freshly inoculated into culture medium will sustain a "lag period" where cell division has not yet begun, followed by a period of "exponential" or "log-phase" growth where all readily available nutrients are rapidly

NATO ASI Series, Vol. H 103
Molecular Microbiology
Edited by Stephen J. W. Busby,
Christopher M. Thomas and Nigel L. Brown

metabolized. As nutrients become exhausted the population begins to starve and enters "stationary phase." Stationary phase is defined as the period when increases in population size, as assayed by cell density or colony forming units (CFU) within the culture, are no longer observed (Kolter et al., 1993). Cells grown under these batch culture conditions can reach densities of 10^9-10^{10} CFU/ml upon entry into stationary phase.

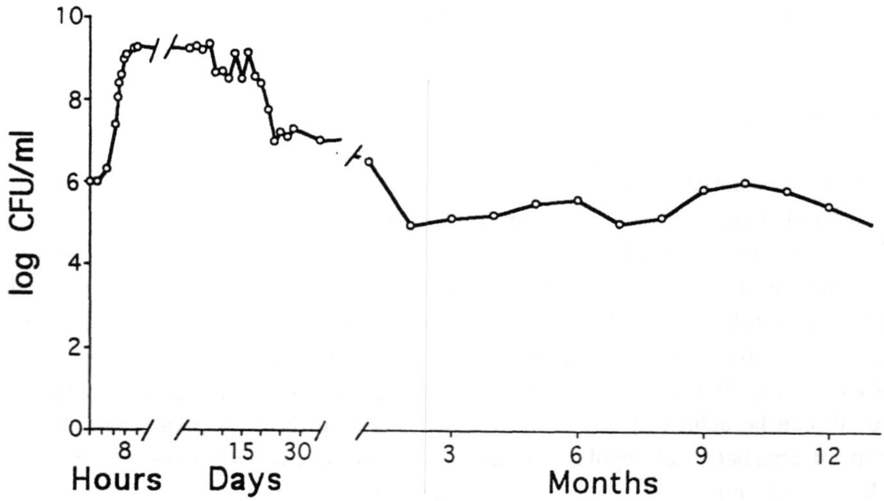

FIG. 1 Long-term incubation of *E. coli*. *E. coli* remains viable for long periods of time without the addition of nutrients. Sterile, distilled water is added periodically to maintain constant volume.

Responses to Starvation

Bacteria have developed different responses to starvation. Some organisms, such as *Bacillus subtilis*, form spores in response to nutrient deprivation (Losick and Youngman, 1984). These spores are highly resistant to desiccation, heat, and radiation and can remain dormant until a nutrient source is encountered. A more complex response to starvation occurs in *Myxococcus xanthus* (Kaiser, 1984). These bacteria normally function as individual cells, however, in response to starvation conditions, millions of bacterial cells will coalesce to form a fruiting body and release spores. Individual cells within the fruiting body enter different developmental programs to enable a sub-population of the cells to survive as spores.

E. coli does not form spores in response to starvation, although a stereotypical pattern of gene expression and morphological changes does occur as cells begin to starve (Huisman et al., 1996). Stationary phase in *E. coli* can be thought of as consisting of at least two distinct stages, an early and a late stage, each accompanied by characteristic changes in cellular physiology, gene expression and population dynamics. Once starvation begins, *E. coli* cells become much smaller and more spherical rather than rod-like, they produce storage compounds such as glycogen, and synthesize enzymes that serve protective functions. Among these protective enzymes are peroxide degrading catalases, and exo- and endo-nucleases involved in DNA damage prevention and repair. Many of the activities induced upon entry into stationary phase are regulated by the stationary phase-specific sigma factor σ^S, encoded by the *rpoS* gene (Hengge-Aronis, 1996). In an early stationary phase culture virtually all the cells have entered this stationary phase program of gene expression. All cells are still metabolically active and capable of re-entering log phase growth.

Once *E. coli* has entered stationary phase, how long do cells remain viable? While the answer to this question is very much dependent on the culture conditions, *E. coli* cultures grown in LB undergo a characteristic growth and death pattern upon prolonged incubation. After about three days approximately 99% of the cells lose viability, the titers of these cultures decrease to 10^7-10^8 CFU/ml and, with continued incubation at 37°C, cultures can be maintained at these levels for many months (Tormo et al, 1990; S. Finkel and R. Kolter, unpublished). Microscopic analysis coupled with the use of vital stains has shown that unlike early stationary phase cultures, most of the cells within these older cultures appear essentially metabolically inactive. However, starting at about 8-10 days after inoculation, the small fraction of the cells which remain viable change morphology and resemble cells in dividing exponential phase cultures (Zambrano et al., 1993). What enables these cells to survive under these nutrient poor conditions?

The GASP Phenotype

Even though cultures which have been starved for many days are thought to lack abundant nutrients, there are still nutrient sources available, namely, the biomass of the culture. In ten-day-old cultures where viable counts are reduced over 100-fold, the detritus consisting of >99% of the initial population that has died is still present. It is believed that the surviving cells can use this biomass as a source of nutrients. Is survival under these conditions merely a stochastic process where, upon the onset of starvation most cells begin to die and release their nutrients, until a point is reached where enough food is available for the survivors to subsist? Or, are there physiological or genetic changes that occur in the surviving population that establish them as the survivors?

These questions can be addressed by analyzing the population dynamics of stationary phase cultures. Our approach has been to use "mixing" experiments in which cells from an aged culture are inoculated into a culture of younger cells (Zambrano et al., 1993). For most of these experiments, initially isogenic cell populations are distinguished from each other using antibiotic resistance gene markers, which themselves confer no competitive advantage.

When cells from an aged (10-day-old) culture are inoculated ~1:10,000 (CFU:CFU) as a minority into a young (1-day-old) culture, several predictions can be made as to what might happen to those aged cells. Initially, we hypothesized one of two outcomes (Fig. 2): (A) the number of minority cells might remain constant and these cells are in some way "death resistant" or (B) these cells would simply parallel the behavior of the majority population, if death were solely a stochastic process. Surprisingly, however, neither of these occurs in practice.

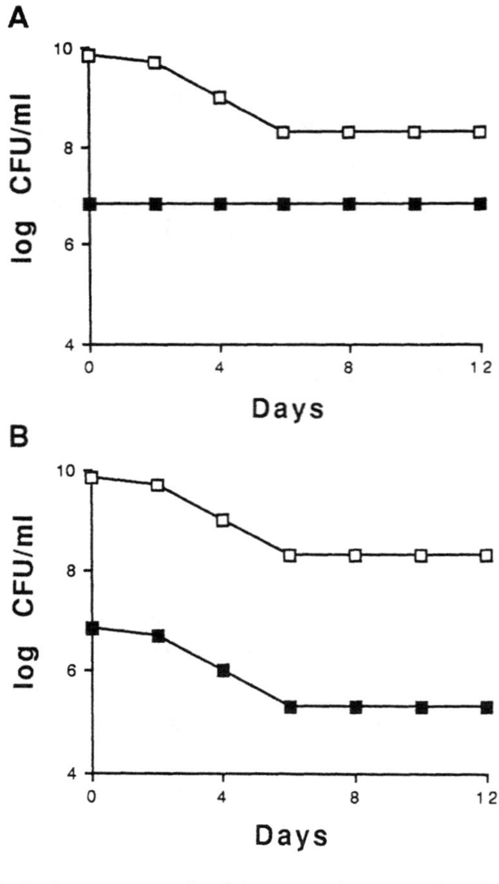

FIG. 2 Hypothetical outcomes of mixing experiments. Aged cells (filled squares) are mixed as a minority with younger cells (open squares). (A) Results expected if minority is "death resistant." (B) Results expected if death is solely a stochastic process.

Instead, the older cells were observed to increase in number, with the concomitant reduction in the younger cell population (Fig. 3). Eventually, the progeny from the aged cells completely displaced the younger cells. These aged cells displayed a phenotype referred to as the Growth Advantage in Stationary Phase, or GASP, phenotype (Zambrano and Kolter, 1993).

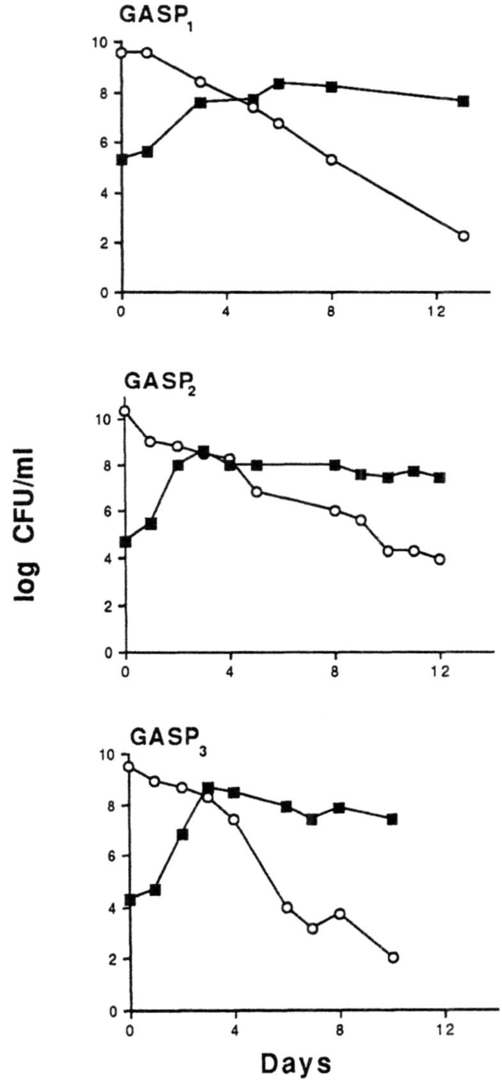

FIG. 3 The GASP phenotype. Cells from aged cultures out compete their parents. $GASP_1$: 10-day-old aged cells as a minority (filled squares) mixed with 1-day-old cells in the majority (open cirlces). Subsequent cycles of 10-day incubations followed by periodic mixing results in $GASP_2$ and $GASP_3$.

The Genetic Basis of GASP and the Role of *rpoS*

Although it is clear that gene expression is altered as cells proceed through stationary phase, accompanied by changes in cellular physiology, the expression of the GASP phenotype is not due simply to physiological adaptation to starvation conditions. In fact, GASP results from a process in which mutant cells able to grow under stationary phase incubation conditions are selected. This has been demonstrated in two ways. First, GASP cells, when allowed to grow for many generations in rich medium without extended incubation in stationary phase, still express the GASP phenotype. That is, they can still outcompete sibling cells which were incubated for only a short time in stationary phase (Zambrano et al., 1993).

Second, evidence for the genetic basis of GASP is shown by the ability to transduce the GASP phenotype from one strain to another without aging the recipient cell. Once a locus is demonstrated to confer the GASP phenotype, it can be moved by bacteriophage P1 transduction into a naive recipient. That recipient cell now displays the GASP phenotype without long-term stationary phase incubation, out-competing a non-transduced population when co-cultured (Zambrano et al., 1993).

We have demonstrated that virtually all wild type *E. coli* K-12 strains which are allowed to incubate in LB medium for greater than 10 days will develop the GASP phenotype (E. Zinser, S. Finkel, S. Gupta, and R. Kolter, unpublished). What are the genes that are being mutated? While it is assumed that mutations in many different loci can confer GASP on a cell, it appears that the earliest GASP mutation acquired during aerobic stationary phase incubation in LB (the $GASP_1$ allele) is consistently a mutation in the *rpoS* gene, encoding the alternative stationary phase-specific sigma factor, σ^S. The σ^S subunit, whose accumulation increases during late-exponential phase, regulates the expression of a large number of genes including those involved in DNA damage prevention and repair, changes in cellular metabolism, biosynthesis, and morphology (Hengge-Aronis, 1996).

Analysis of the earliest $GASP_1$ mutants indicated that these strains had suffered a reduction, but not an elimination, of σ^S activity. The first GASP allele of *rpoS* to be identified had a small duplication near the 3' end of the gene, resulting in a frame-shift which replaced the last four residues with 39 new amino acids (Zambrano et al., 1993). Other *rpoS* alleles have now been identified from GASP mutants and show a variety of changes including missense and frameshift mutations (S. Gupta and R. Kolter, unpublished). These mutations can affect both the stability of the protein as well as its ability to interact with core polymerase or DNA. Although the nature of each of these mutations is different, each of these mutant proteins displays a reduced activity or "attenuated" phenotype, never a total loss of activity. It is believed that this reduction, but not elimination, of σ^S activity is required to express the GASP phenotype since null mutants of *rpoS* are not able to compete with wild type strains during stationary phase (Zambrano et al., 1993).

It is interesting to note that many strains exist both in the laboratory and in nature that are attenuated for σ^S activity. Laboratory strains can be maintained indefinitely as *rpoS*^{att} if they are not mixed with *rpoS*$^{+}$ cells. A simple assay for the activity of σ^S is the determination of the degree of "bubbling" observed within a colony of *E. coli* when exposed to hydrogen peroxide since the stationary phase-specific catalase, HPII, is under the regulation of σ^S (Sak et al., 1989). If σ^S activity is reduced, so is the level of catalase. This is reflected in a delay in the advent and a decrease in the vigor of the bubbling.

E. coli samples found in both soil and fresh water are growing under nutrient limited conditions and can be expected to persist for long periods in stationary phase. Similar conditions can be rationalized for those strains isolated from humans; while there may be an increase in total food passaging through the gut, as compared to fresh water or soil, there is also a larger number of total organisms competing for those nutrients. Therefore it is not surprising that clinical and environmental isolates of *E. coli* typically display the GASP phenotype when competed with laboratory strains and many have an *rpoS*^{att} allele (E. Zinser, S. Gupta, and R. Kolter, unpublished). This is thought to be the consequence of the specific selective conditions in which these strains are found. Thus in some cases, the conditions encountered may be similar to the conditions *E. coli* finds during prolonged incubation in LB. In addition, clinical strains have been shown to exhibit the GASP phenotype when aged and then competed with their parental strain (E. Zinser and R. Kolter, unpublished). This observation also raises of the question of why all strains are not *rpoS*^{att}. The allelic variation observed for *rpoS* suggests that in different environmental niches different *rpoS* alleles are selected and that under some conditions it is more advantageous to have a wild type allele.

Stationary Phase Cultures Are Dynamic

The acquisition of the GASP phenotype by cells within a culture indicates that stationary phase cultures are dynamic. Even though the total viable counts within a culture remain constant after approximately 3 days in batch culture, we can see that by ten days mutants within the population can take over these cultures. However, the process does not end there. If a cell carrying a GASP allele is isolated, regrown and allowed to incubate for another ten days to 2 weeks, another GASP allele will arise, able to outcompete the first. This process can be repeated such that a succession of GASP-conferring mutations can be isolated, each enabling the new mutant to outcompete its parent strain (Fig. 3; Zambrano, 1993; Zambrano et al., 1993).

This can also be observed if cultures are allowed to continue to incubate with periodic sampling of the cells during stationary phase. Cells sampled at Day 1 will be out competed by cells sampled at Day 10. Twenty-day-old cells will outcompete those 10-day-old cells, and so forth (Fig. 4; S. Finkel and R. Kolter, unpublished).

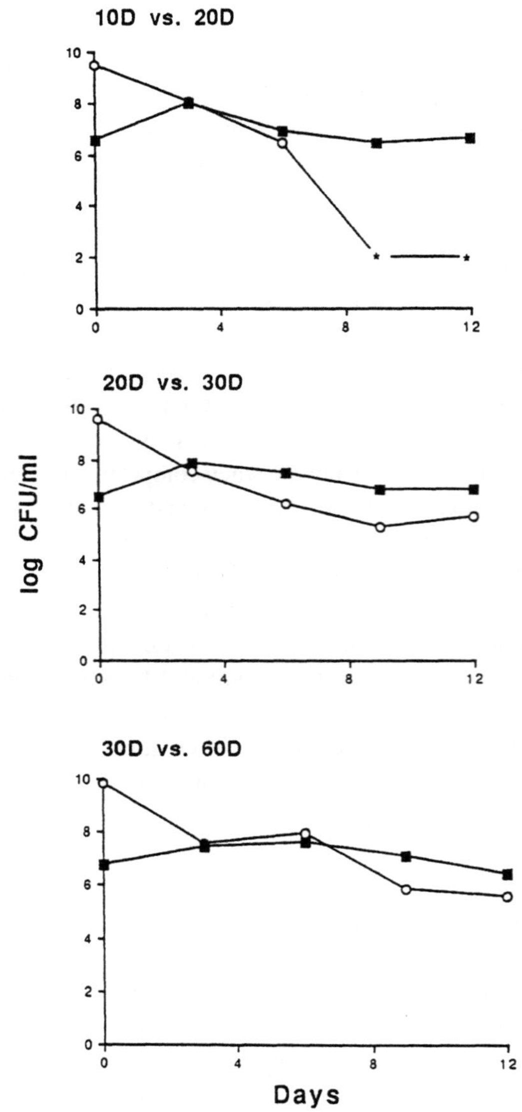

FIG. 4 GASP mutants continue to arise with prolonged incubation. Cells are sampled periodically and mixed with cells from younger cultures. Older cells (filled squares) are mixed as a minority with younger cells (open circles). (A) 20-day-old vs. 10-day-old, (B) 30-day-old vs. 20-day-old, (C) 60-day-old vs. 30-day-old. Asterisks indicate no detectable CFU.

Hence, new GASP-conferring alleles appear within the population periodically which confer on the mutant the ability to outcompete its siblings, allowing that mutant to take over the culture; until it is replaced by more fit mutants (Fig. 5).

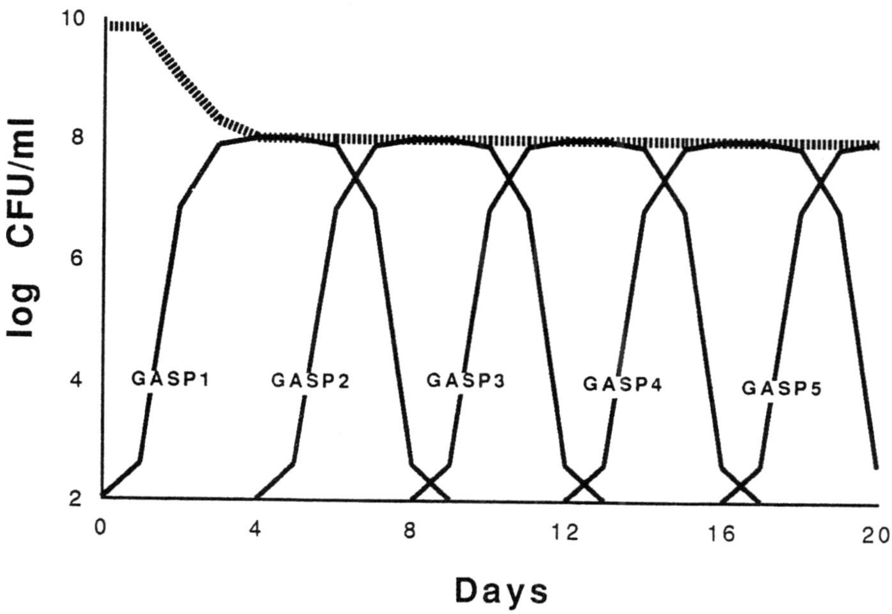

FIG. 5 Waves of GASP takeovers occur constantly.

Other Genes Required to Express the GASP Phenotype

In addition to mutations in *rpoS*, it is clear that other mutations occur which confer the GASP phenotype. Active investigation is currently underway to identify subsequent GASP mutations. However, in addition to new or altered activities that the cell may develop by the acquisition of beneficial mutations, there are other activities which much be present in order for the cell to survive during stationary phase.

Mutational screens have been utilized to identify activities required for survival during competition in stationary phase (Zambrano, 1993). Using transposon-insertion mutagenesis, mutations were found that do not affect the viability of the mutant strain when grown in pure culture, but compromise the ability of the mutant strain to compete when co-cultured with the wild type parent. Identification of these mutants is possible since cells expressing the GASP phenotype are able to grow upon a lawn of wild type cells (Fig. 6).

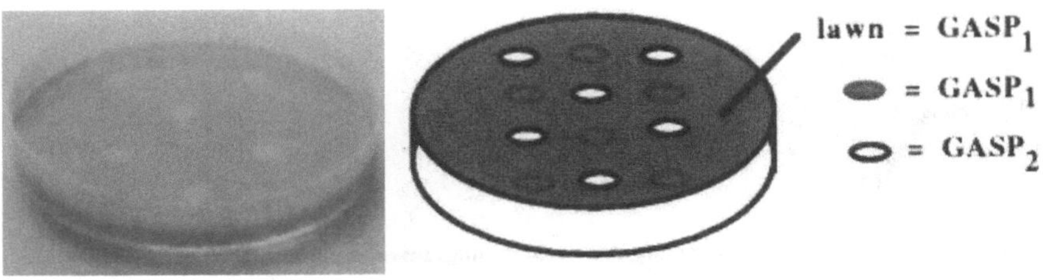

FIG. 6 Cells expressing the GASP phenotype can grow on a lawn of younger cells. A lawn of GASP$_1$ cells is spread on nutrient agar. GASP$_1$ and GASP$_2$ cells are then spotted onto the lawn. GASP$_2$ cells can grow on the lawn, while GASP$_1$ cells cannot.

Mutations were found in *nuoF* and *nuoM* which encode the two subunits of the bacterial NADH dehydrogenase I (NDH-1) (Zambrano and Kolter, 1993), *ompR* which encodes a regulator of outer membrane porins, and *nhaA* which encodes a sodium-proton antiporter (Zambrano, 1993). Mutations in NDH-1 and the antiporter might be expected to affect the ability of the cell to modulate energy flux within the cell and a loss of OmpR activity may affect the cell's ability to regulate transport.

The Generality of the GASP Phenotype

So far we have discussed the ability of *E. coli* to express the GASP phenotype while growing in rich LB medium; a well-characterized organism growing under standard laboratory conditions. How generalizable is the GASP phenomenon? Does it only happen under a narrow set of circumstances to a particular organism, or can it be observed more broadly?

The ability of *E. coli* to express GASP under different culture conditions was assessed. In addition to LB, *E. coli* has been shown to express the GASP phenotype when grown in minimal salts medium supplemented with glucose or casamino acids (Zambrano, 1993), as well as in LB under anaerobic conditions (E. Zinser and R. Kolter, unpublished). GASP has also been observed when *E. coli* is incubated for long periods on agar plates as well as in soft agar stabs (S. Finkel and R. Kolter, unpublished). However, the kinetics of GASP and the population dynamics observed under these different culture conditions differ from those observed for growth in aerobic LB culture, as described above. For example, the population takeovers are much slower in minimal glucose medium, the kinetics of death within the culture are altered (the characteristic 99% death of the majority population does not occur until after 30-60 days of incubation) and the rates of takeover are reduced (Fig. 7; Zambrano, 1993). Differences are also observed for the kinetics of GASP in anoxic LB medium (E. Zinser and R. Kolter, unpublished).

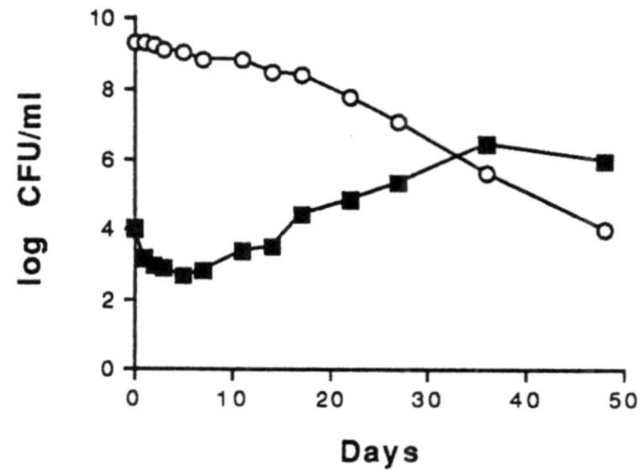

FIG. 7 Cells express the GASP phenotype in minimal glucose medium. Cells aged for 62 days in minimal glucose medium (filled squares) are mixed as a minority into an overnight culture of young cells (open circles).

Even more striking however, is the apparent lack of universality of the GASP mutations which are selected. For example, cells which are aged under anaerobic conditions in LB, which display the GASP phenotype when mixed with younger anaerobic cultures, do not express GASP when mixed with young aerobic cultures, and vice-versa (E. Zinser and R. Kolter, unpublished). That is, the conditions under which these strains are incubated are so different that the apparent advantages gained by these strains during prolonged incubation in stationary phase do not confer the advantage when the environment is greatly altered.

Differences are also observed between the ability of natural isolates to GASP as compared to laboratory strains. Most natural strains, including the environmental and clinical isolates, display the GASP phenotype (Fig. 8; E. Zinser, S. Gupta, and R. Kolter, unpublished). In fact, some clinical isolates can out compete laboratory strains much more quickly than aged laboratory strains against their parents. As stated above, many of these strains already have *rpoS*^{att} mutations which may be partially responsible for their increased ability to compete. Undoubtedly, these strains carry other alleles which confer a competitive advantage when competing for nutrients.

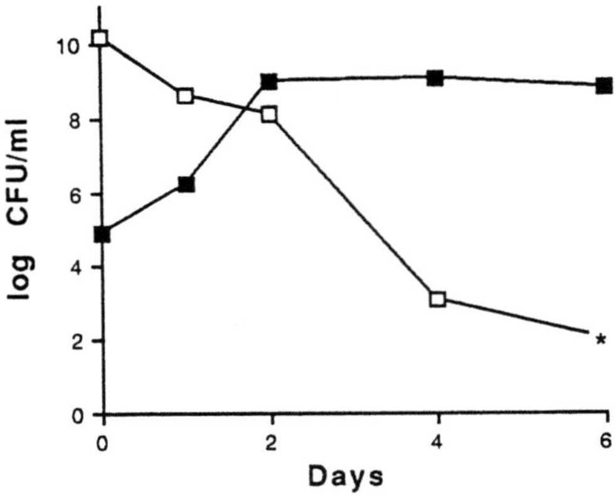

FIG. 8 GASP occurs in clinical isolates of *E. coli.* Cells from a 10-day-old culture of a clinical *E. coli* isolate (filled squares) is inoculated as a minority into an overnight culture of the same clinical isolate (open circles). Asterisks indicate no detectable CFU.

Finally, the GASP phenotype has been observed in other micro-organisms (M. Espinosa-Urgel, 1995, Ph.D. Thesis, Universidad Complutese de Madrid; Zinser and Kolter, unpublished). Experiments similar to those performed on *E. coli* have been used to show that GASP is expressed both by other gram-negative (*Shigella dysenteriae*, *Enterobactercloacae*) and gram-positive bacteria (*Staphylococcus aureus, Bacillus globigii, and Enterococcus faecalis*), as well as by yeast (*Saccharomyces cerevisiae*).

Conclusion: The Role of GASP in Adaptation and Evolution

Previous studies directed toward understanding the ability of a bacterial population to evolve over time have been performed either in chemostats or by constant serial passage with dilution into fresh medium (Dykhuisen, 1990; Elena et al., 1996). Under either condition, there is a selective pressure to increase fitness relative to all other cells in the population under a constant environment. Under these conditions, relative fitness will indeed increase. For example, Elena et al. (1996) have demonstrated almost 50% increases in relative fitness for *E. coli* which has been passaged through 20,000 generations in minimal glucose medium. These fitter cells have acquired mutations which increase their ability to utilize the single carbon source. The increases in fitness observed are not gradual, but appear to be sporadic.

This is consistent with a model of punctuated equilibrium or periodic selection in which relative fitness remains roughly constant for long periods until an advantageous mutation is selected. Then, those cells with the advantageous mutation will rapidly take over the population at which time equilibrium is re-established because all cells are now of equal fitness. Then the cycle repeats, with a concomitant increase in fitness.

The basic concepts of increased fitness and population fluctuation described for constant culture conditions (either chemostats or serial passage) also apply to the experiments conducted to explore the GASP phenotype. Yet, striking differences in the dynamics of the population changes between the two systems are observed. The batch culture conditions result in environments which are constantly changing: nutrients are used up, waste products accumulate, and cellular debris is released. The rate at which GASP mutations can take over appears to be faster than those observed in the constant culture environments. This results in an environment where the selective pressures are constantly changing, therefore a GASP mutation which confers a competitive advantage at 10 days, may not confer the same advantage many weeks or months later. In other words, a ten-month-old GASP allele may not confer increased fitness on a cell compared to a five-month-old GASP allele if the environmental conditions which gave rise to the older GASP allele are not present.

The highly dynamic nature of stationary phase cultures has not always been appreciated, but it is essential to an understanding of the events that occur when bacterial cells become nutrient deprived. There is little doubt that continued studies of stationary phase events in *E. coli* and other organisms will identify new genes and functions that will have implications for our understanding of mutational mechanisms and evolutionary processes, bacterial persistence in pathogenesis and in the environment, and new insights into basic metabolic systems.

Acknowledgements

This work is funded by the National Science Foundation (MCB9207323). S.E.F. is the recipient of a Helen Hay Whitney Postdoctoral Fellowship. We thank Heidi Goodrich-Blair, Manolo Espinosa-Urgel and Leslie Pratt for stimulating discussions and critical reading of this manuscript.

References

Dykhuisen DE (1983) Experimental studies of natural selection in bacteria. Annu Rev Ecol Syst 21, 373-398

Elena SF, Cooper VS, Lenski RE (1996) Punctuated evolution caused by selection of rare beneficial mutations. Science 272, 1802-1804

Hengge-Aronis R (1996) Regulation of gene expression during entry into stationary phase. In *Escherichia coli* and *Salmonella typhimurium*: Cellular and Molecular Biology, Neidhardt FC et al (eds.) American Society for Microbiology, Washington, DC, pp. 1497-1512

Huisman GW, Siegele DA, Zambrano MM, Kolter R (1996) Morphological and physiological changes during stationary phase. In *Escherichia coli* and *Salmonella typhimurium*: Cellular and Molecular Biology, Neidhardt FC et al (eds.) American Society for Microbiology, Washington, DC, pp. 1672-1682

Kaiser D (1984) Regulation of multicellular development in *Myxobacteria*. In Microbial Development, Losick R, Shapiro L (eds.) Cold Spring Harbor Laboratory Press, Cold Spring Harbor, NY, pp. 197-218

Kolter R, Siegele DA, Tormo A (1993) The stationary phase of the bacterial life cycle. Annu Rev Microbiol 47, 855-874.

Losick R, Youngman P (1984) Endospore formation in *Bacillus*. In Microbial Development, Losick R, Shapiro L (eds.) Cold Spring Harbor Laboratory Press, Cold Spring Harbor, NY, pp. 63-88

Sak BD, Eisenstark A, Touati D (1989) Exonuclease III and the catalase hydroperoxidase II in *Escherichia coli* are both regulated by the *katF* gene product. Proc Natl Acad Sci USA 86, 3271-3275

Tormo A, Almirón M, Kolter R (1990) *surA*, an *Escherichia coli* gene essential for survival during stationary phase. J Bacteriol 172, 4339-4347.

Zambrano MM (1993) *Escherichia coli* mutants with a genetic advantage in stationary phase. Ph.D. Thesis, Harvard University, Cambridge, MA

Zambrano MM, Kolter R (1993) *Escherichia coli* mutants lacking NADH dehydrogenase I have a competitive disadvantage in stationary phase. J Bacteriol 175, 5642-5647

Zambrano MM, Siegele DA, Almirón M, Tormo A, Kolter R (1993) Microbial competition: *Escherichia coli* mutants that take over stationary phase cultures. Science 259, 1757-1760

THE CITRIC ACID CYCLE AND OXYGEN-REGULATED GENE EXPRESSION IN *ESCHERICHIA COLI*

Jeffrey Green and John R Guest

The Krebs Institute for Biomolecular Research, Department of Molecular Biology and Biotechnology, University of Sheffield, Western Bank, Sheffield S10 2TN, UK

1. Introduction

Escherichia coli is a metabolically versatile organism which uses a variety of substrates and three different metabolic modes to support growth under aerobic or anaerobic conditions. The choice of metabolic mode depends on the supply of electron acceptors (Table 1). Aerobic respiration is the most productive because carbon sources can be completely oxidised to carbon dioxide via the citric acid cycle and as many as 24 reducing equivalents per mol of glucose can be processed via the proton-translocating aerobic respiratory chain. Anaerobic respiration with e.g. nitrate or fumarate in place of oxygen, is less productive because the citric acid cycle is repressed under anaerobic conditions. As a result, the substrate is only partially oxidised (mainly to acetate) and fewer reducing equivalents enter the corresponding anaerobic respiratory chain. Then, in the absence of an exogenous electron acceptor, the mixed acid fermentation can be used with appropriate substrates. This is the least productive metabolic mode, involving a redox-balanced dismutation of the substrate to form acetate, ethanol, hydrogen, carbon dioxide, and traces of formate, succinate and lactate. Acetyl-CoA serves as an endogenously-generated electron acceptor and the energy is derived from substrate-level phosphorylation rather than membrane-associated proton translocation.

Table 1. Metabolic modes of *Escherichia coli*

	Metabolic mode	Electron acceptor	E_0' (mV)	ΔG_0' (kJ/mol gluc)	Energy yield (per mol gluc)
1	Aerobic respiration	Oxygen	+820	-2830	4ATP + 24[H]
2	Anaerobic respiration	Nitrate	+430	-858	4ATP + 8[H]
	" "	Nitrite	+374	?	4ATP + 8[H]
	" "	DMSO	+160	-650	4ATP + 8[H]
	" "	TMAO	+130	-625	4ATP + 8[H]
	" "	Fumarate	+30	-550	4ATP + 8[H]
3	Fermentation	Acetyl-	-412	-218	3ATP

Given the choice, *E. coli* adopts the most energetically favourable metabolic mode. Thus, aerobic respiration is preferred to anaerobic respiration, which in turn is preferred to fermentation. This reflects the energy yields (ATP + H), the mid-point potentials of the electron acceptors and the free energies of the corresponding

NATO ASI Series, Vol. H 103
Molecular Microbiology
Edited by Stephen J. W. Busby,
Christopher M. Thomas and Nigel L. Brown

metabolic processes (Table 1). Switching between different modes involves significant changes in enzyme synthesis and these are largely controlled at the transcriptional level by five regulatory systems which respond to: the absence of oxygen or onset of anaerobiosis (ArcBA and FNR); the presence of nitrate or nitrite (NarXL and NarPQ); and the accumulation of formate, which signals the absence of exogenous electron acceptors and the need to adopt the fermentative mode (FhlA). The observed preference for nitrate and nitrite as anaerobic electron acceptors rather than fumarate, dimethylsulphoxide or trimethylamine-N-oxide, is also mediated by the NarXL/PQ regulatory systems acting positively or negatively on the expression of different anaerobic respiratory processes.

The discovery of the oxygen-dependent regulators (ArcA and FNR) emerged from parallel studies on the expression of two membrane-bound flavoenzymes, succinate dehydrogenase (SDH) and fumarate reductase (FRD). Both catalyse the interconversion of succinate and fumarate, but whereas SDH is expressed aerobically to function in the citric acid cycle and repressed anaerobically, FRD is repressed aerobically but derepressed anaerobically to serve as the terminal reductase in fumarate respiration and to replace SDH in the anaerobic citric acid cycle. The two-fold aim of this chapter is to highlight recent advances in (i) identifying genes encoding citric acid cycle enzymes and (ii) elucidating the mechanisms controlling their oxygen-dependent expression. Space constraints demand that the coverage is selective, reflecting the authors interests, rather than comprehensive. Other reviews containing similar and related material can be found in this volume and elsewhere: Cronan and LaPorte (1996); Guest (1995); Guest *et al.* (1996a); Lynch and Lin (1996a, b).

2. The Citric Acid Cycle of *Escherichia coli*

The citric acid cycle catalyses the total oxidation of acetyl units derived from glycolysis (pyruvate) or other catabolic pathways (Fig. 1). It is the major energy-generating pathway in aerobic heterotrophs and it is also an important source of precursors for cellular biosynthesis. The citric acid cycle of *E. coli* is an inducible pathway that is most highly induced during aerobic growth on non-fermentable substrates in media lacking amino acids and other biosynthetic intermediates. It is subject to catabolite repression, anaerobic repression and end-product repression. The major effect of anaerobiosis or excess glucose is to repress the conversion of 2-oxoglutarate to fumarate via the 2-oxoglutarate dehydrogenase complex (ODHC), succinyl-CoA synthetase (SCS) and succinate dehydrogenase (SDH). This transforms the aerobic cycle into its branched or non-cyclic form and as a result, metabolic carbon flows at a much reduced rate through oxidative and reductive routes to 2-oxoglutarate (for biosynthetic purposes) and succinate (a minor fermentation product), respectively (Fig. 1). Under such conditions most of the glycolytic carbon is converted to acetate + CO_2 (anaerobic respiration) or acetate, ethanol + CO_2 (mixed acid fermentation). This transformation is accompanied by significant changes in enzyme synthesis, such that some of the aerobic enzymes are replaced by alternative enzymes that catalyse the same or related reactions.

For example, *E. coli* uses three enzymes to oxidise pyruvate under different physiological conditions. The pyruvate dehydrogenase complex (PDHC) operates during aerobic respiratory growth but its function is partially replaced by pyruvate formate-lyase (PFL) during anaerobic respiration and completely replaced by PFL

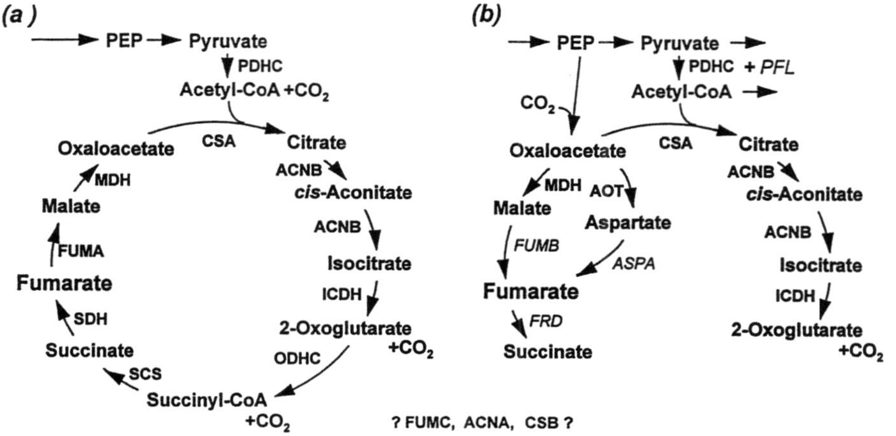

Fig. 1. Cyclic and non-cyclic forms of the citric acid cycle of *Escherichia coli* that operate under (a) aerobic conditions with limiting glucose and (b) anaerobic conditions or excess glucose.

during fermentation. The role of the third enzyme, pyruvate oxidase (PoxB) is less clear. It is an aerobic stationary phase enzyme which converts pyruvate directly and wastefully to acetate rather than acetyl-CoA. Other significant changes include the replacement of fumarase A (FumA) and SDH by their anaerobic counterparts, FumB and fumarate reductase (FRD), and the induction of an alternative reductive route from oxaloacetate to fumarate via aspartate, under anaerobic conditions (Fig. 1).

In order for the cycle to fulfil its anabolic role of providing precursors for cellular biosynthesis, intermediates have to be replenished by the so-called anaplerotic routes. Different routes are used depending on the carbon source. Thus with glucose or glycerol as substrates, oxaloacetate is replenished by a phosphopyruvate carboxylase-mediated carboxylation. With pyruvate or lactate, oxaloacetate is replenished by a similar route once pyruvate has been converted to phosphopyruvate by phosphopyruvate synthetase. In the case of acetate, intermediates are replenished by the glyoxylate cycle which uses isocitrate lyase and malate synthase to catalyse the net synthesis of oxaloacetate from two acetyl-units. No anaplerotic route is needed for growth on citric acid cycle intermediates, but the metabolic cycle has to be complemented by the production of acetyl-CoA from oxaloacetate or malate by phosphopyruvate carboxykinase or one of two malic enzymes, pyruvate kinase and PDHC, in order to complete the oxidation of these substrates.

2.1 Gene-enzyme relationships

Genes encoding all of the enzyme activities of the citric acid cycle have been cloned, sequenced and located in the *E. coli* linkage map (Fig. 2). A non-uniform set of symbols has evolved reflecting the nutritional requirements imposed by their inactivation (*glt*, glutamate; *ace*, acetate; *suc*, succinate), the enzyme name (*lpd*, lipoamide dehydrogenase; *fum*, fumarase, *sdh*, succinate dehydrogenase) or their downstream location in a previously named operon (*suc*, for the specific components of the 2-oxoglutarate dehydrogenase complex and the succinyl-CoA synthetase subunits. In many cases the transcriptional relationships have been defined (Fig. 2).

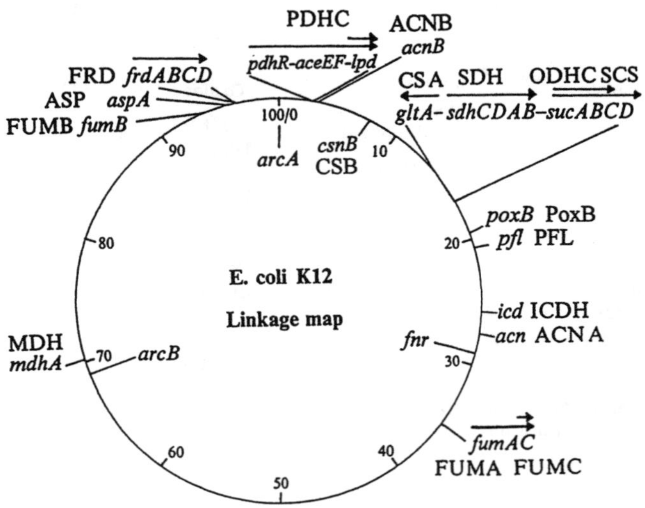

Fig. 2. Circular linkage map of *E. coli* showing the positions and some transcriptional relationships of the citric acid cycle genes and the locations of the *arc* and *fnr* genes.

The genes encoding four of the enzymes (CSA, SDH, ODH and SCS) are arranged in a major cluster of adjacent transcriptional units. The operon encoding the PDHC (*pdhR-aceE-aceF-lpdA*) represents a minor cluster which is interesting because it encodes a pyruvate-sensitive repressor (PdhR) as well as the enzymatic pyruvate dehydrogenase, lipoate acetyltransferase and lipoamide dehydrogenase subunits. This operon is also interesting because the *lpd* gene can be transcribed from the *pdh* promoter and from an *lpd* promoter, which is co-regulated with the *suc* and *gcv* operons, in order to provide subunits for assembly into the respective PDHC, ODHC and glycine cleavage system. The other genes are scattered in the linkage map (Fig. 2).

2.2 Multiple enzymes catalysing individual steps in the citric acid cycle

One completely unexpected finding that has emerged from recent molecular-genetic studies and the application of 'reverse genetic' procedures, is the fact that several steps in the cycle are catalysed by more than one genetically-distinct enzyme. The existence of analogous flavoprotein complexes, SDH and FRD, catalysing the aerobic and anaerobic interconversion of fumarate and succinate, has already been mentioned. There are also three fumarases , at least two (possibly three) aconitases (Gruer *et al.*, 1997), and a second citrate synthase (CSB) which resembles the dimeric mitochondrial enzymes rather than the major hexameric enzyme encoded by the *gltA* gene, CSA (Patton *et al.*, 1993). The second citrate synthase was originally detected in a second-site revertant of a *gltA* deficient strain, and it seemed that its appearance might be due to the reactivation of a cryptic gene or the derepression of a gene with an overlapping function but requiring specific conditions for induction. The CSB gene, here designated *csnB*, was recently located by screening the complete *E. coli* genome (Fig. 2) and it would appear that CSB is an N-terminally truncated product of a homologue of the *prpC* gene involved in propionate catabolism in *Salmonella typhimurium* (Guest *et al.*, unpublished).

Two of the fumarases (FumA and FumB) are almost identical dimeric, oxygen-labile enzymes containing one [4Fe-4S] cluster per subunit. Due to their instability, these enzymes were not detected until their genes had been isolated and characterised. Of the two, FumA is the aerobic citric acid cycle enzyme, whereas FumB is induced under anaerobic conditions, presumably to function in the reductive branch of the anaerobic cycle and to generate fumarate for use as an anaerobic electron acceptor (Fig. 1). In contrast, the third fumarase (FumC) is an entirely different stable tetrameric enzyme that very closely resembles the mitochondrial enzymes. Its expression is specifically induced by oxidative stress, which could mean that FumC either maintains a limited level of citric acid cycle activity under conditions that are likely to inactivate the other fumarases or alternatively, it accelerates the restoration of normal metabolic activity after exposure to such stress.

Evidence for the existence of multiple aconitases came from the detection of residual aconitase activities first in single then double mutants constructed by sequentially cloning and inactivating two chromosomal genes (*acnA* and *acnB*), only to be left with an activity ($\leq$ 4% wild-type) that may be the product of a third gene, *acnC* (Gruer *et al.*, 1997). Clues about the relative roles of the AcnA and AcnB have come from studying the expression of the corresponding *acn-lacZ* fusions during the growth cycle, under different physiological conditions, and in strains in which specific global regulatory genes are mutated or amplified (Bradbury *et al.*, 1996). As a result it is apparent that AcnB is the major citric acid cycle enzyme synthesised during exponential phase and anaerobically repressed via ArcA control, whereas AcnA is a stationary phase enzyme which resembles both FumC in being induced in response to oxidative stress and FumA in its Fur-activated response to iron. Whether AcnC is a genetically distinct aconitase of specific function rather than the activity of a related enzyme possessing a broad substrate specificity that includes citrate, isocitrate and cis-aconitate, has still to be established.

Other interesting features of the bacterial aconitases are their structural and phylogenetic relationships to mammalian mitochondrial aconitases, the cytoplasmic iron regulatory proteins (IRP), and other members of the aconitase family, isopropylmalate isomerase (IPMI) and homoaconitase (Gruer *et al.*, 1997). Whereas most of the proteins are predicted to have a four-domain structure similar to that defined crystallographically for the mitochondrial enzyme, the bacterial IPMI and AcnB proteins exhibit two architecturally distinct variations within the same overall structure (Fig. 3). Thus domain-4 of bacterial IPMI (but not fungal IPMI) exists as an independent subunit rather than being at the C-terminal end of a monomeric protein, connected to domain-3 via a long linker peptide. Then in AcnB, domain-4 is located

N-terminally and linked directly to domain-1. The cyclically permuted domain orders exhibited by AcnA and AcnB suggest that the ancestral coding regions were rearranged or fused in different ways during the course of evolution.

The iron regulatory proteins control the iron-dependent translation or stability of relevant mRNA transcripts in vertebrates (Hentze and Kuhn, 1996). Some of these proteins designated IRP1, function alternately as cytoplasmic aconitases under iron-sufficient conditions or as regulators after reversible disassembly of their iron-sulphur clusters during iron-deprivation or redox stress. They operate by binding site-specifically to stem-loop structures via the active-site clefts of the apo-proteins. It is interesting to note that AcnA is 53 % identical to mammalian IRP and 29 % identical to the mitochondrial aconitases, whereas AcnB is more distantly related to IRP, the mitochondrial aconitases and AcnA (15-17 % sequence identities). This raises the possibility that one of the *E. coli* aconitases, particularly AcnA, may perform a redox-stress or iron-responsive regulatory role in gene or transcript expression, in addition to its enzymatic function.

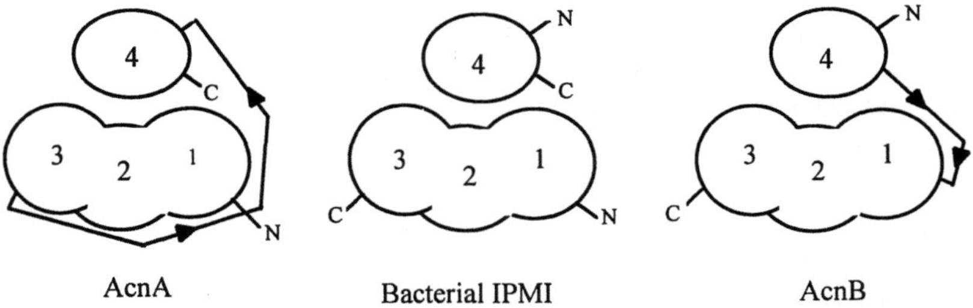

Fig. 3. Predicted structural variations in the domain organisations of bacterial IPMI and AcnB relative to AcnA and other members of the aconitase protein family.

2.3 Transcriptional regulation

The regulation of citric acid cycle gene expression has not been studied systematically, but quite a lot of information has accumulated for some of the genes and operons. It comes mainly from direct measurements of enzyme specific activities and studies with *lacZ* fusions (which enable the expression of one enzyme to be quantified in the presence of one or more iso-functional enzymes) with cultures grown under different physiological conditions, or cultures in which specific global regulators are inactivated or amplified. In many cases the transcripts have been identified but relatively few promoters have been analysed in detail at the molecular level, in order to define the binding sites for RNA polymerase and relevant transcription activators or repressors. The effects of relevant global regulators are summarised in Table 2. In view of the dual role of the cycle and the multiple roles of some of the enzymatic steps, it is not surprising that each gene seems to be controlled by specific regulatory networks of differing complexity. There are some common themes relating to both anaerobic repression by ArcA (and FNR) and anaerobic activation by FNR, but catabolite repression appears not to be mediated simply or exclusively by CRP. It is also clear that some of the promoters respond to specific regulators that can be rationalised in functional terms. Ultimately it should be possible derive a complete molecular formulation for the mechanism by which this seemingly highly co-ordinated metabolic cycle and all of its related activities, are expressed to meet the changing needs of cells growing in a diverse range of environments and at different stages of the growth cycle.

Table 2 Responses of citric acid cycle enzymes and the corresponding promoters to global regulators: +, activation; −, repression; O, little or no effect; . not tested; −g, repression by glucose; Og, no effect of glucose; aer, only tested aerobically.

Promoter	***Regulatory gene***						
	arcA	***fnr***	***crp***	***fruR***	***soxRS***	***fur***	***rpoS***
pdh	O	±?	+?	.	.	.	.
lpd	−	.	+	.	.	.	.
gltA	−	O	O	O	.	.	.
acnA	−	−	+	+	+	+	+
acnB	−	O	+	−	O	O	−
icd	−	.	−g	+	.	.	.
sucA	−	.	+	.	.	.	.
sdhC	−	−	O	O	.	O	.
fumA	−	−	.	.	O	+	O
fumC	−	O	O	.	+	O	+
fumB	.	+	Og	.	O aer	O aer	.
mdh	−	O	. −g	.	.	O	.

2.4 Evolutionary aspects

The central position of the citric acid cycle in all metabolic schemes suggests that it arose at an early stage in metabolic evolution. This view is further strengthened by the fact that enzymes catalysing steps in the cycle, or close relatives of these enzymes, are ubiquitous throughout all forms of life whether *Prokarya*, *Eukarya* or *Archaea*, and irrespective of whether they are aerobes or anaerobes. It would therefore seem highly likely that the cycle had its origins in the early anaerobic environment of our planet. The earliest cellular organisms probably obtained metabolic energy by fermenting the rich mixture of organic compounds, including di- and tri-carboxylic acids, that had accumulated in the prebiotic era. The reactions currently operating in the oxidative and reductive routes of the anaerobic citric acid cycle may thus represent some of the earliest metabolic transformations performed by primitive living organisms. Then, with the oxygenation of the earth's atmosphere, these routes might have been linked by the PDH and ODH reactions and the metabolic flow adjusted to generate an oxidative metabolic cycle. It could thus be speculated that the citric acid cycle recapitulates its evolutionary history during the anaerobic-aerobic transition of *E. coli*. Equally plausible is the view that the cycle evolved from the same or similar ancestral reactions, as a cyclic mechanism for CO_2 fixation driven in the reverse direction by pyrite formation (Maden, 1995). Such a mechanism is analogous to the reductive citric acid cycle operating in some bacteria and archaebacteria, where the reductive carboxylation of succinyl-CoA and acetyl-CoA are driven by reduced ferredoxin. In this case the transition to an oxidative cycle would be achieved by replacing the reductive carboxylation steps by the oxidative decarboxylation activities of the PDH and ODH complexes. It is of course conceivable that the cycle has evolved more than once, each time taking advantage of contemporary catalytic components. As more genes from a wide range of existing organisms are sequenced and compared, the potentially tortuous course of citric acid cycle evolution may emerge.

With respect to the multiplicity of citric acid cycle genes in *E. coli*, it is clear that gene duplication has provided opportunities for the functional specialisation of structurally analogous pairs of enzymes and the acquisition of independent regulatory mechanisms tailored to specific roles in aerobic and anaerobic metabolism, e.g. SDH/FRD, FumA/FumB. In other cases it would appear that a distantly related gene encoding an iso-functional enzyme has been recruited provide specific catalytic or regulatory advantages, e.g. the redox-stress induced FumC, the iron and redox-stress induced AcnA, and the ill-characterised CSB. Significantly, the latter enzymes are more closely related to their mammalian or Gram-positive counterparts as if they may have been imported by an ancestral *E. coli*. In both scenarios it would appear that the retention of two or more differentially regulated genes is preferred to the constitutive expression or to the multi-factorial regulation of a single all-purpose gene, and that the genome expands to accommodate the extra genes.

2.5 Engineering pyruvate metabolism

As mentioned above, *E. coli* expresses three enzymes (PDHC, PoxB and PFL) for oxidising pyruvate under different conditions. These enzymes control the flux of glycolytic carbon into the citric acid cycle and hence into growth-related products including the products of heterologous gene expression. The ability to manipulate the synthesis or activity of these enzymes could be used to redirect substrate carbon into biotechnologically useful products and away from acetate (which accumulates at high cell densities and is a major cause of poor productivity and cell death). Two strategies have been used to 'engineer' aerobic pyruvate metabolism: first, by altering the transcriptional control of PDHC or PoxB expression; and second, by genetically manipulating the catalytic activity of PDHC (Guest *et al.*, 1996b, 1997).

The first strategy was to replace the *pdh* promoter (and *pdhR* gene) in the wild-type *E. coli* chromosome, or the *poxB* promoter of a strain lacking PDHC activity, by the LacI-regulated P_{tac} promoter. This allowed the synthesis of the corresponding enzymes to be controlled by the concentration of exogenously supplied IPTG. Both strains required either acetate or IPTG for aerobic growth in glucose minimal medium. Furthermore, as shown in Fig. 4a, growth rates and hence the metabolic flux to acetyl-CoA or acetate, could be controlled over a relatively wide range by the non-metabolisable inducer, IPTG. These results also demonstrate that in the absence of PDHC, the energetically wasteful stationary-phase enzyme (PoxB) can support the growth of *E. coli* provided that it is induced early in the growth cycle.

The second strategy was to alter the number and type of lipoyl domains associated with the lipoate acetyltransferase (E2p) subunits of the PDHC. In the PDHC multiple copies of two other subunits (E1p and E3) are assembled around a core of 24 E2p subunits with an overall polypeptide chain stoicheiometry of approximately 1.0:1.0:0.5 (E1p:E2p:E3). The E2p subunit contains three similar lipoyl domains (lip) connected to an E3-binding domain (E3bd) by mobile ≈ 20-residue linkers (xxxx) and a catalytic and inner-core-forming domain (CAT). Each lipoyl domain is post-translationally modified by a lipoyl cofactor that is covalently bound to a

lip-xxxx-**lip**-xxxx-**lip**-xxxxx-**E3bd**-xxx-**CAT**

prominent lysine residue. The lipoyl domains protrude from the inner core to permit the sequential reductive acetylation, transacetylation and reoxidation of the lipoyl cofactors at the three types of active site, during the cycle of reactions catalysed by the complex. A curious feature of the PDHC of *E. coli* and other Gram-negative bacteria is the retention of three lipoyl domains compared with the E2 subunits of other 2-oxo acid dehydrogenase complexes which have only one or two lipoyl domains per E2p chain. Biochemical studies with genetically modified PDH complexes containing less than three lipoyl domains or complexes in which some but not all of the lipoyl domains are rendered unlipoylatable by a lysine-glutamine substitution, indicated that three lipoyl domains are not essential for maximum enzymatic activity. However, recent physiological studies with isogenic strains containing genetically engineered *pdh* operons have indicated that PDHC function is impaired if the number of lipoyl domains is lower or higher than three per E2p chain.

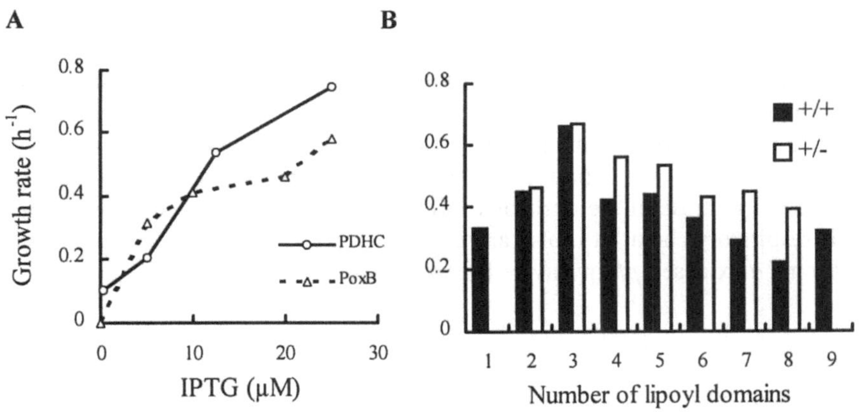

Fig. 4. Maximum specific growth rates of genetically engineered bacteria on glucose minimal media with: **A**, strains containing P_{tac}-regulated *pdh* operon (o) or *poxB* gene and no PDHC (Δ); **B**, strains containing PDHC with altered numbers and types of lipoyl domains per E2p chain, all lipoylated (■ +/+) or outermost domains lipoylated rest unlipoylated (□ +/-).

The growth rates reflected by 1 to 9 wild-type lipoyl domains or 1 wild-type domain at the N-terminus and up to 7 unlipoylated inner domains, are shown in Fig. 4b. They clearly indicate that 3 lipoyl domains per E2p chain is optimal and that bacteria having a graded range of inherently different capacities for metabolising pyruvate can be engineered by altering the catalytic proficiency of the PDHC. The results further indicate that an optimal growth rate can be achieved when only the outermost domain of three is lipoylated, suggesting that the retention of three domains is to facilitate optimal interactions with the various active sites rather than providing extra cofactors to participate in catalysis. It would also appear that measuring the specific growth rate of an engineered strain is a more sensitive indicator of the catalytic competence than measuring the specific activity of an altered PDHC under substrate saturating conditions.

3. Anaerobic Gene Expression in *Escherichia coli*

Oxygen and aerobic respiration confer enormous energetic benefits on a facultative anaerobe by allowing complete oxidation the growth substrate, compared with anaerobic respiration or fermentation (Table 1). The ability to adapt to the availability of oxygen and other potential terminal electron acceptors by expressing appropriate groups of genes is largely controlled at the level of transcription by four global regulators: the oxygen-responsive ArcBA and FNR regulators; and the nitrate/nitrite responsive NarXL and NarQP regulators, which act in concert to impose the metabolic priorities of *E. coli.* The remainder of this chapter will focus on the oxygen-responsive transcription regulators ArcBA and FNR.

3.1 The ArcBA system

The ArcBA regulatory system has been characterised by Lin and co-workers following the isolation of two classes of mutant that failed to anaerobically repress the expression of succinate dehydrogenase and other aerobic enzymes. The corresponding genes were designated *arcB* and *arcA* (Fig. 2) to denote the lack of aerobic respiration control. They encode a membrane-bound sensor (ArcB) and a cytoplasmic regulator (ArcA) that form a typical two component signal-transducing system (Fig. 5). The *arc* system was subsequently shown to control the expression of at least 30 transcriptional units, primarily by acting as an anaerobic repressor (Lynch and Lin, 1996). Thus it represses the expression of genes associated with aerobic metabolism e.g. those of the citric acid cycle (Table 2), the glyoxylate cycle and the aerobic respiratory chain. It is also an activator of *cydAB* (encoding the microaerobic cytochrome *d* oxidase), *pfl* (pyruvate formate-lyase) and *arcA* itself.

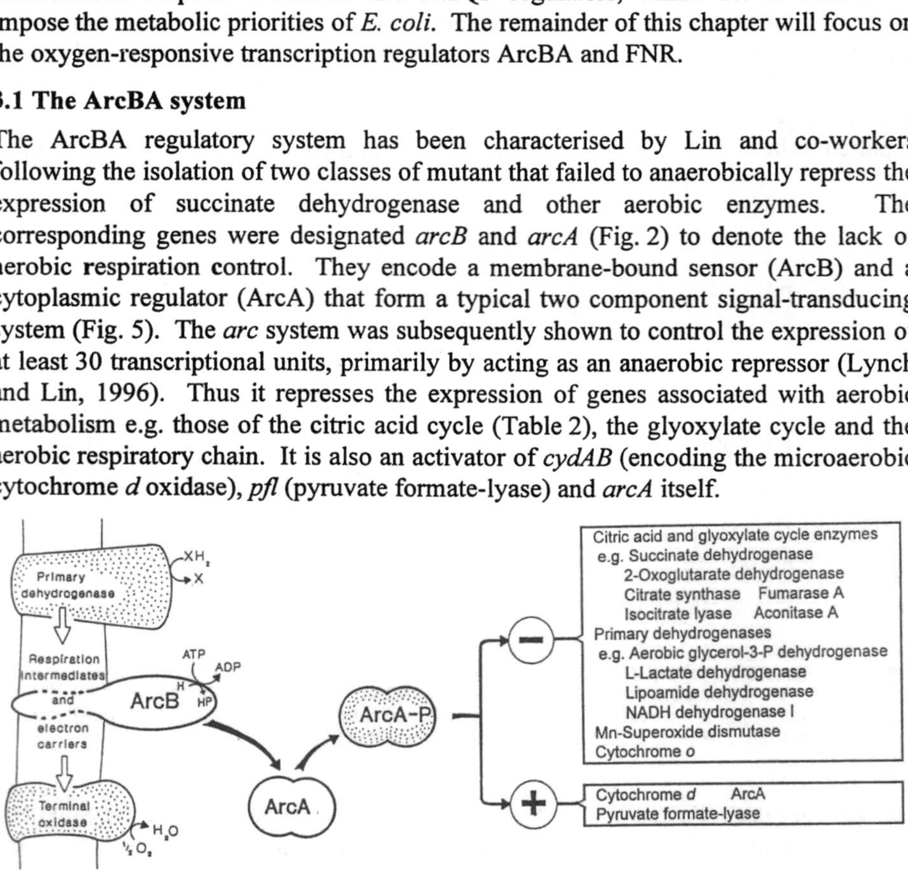

Fig. 5. Schematic representation of the Arc two component sensor-regulator system.

3.1.1 Properties of the ArcBA proteins

Two component sensor-regulator couples are used in many organisms for regulating gene expression in response to a variety of environmental stimuli. The sensors are characterised by the presence of an N-terminal input domain linked to a histidine autokinase transmitter that is phosphorylated in response to the primary signal. The regulator is then activated by phosphorylation of a conserved aspartyl residue in its N-terminal receiver module.

The sensor (ArcB) is an 174,000 Da dimeric protein that is associated with the cytoplasmic membrane. Its amino acid sequence indicates that it belongs to a subgroup of sensors known as hybrid kinases, which possess a receiver domain in addition to a histidine kinase transmitter domain (Fig. 6). ArcB has only two membrane spanning units indicating that only seven amino acid residues are exposed to the periplasm, the bulk of the protein being in the cytoplasm. Such an organisation suggests that ArcB is stimulated from within the membrane or the cytoplasm, because other sensors of this type possess more extensive periplasmic domains.

The regulator (ArcA) is a monomer of 27,000 Da that is similar to the response-regulators of two-component systems that have an N-terminal receiver domain and a helix-turn-helix DNA-binding motif near the C-terminus (Fig. 6).

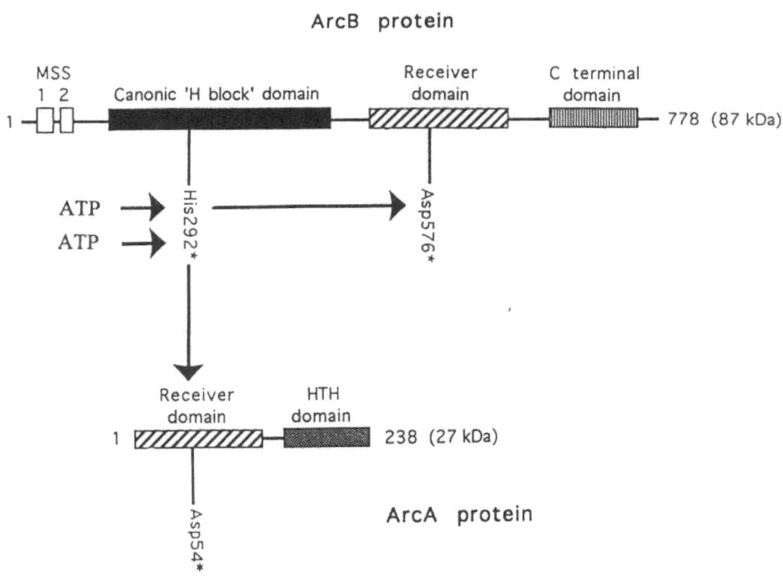

Fig. 6. Schematic representation of the domain structure of ArcA and ArcB and the corresponding signal transduction pathway. The membrane-spanning segments (MSS) and the helix-turn-helix (HTH) DNA-binding motif are indicated. The amino acid residues involved in the signal transduction pathway are highlighted. (Adapted from Lynch and Lin, 1996).

3.1.2 The Arc signal transduction pathway

The precise nature of the stimulus sensed by ArcB is unclear. An early suggestion was that it responds to anaerobic stress in the aerobic respiratory chain. However, the Arc system is unaffected by mutations that block the synthesis of a likely mediator, ubiquinone. The accumulation of the products of anaerobic metabolism such as

NADH, D-lactate, acetate, and pyruvate, could also signal the onset of anoxia, especially as these compounds increase the level of ArcB phosphorylation *in vitro* by inhibiting an autophosphatase activity, and may thus in turn increase the level of active ArcA. Most recently, the lowering of the electrochemical proton potential that is associated with the switch to anaerobic growth, has been offered as a potential signal.

Whatever the primary signal(s) may be, it is well established that ArcA activation proceeds through a cascade of phosphorylated intermediates. The initial response to anaerobiosis is the autophosphorylation of H292 of ArcB followed by internal transfer to D576 of the receiver domain. This facilitates the re-phosphorylation H292 and subsequent phosphoryl transfer from H292 of ArcB to D54 of ArcA (Fig. 6). Phosphorylation of ArcA promotes dimerisation and allows site-specific binding to the promoter regions of target genes.

3.1.3 DNA binding by ArcA

Recent *in vitro* studies have shown that phosphorylated ArcA binds site-specifically to relevant promoters in regions containing an ArcA-P box, [A/T]GTTAATTA[A/T]. Indeed, sequences conforming to this consensus are found in all known Arc-regulated genes (Lynch and Lin, 1996). The sizes of the protected regions in DNase I footprints vary from 25 to 105 bp depending on the promoter, indicating that ArcA-P dimer aggregation occurs at some sites. This may be significant if as proposed, multi-protein complexes are formed at some Arc-regulated promoters (see below). High resolution footprints are now required and site-directed mutagenesis within the ArcA-P box to confirm that this box is necessary for Arc-dependent regulation *in vivo*.

3.2 The FNR protein

The *fnr* gene of *E. coli* was identified following the isolation of pleiotropic mutants that lack the ability to use fumarate or nitrate for anaerobic respiratory growth (Guest *et al.*, 1996). The mutants were designated *fnr* to signify their defects in fumarate and nitrate reduction (Fig. 2). Unlike the ArcBA system, FNR acts as its own sensor and regulator. The FNR modulon currently contains 29 transcriptional units (70 genes) that are mainly associated with anaerobic energy generation. Indeed, all but one is concerned with anaerobic metabolism or with the transport or synthesis of substrates and cofactors required by other members of the modulon (Guest *et al.*, 1996). One particularly interesting member is *arcA*, which could be regarded as putting FNR at the highest level of oxygen-responsive transcription regulation in *E. coli*.

3.2.1 Relationship to CRP (CAP)

Comparison of the amino acid sequences of FNR and CRP indicates that all of the secondary structural elements of CRP are retained in FNR, including the helix-turn-helix motif in the DNA-binding domain, the large β-roll in the cyclicAMP-binding domain, and the major helix at the dimer interface (Fig. 7). However, there are some important differences between the two proteins. Aerobically isolated FNR protein is monomeric (M_r 28,000 by gel filtration, 30,000 by SDS-PAGE) rather than dimeric, although it appears to be dimeric when bound to DNA (see below). The residues that interact with cyclicAMP in CRP are not conserved, and there is no evidence for an interaction between cyclicAMP (or any other nucleotide) and FNR. But the major difference between the two proteins is the presence of a cysteine-rich N-terminal extension in FNR, which contains three of the four essential cysteine residues (C20, C23, C29 and C122). Moreover, aerobically isolated FNR contains a variable amount of iron (0.02-1.10 atoms per monomer), the iron content being inversely-related to cysteine sulphydryl reactivity. This indicates that the cysteine residues serve as iron ligands, and for some considerable time, it has been inferred that the N-terminal region of FNR contributes to an iron-binding sensory domain, which initiates an oxygen- (redox-) mediated conformational change in FNR resulting in the activation or repression of transcription.

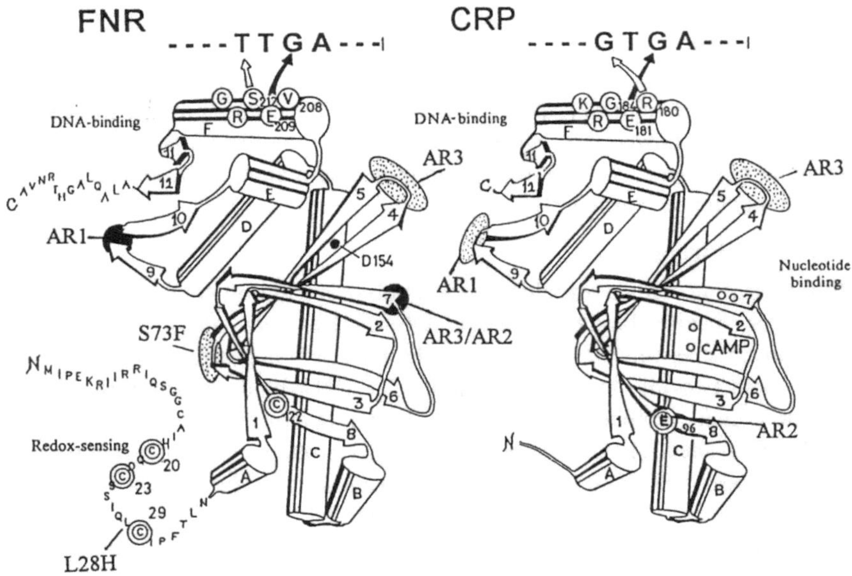

Fig. 7. The predicted structure of the FNR monomer based on the structure of CRP (CAP). The essential cysteine residues that ligand the oxygen- or redox-sensitive iron-sulphur cluster are ringed, the discriminatory interactions between the DNA binding helix (α_F) and specific base pairs in the FNR-site are indicated by arrows. Regions of FNR that are involved in making activating contacts with RNA polymerase are: AR1 which is thought to contact the α_{CTD} at both Class I and Class II promoters; AR2 and AR3 which are important at Class II promoters were they may contact α_{NTD} and σ respectively.

3.2.2 FNR:DNA interactions

The relationship between FNR and CRP extends to their cognate DNA-binding sites. FNR recognises a 22-bp sequence (the FNR box) which consists of the inverted repeat (----**TTGA**T----A**TCAA**----) deduced from comparisons of FNR-regulated promoters and confirmed by *in vitro* footprinting. The FNR box is related to the CRP-site consensus such that the **GTGA** half-site motif of the latter is replaced by **TTGA**. The specificity of FNR and CRP binding can be interconverted by replacing T(A) with G(C) and *vice versa*, at the critical symmetrical positions in the half-site motifs. The DNA-recognition specificities of the two regulators can also be switched by appropriate amino acid substitutions in their DNA-binding helices (α_F). Studies of this type amply confirm that there is a structural and functional relationship between FNR and CRP and that their modes of action are likely to be closely related, despite the differences in primary signal detection. Specificity for each binding site is through a common interaction between $E209_{FNR}$ ($E181_{CRP}$) and the **G-C** base pair in each half-site, and a discriminatory interaction between $S212_{FNR}$ and the unique **T-A** base pair in the FNR half-sites, which replaces that between $R180_{CRP}$ and the corresponding **G-C** base pair in the CRP half-sites (Fig. 7). The role of the conserved interaction between $R213_{FNR}$ ($R185_{CRP}$) and the common **G-C** base pair in sequence discrimination is uncertain. The corresponding binding-site motifs can therefore be denoted -**E**--**SR** (FNR) and **RE**---R (CRP).

A variety of footprinting techniques have been used to study FNR:DNA interactions *in vitro*. Several positively-regulated (*FFpmelR*, *pflP6*, *fdn*, *narGHJI*, *nirB*) and negatively-regulated (*ndh* and *fnr*) promoters have been investigated. In all cases, the protected regions (24-33 bp) overlapped the sites predicted from sequence comparisons and showed that two FNR monomers (one dimer) are (is) bound. The presence of hypersensitive sites within the protected regions indicate that FNR bends DNA in a manner similar to CRP.

3.2.3 The FNR trigger

Early footprinting studies revealed that anaerobiosis and iron containing FNR are not essential for specific binding to target DNA. However, although the protection patterns are independent of iron content, the DNA-binding affinities (K_d 10^{-7} M) are reduced about two-fold by iron-depletion. The DNA-binding affinities are likewise lower for mutant proteins lacking the essential N-terminal cysteine residues, and binding is abolished for the iron-deficient C122A protein. The observation that iron-containing FNR (holo-FNR) activates and represses transcription *in vitro* represented a significant advance, because it showed that the FNR protein could be isolated in an active form (Sharrocks *et al.*, 1991). A further milestone was provided by the observation that inactive apo-FNR could be converted to active holo-FNR by treatment with ferrous ions and β-mercaptoethanol (Green and Guest, 1993). This posed important questions concerning the nature of the bound iron (the redox-sensing cofactor).

Another fundamentally important advance in studying the mechanism of oxygen- or redox-sensing by FNR came with the isolation of the FNR* mutants by Kiley and co-workers. The FNR* proteins retain some ability to regulate transcription *in vivo* even during aerobic growth (Lazazzera *et al.*, 1993). Two classes of FNR* mutants were identified with respect to the locations of the corresponding amino acid substitutions, which affected the dimer interface (D154A) or the N-terminal redox-sensing region (L28H). FNR* proteins show enhanced DNA binding compared to wild-type protein and the D154A substitution produces an FNR* protein that is substantially dimeric under aerobic conditions whereas wild-type FNR protein is monomeric. These observations strongly suggest that dimerisation is a key component in FNR activation.

Combining the two classes of substitution (L28H, D154A) in one FNR** protein yields an FNR protein that is not only dimeric but contains two [3Fe 4S] clusters per dimer (Khoroshilova *et al.*, 1995). The knowledge gained from studies with the FNR** protein has now been applied to wild-type FNR. Advantage has been taken of new purification and reconstitution procedures and it is now apparent that anaerobic FNR contains one [4Fe-4S] cluster per monomer and that the presence of the iron-sulphur cluster promotes dimerisation and enhances site-specific DNA binding (Green *et al.*, 1996; Lazazzera *et al.*, 1996). FNR having a full complement of two [4Fe-4S] clusters per dimer binds target DNA with a K_d of 10^{-8} M, that is a 7 to 10 fold higher affinity than that of untreated protein, and 5 times better than chemically reconstituted FNR.

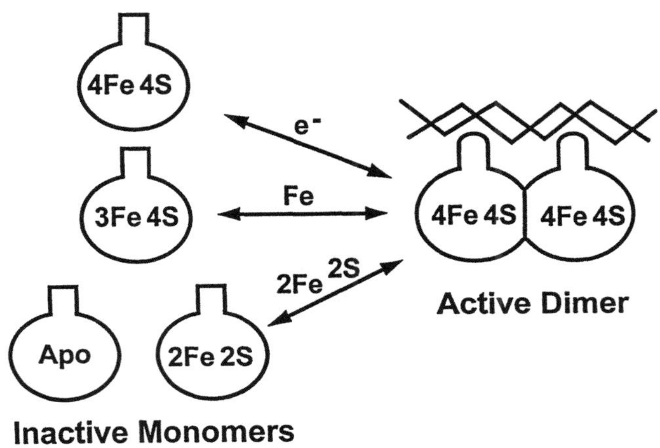

Fig. 8. Routes to the active FNR dimer including: reduction of pre-existing [4Fe-4S] clusters; the incorporation of a single ferrous ions into 3[Fe-4S] clusters; the reversible conversion of [2Fe-2S] to [4Fe-4S] clusters; and the complete assembly of [4Fe-4S] clusters from apo-FNR.

The presence of the [4Fe-4S] cluster is associated with a characteristic absorbance in the visible region around 420 nm which is rapidly bleached upon exposure to air. Anaerobically prepared FNR is epr silent, consistent with the presence of a $[4Fe\text{-}4S]^{2+}$ cluster but it is rapidly destroyed upon exposure to air. A signal characteristic of a $[3Fe\text{-}4S]^{+}$ cluster appears transiently, and a more stable but inactive $[2Fe\text{-}2S]^{2+}$ form of FNR is produced (Khoroshilova *et al.*, 1997). This 3Fe FNR protein is the form observed with the FNR** mutant. Thus the loss of the [4Fe-4S] cluster upon exposure to air provides the switch for the inactivation of FNR during aerobic growth.

The mechanism of oxygen sensing by FNR is thus becoming clearer (Fig. 8). However, substantial questions remain in defining the exact nature of the switch (is complete destruction of the cluster necessary?), the signal (does FNR sense oxygen *per se* or redox state?), and the interaction between FNR and RNA polymerase (how is transcription activation or repression effected?).

3.3 Communication with RNA polymerase

The initial steps of FNR-mediated transcription regulation can be defined in terms of the anaerobic acquisition of the iron-sulphur cluster leading to enhanced site-specific binding to DNA. But how does this modulate the activity of RNA polymerase? Studies with FNR, CRP and FNR homologues are providing some clues. The basic transcription machinery of *E. coli* consists of a multi-subunit ($\alpha_2\beta\beta'\sigma$) RNA polymerase (Fig. 9). The commonly used σ subunit is σ^{70} which recognises and binds to two promoter elements, the -10 and -35 boxes. The α subunit is also involved in DNA interactions at some promoters. It binds via its C-terminal domain to UP elements, regions of A-T rich DNA located upstream of the -35 element. The strength of any promoter is governed by the presence (or absence) and quality of the -10, -35 and UP elements. Variations and additions to this basic model provide mechanisms for regulating transcription in response to different environmental stimuli.

3.3.1 Transcription activation

In common with most positively-regulated promoters, FNR-activated promoters tend to have -10 sequences but a poor -35 sequences. However, there is usually an FNR-site centred at about -41.5 that compensates for the weak -35 element and to permit anaerobic FNR to activate transcription. Thus the architecture of most FNR-activated promoters resembles that of Class II CRP-dependent promoters (Busby and Kolb, 1996). This type of organisation forms the basic transcriptional unit for FNR-mediated activation which can be further modulated by adding sites for other regulatory factors upstream of the basic unit (Fig. 10) or by the presence of additional upstream FNR-sites.

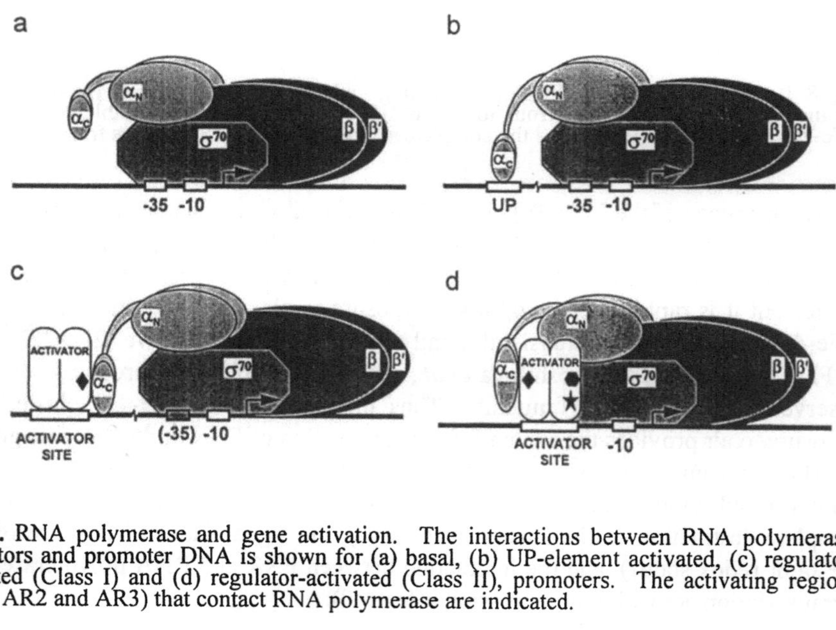

Fig. 9. RNA polymerase and gene activation. The interactions between RNA polymerase, regulators and promoter DNA is shown for (a) basal, (b) UP-element activated, (c) regulator-activated (Class I) and (d) regulator-activated (Class II), promoters. The activating regions (AR1, AR2 and AR3) that contact RNA polymerase are indicated.

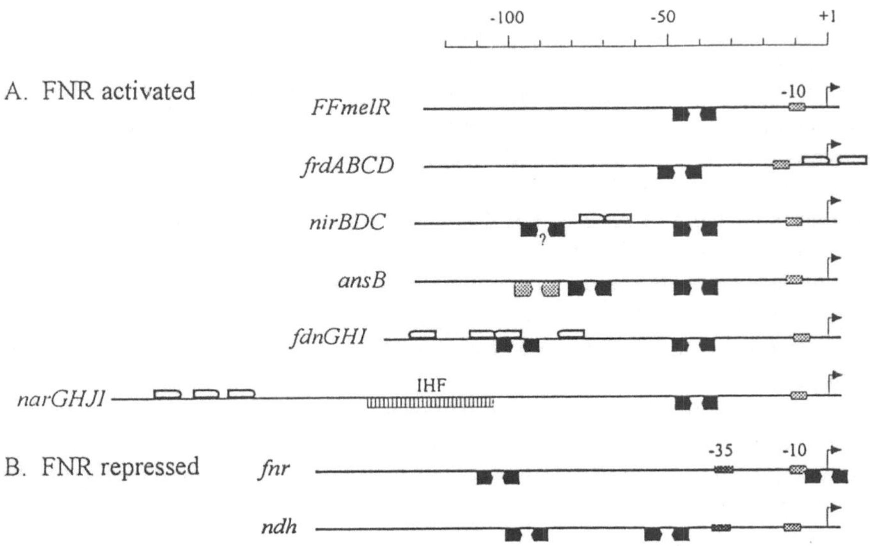

Fig. 10. Architecture of FNR regulated promoters. The positions of regulator binding sites (FNR, ■ ; CRP, ▩ ; NarL ▭), the -10 and -35 elements are located relative to the transcription start point (arrows).

Studies on transcription regulation by FNR have been greatly facilitated by equivalent investigations with CRP. Transcription is activated from CRP-dependent Class II promoters by direct contacts between three Activating Regions (AR1, AR2 and AR3) of CRP and the α and σ subunits of RNA polymerase. The AR1 of the upstream CRP monomer contacts the C-terminal domain of the α subunit of RNA polymerase whereas AR2 and AR3 of the downstream subunit contact the N-terminal domain of the α subunit of RNA polymerase and region 4 of σ^{70} respectively. The AR2 and AR3 contacts are not possible at Class I activated promoters (regulator site at -61.5 or beyond) and a single contact between the C-terminal domain of α and AR1 of the downstream CRP subunit is sufficient (Busby and Kolb, 1996).

Analysis of FNR mutants generated by low fidelity PCR has identified a region between β_4 and β_5 corresponding to AR3 characterised by the mutation G85A (Fig. 7). This approach assigned AR1 of FNR to the loop between β_3 and β_4 characterised by the mutation S73F. However, the substitution S73A did not affect the ability of FNR to activate transcription, therefore it seemed that FNR possesses an AR1 equivalent but that S73 is not part of it. Further evidence for the presence of an AR1 in FNR comes from the observation that FNR activates expression of synthetic Class I promoters with FNR boxes 61, 71, 82 and 92 base pairs upstream of the transcription start point. However, for efficient FNR-dependent activation an improved -35 hexamer (TTAAAG) obtained by inserting a single base into the parental sequence (TAAAGA) was required (Wing *et al.*, 1995). Another approach has been to study HlyX, an FNR homologue of *Actinobacillus pleuropneumoniae*, which as well as complementing *fnr* mutations, can anaerobically activate several genes that are not activated by FNR. The amino acid sequence of HlyX is 73% identical to that of FNR and it was reasoned that the distinct but overlapping specificities of the two regulators may derive from differences in the quality of their respective activating contacts with RNA polymerase. Site-directed mutagenesis of HlyX has indicated that AR3 may include a region of β_7 as well as the β_4-β_5 loop, or that β_7 may contribute to AR2 of HlyX (FNR). It has also indicated that AR1 is a surface exposed region that includes A187 (Fig. 7). Thus it would appear that AR1 is located, at least in part, in the same loop between β_9 and β_{10} in FNR. Partial proteolysis of FNR with trypsin releases a polypeptide initiating at G185 indicating that this region is accessible and can therefore participate in protein-protein interactions.

These observations confirm that FNR activates transcription in a similar manner to CRP, by forming specific activating contacts with RNA polymerase. These contacts serve to compensate for the weak basal promoter elements in activated promoters, by recruiting RNA polymerase to the promoter and facilitating promoter clearance.

3.3.2 Transcription repression

FNR repressible promoters possess good -10, -35 elements and multiple FNR sites. The locations of the binding sites for global regulators are generally more variable in repressed promoters (operator sequences) than in activated promoters (Collado-Vides *et al.*, 1991). This appears to be true for promoters that are repressed FNR, where FNR binding sites are found at: -50.5 and -94.5 in *ndh*; -0.5 and -103 in *fnr*; -35.5 in *sodA*; and -106.5, -75.5 and +107.5 in *narX* (Guest *et al.*, 1996). One reason for this variability may be to allow each promoter to be regulated by FNR but to different extents. Model systems predict that the greatest degree of repression is observed when the operator binds to the spacer region between the -35 and -10 elements thereby occluding both promoter elements. Intermediate repression occurs when the operator sequence overlaps both the -10 element and the transcription start point, such that the repressor occludes the initiation region but RNA polymerase can still form productive complexes by interacting with the -35 element. The weakest repression should occur when the operator is located immediately upstream of the -35 element, where the repressor least affects RNA polymerase binding because it leaves exposed the crucial -10 element and transcript start site (Collado-Vides *et al.*, 1991).

The FNR-repressed *ndh* gene encodes a non-proton translocating NADH dehydrogenase. It seems to fall into the latter category by having a high affinity FNR binding site centred at -50.5 and a low affinity site at -94.5. The presence of tandem FNR sites in the *ndh* promoter resembles the arrangement of CRP sites in promoters of the CytR regulon where repression occurs when CytR binds by direct protein-protein interactions to tandemly bound CRP dimers, thus stabilising the 2CRP-CytR-DNA complex (Busby and Kolb, 1996). However, this mechanism is unlikely to operate at the *ndh* promoter because repression is mediated solely by FNR.

A model for FNR-mediated repression of *ndh* expression has been developed from *in vitro* studies with a series of altered *ndh* promoters. During aerobic growth FNR is inactive and not bound to the *ndh* promoter thus allowing RNA polymerase free access to the -10 and -35 promoter elements (Fig. 11a). RNA polymerase binds to a region spanning -60 to +20 in which the C-terminal domain of the α subunit (α_C) occupies the region between -37 and -55. This binary RNA polymerase-*ndh* complex is competent for the initiation of transcription. Adding FNR does not prevent RNA polymerase binding to the *ndh* promoter so FNR-mediated repression is not simply due to promoter occlusion. On the contrary, it would seem that low concentrations of active FNR, i.e. under microaerobic conditions, the high affinity FNR site at -50.5 (FNR I) is occupied but the α_C subunit of RNA polymerase is displaced. Then as a consequence of FNR-induced DNA bending, α_C can contact a weaker alternative binding site which overlaps the FNR site centred at -94.5 (FNR II), and some transcription can still occur (Fig. 11b). Under strictly anaerobic conditions the cellular concentration of active FNR is sufficient to fully occupy both FNR sites, thus blocking both α_C binding sites but still allowing the bulk of RNA polymerase to interact with the *ndh* promoter (Fig. 11c). As a result, α_C is 'orphaned' (lacks a binding site) and in the unbound state inhibits transcription initiation and destabilises RNA polymerase-DNA interactions. The presence of RNA polymerase in repression complexes is not unprecedented because it can form a ternary complex with DNA-bound *lac* repressor.

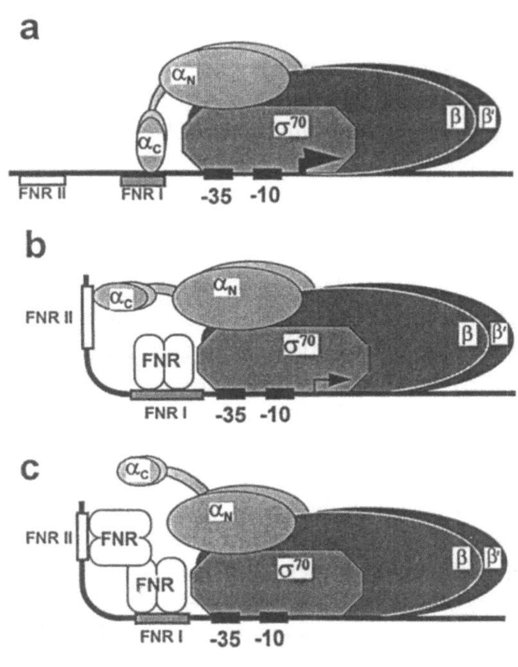

Fig. 11. A model of FNR-mediated repression of *ndh* expression (see text for details).

Interestingly, this postulated mechanism offers a direct link between transcription activation and repression. A key component of transcription activation from Class II CRP-dependent promoters is the formation of 'anti-inhibition' contacts between CRP and α_C which relieve the inhibitory effects of the otherwise unbound α_C. On the other hand, FNR-mediated *ndh* repression appears to involve the formation of unbound 'orphan' α_C by displacing it from its functional binding sites. Presumably the promoter architecture, particularly with respect to the positions of the FNR binding sites, precludes FNR from making the necessary 'anti-inhibition' contacts so that transcription is repressed. Activation of transcription may thus require the suppression of the inhibitory effects of an unbound α_C, whereas repression requires the generation of an inhibitory α_C through regulator-DNA contacts which prevent α_C-DNA contacts.

3.3.3 Multi-factorial regulation

Many FNR-regulated genes are members of other modulons and as such the basic pattern of FNR-mediated activation and repression is modified. The impact of individual regulators responding to different stimuli upon expression from a single promoter raises many questions concerning their interaction with RNA polymerase and with each other. It is envisaged that these specific interactions provide the molecular mechanism for establishing the metabolic priorities of *E. coli*, ensuring that the most energetically efficient metabolic process operates. From the arrangement of different regulatory sites in FNR-dependent promoters (Fig. 10) several mechanisms are possible, including: competition for overlapping binding sites; co-operative DNA-binding; independent interactions between each regulator and RNA polymerase; and activity modulation one regulator that contacts RNAP by another that makes no direct contact with RNA polymerase.

3.3.4 Dual regulation by FNR and NarL

Most of the anaerobic respiratory genes are regulated by both FNR and the NarXL two component sensor-regulator system (or its counterpart NarQP) in response to anaerobiosis and nitrate, respectively. NarL can act as a classical repressor by binding close to the transcription initiation site, e.g. in the *frdA* (Fig. 10) and *dmsA* promoters. However, there is much still to be learned about the mechanism of FNR- and NarL- activated transcription. It is not unequivocally established whether there is a need for direct interaction between FNR and NarL. Such interactions could explain the lack of nitrate-inducible *narG* expression in an *fnr* mutant and the lowered anaerobic *nirB* expression observed in a *narL* mutant in the absence of nitrate or when the NarL-site is deleted or altered. Also, studies with some *fnr** mutants have indicated that an interaction between FNR and NarL is needed for maximal aerobic expression of a *narG-lacZ* reporter. In contrast, FNR is not required for *narG* expression when the -35 element is replaced by the consensus sequence and the resulting activity is enhanced by NarL, indicating that a FNR:NarL interaction is not essential. The impact of other regulators such as IHF on *narG* expression should not be overlooked: by bending DNA it may facilitate the formation of multi-protein complexes and hence direct contact between FNR, NarL and RNA polymerase.

These contradictory reports highlight the need for further experimentation which should be guided by the recently reported NarL structure.

3.3.5 FNR and CRP

The FNR and CRP binding sites are sufficiently similar to envisage that there may be limited competition for alternate binding sites. The colicin E1 gene (*cea*) is both FNR activated and catabolite repressed, and the *cea* promoter contains several potential FNR-sites, the best of which (centred at -64.5) is also a potential CRP-site. The anaerobic induction of *ansB* (asparaginase II) is mediated by an FNR-site at -41.5 (Fig. 10) and catabolite repression is mediated by a CRP at -91.5. Both FNR and CRP are required for maximum expression and this is achieved by co-dependent activation in which both regulators interact with RNA polymerase. It appears that the AR1 of the downstream subunit of CRP and the AR3 of the downstream subunit of FNR make the important contacts. Interestingly the *ansB* promoter of *Salmonella typhimurium* is regulated by two CRP dimers. In this case the activation is synergistic rather than co-dependent. Thus it appears that co-dependent activation is determined by the nature of the regulators and the context in which they exist.

3.3.6 FNR and ArcA interactions

Finally, the interaction between the two oxygen responsive regulators, FNR and ArcA, is important because they co-ordinate the response to anoxia by reciprocal anaerobic activation and repression mechanisms. They are intimately linked via the anaerobic activation of *arcA* by FNR, which presumably tunes the co-ordinated regulation of aerobic and anaerobic metabolism. The overall physiological consequence of this link would be for FNR to intensify the repressing effects of ArcA on 'aerobic' gene expression as anaerobiosis deepens. The dual action of FNR and ArcA on the expression of some genes (*sodA* in the absence of Fur, *cyo* and *cyd*) may be explained by FNR regulation of *arcA* but in other cases (*focA-pfl* and *arcA*) direct FNR:ArcA interaction is indicated. The nature of the interaction has not been defined but a model involving IHF-induced DNA-bending and a transcription complex containing two FNR dimers, ArcA and RNA polymerase, has been proposed for *focA-pflP6* and *P7*.

General

The molecular-genetic approach has revealed fascinating insights into the structural and regulatory diversity of the genes and enzymes that operate at the central core of metabolism. Predictions about their evolutionary origins should become clearer as more genes from a wider range of organisms are characterised and compared. It should likewise soon be possible to combine molecular-genetic and physiological approaches in order to engineer a range of highly efficient bacterial biocatalysts designed for specific purposes. Our understanding of transcription and transcription regulation will make a major contribution to the construction of 'designer bacterial factories', offering prospects for externally controlling specific gene expression to maximise product yields. With more information, particularly relating to the structures and interactions between the various components of transcriptional complexes, the approach should become directed rather than empirical.

Acknowledgements. We are grateful to our colleagues for information used in compiling this chapter: Ahmed M Abdel-Hamid, Mandy Baldwin, Barrie Cassey, Louise Cunningham, Margaret M Attwood, Wenmao Meng and Justin T Evans. We also thank The Wellcome Trust and the Biotechnology and Biological Sciences Research Council for financial support.

References

Bradbury AJ, Gruer MJ, Rudd KE, Guest JR (1996) The second aconitase (AcnB) of *Escherichia coli*. Microbiology 142: 389-400

Busby S, Kolb A (1996) The CAP modulon. In Regulation and Gene Expression in *Escherichia coli* (Lin ECC and Lynch AS eds) RG Landes & Co Austin Texas pp 255-279

Collado-Vides J, Magasanik B, Gralla J (1991) Control site location and transcriptional regulation in *Escherichia coli*. Microbiol Rev 55: 371-394

Cronan JE JR, LaPorte D (1996) Tricarboxylic acid cycle and glyoxylate bypass. In *Escherichia coli* and *Salmonella* (Neidhardt FC *et al* eds) ASM Press Washington DC pp 206-216

Green J, Bennett B, Jordan P, Ralph ET, Thomson AJ, Guest JR (1996) Reconstitution of the [4Fe-4S] cluster in FNR and demonstration of the aerobic-anaerobic transcription switch *in vitro*. Biochem J 316: 887-892

Green J, Guest JR (1993) Activation of FNR-dependent transcription by iron: an *in vitro* switch for FNR. FEMS Microbiol Lett 113: 219-222

Gruer MJ, Artymiuk P, Guest JR (1997) The aconitase family: three structural variations on a common theme. Trend Biochem Sci 22: 3-6

Gruer MJ, Bradbury AJ, Guest JR (1997) Construction and properties of aconitase mutants of *Escherichia coli*. Microbiology 143: 1837-1846

Guest JR (1995) Adaptation to life without oxygen. Phil Trans R Soc Lond B 350: 189-202.

Guest JR, Attwood MM, Machado RS, Matqi KY, Shaw JE, Turner SL (1997). Enzymological and physiological consequences of restructuring the lipoyl domain content of the pyruvate dehydrogenase complex of *Escherichia coli*. Microbiology 143: 457-466

Guest JR, Green J, Irvine AS, Spiro S (1996a) The FNR modulon and FNR-regulated gene expression. In Regulation and Gene Expression in *Escherichia coli* (Lin ECC and Lynch AS eds) RG Landes & Co Austin Texas pp 317-342

Guest JR, Quail MA, Davé E, Cassey B, Attwood MM (1996b) Regulatory and other aspects of pyruvate dehydrogenase complex synthesis in *Escherichia coli*. In Biochemistry and Physiology of Thiamin Diphosphate Enzymes (Biswanger H and Schellenberger A eds) A u C Intemann Prien pp 326-333

Hentze MW, Kuhn LC (1996) Molecular control of vertebrate iron metabolism: mRNA-based regulatory controls operated by iron, nitric oxide, and oxidative stress. Proc Natl Acad Sci USA 93: 8175-8182

Khoroshilova N, Beinert H, Kiley PJ (1995) Association of a polynuclear iron-sulfur center with a mutant FNR protein enhances DNA-binding. Proc Natl Acad Sci USA 92: 2499-2505

Khoroshilova N, Popescu, C. Munck, E. Beinert H, Kiley PJ (1997) Iron-sulfur cluster disassembly in the FNR protein of *Escherichia coli* by O_2: [4Fe-4S] to [2Fe-2S] conversion with loss of biological activity. Proc Natl Acad Sci USA 94: 6087-6092

Lazazzera BA, Bates DM, Kiley PJ (1993) The activity of the *Escherichia coli* transcription factor FNR is regulated by a change in oligomeric state. Genes Dev 7: 1993-2005

Lazazzera BA, Beinert H, Khoroshilova N, Kennedy MC, Kiley PJ (1996) DNA-binding and dimerization of the Fe-S containing FNR protein *Escherichia coli* are regulated by oxygen. J Biol Chem 271: 2762-2768

Lynch AS, Lin ECC (1996a) Responses to molecular oxygen. In *Escherichia coli* and *Salmonella* (Neidhardt FC *et al.* eds) ASM Press Washington DC pp 1526-1538

Lynch AS, Lin ECC (1996b) Regulation of aerobic and anaerobic metabolism by the ARC system. In Regulation and Gene Expression in *Escherichia coli* (Lin ECC and Lynch AS eds) RG Landes & Co Austin Texas pp 361-381

Maden BE (1995) No soup for starters? Autotrophy and the origins of metabolism. Trend Biochem Sci 20: 337-341

Patton A J, Hough DW, Towner P, Danson MJ (1993) Does *Escherichia coli* possess a second citrate synthase gene? Eur J Biochem 214: 75-81

Sharrocks AD, Green J, Guest JR (1991) FNR activates and represses transcription *in vitro*. Proc Roy Soc Lond B 245: 219-226

Wing HJ, Williams SM, Busby SJW (1995) Spacing requirements for transcription activation by *Escherichia coli* FNR protein. J Bacteriol 177: 6704-6710

Part 2

Genomes and their Survival

GENE TRANSFER BY BACTERIAL CONJUGATION: ESTABLISHMENT OF THE IMMIGRANT PLASMID IN THE RECIPIENT CELL

Brian M Wilkins and Steven Bates

Department of Genetics, University of Leicester, Leicester LE1 7RH, UK

Introduction

Bacterial plasmids are extrachromosomal DNA elements that replicate autonomously in their host cells. In addition to containing sectors for replication and maintenance stability, plasmids collectively carry diverse cargoes of specialised genes for functions of environmental, medical and commercial importance. Examples are degradative enzymes, antibiotic-resistance and virulence factors, and DNA restriction enzymes.

Natural plasmids also encode systems for their lateral transfer between bacteria by the process of conjugation. These systems are remarkable for their broad transfer range or 'promiscuity'. As a consequence, conjugation is viewed as a major route for disseminating bacterial genes and for generating genetic diversity in natural populations. Even if the transferred plasmid cannot be maintained autonomously in the new bacterium, the incoming genes may integrate into the genome of the new cell by transposition or some other type of recombination. In addition, some plasmids can insert into the bacterial chromosome to form a covalently continuous DNA structure called an Hfr chromosome. Formation of such cointegrates potentiates conjugative transfer of chromosomal genes and their subsequent incorporation into the resident chromosome of the recipient cell by homologous recombination.

Bacterial conjugation is defined as a plasmid-encoded process in which DNA is transferred from a donor to a recipient bacterium by a specific mechanism requiring cell to cell contact. Considerable progress has been made in the last few years in elucidating mechanistic aspects of the process, with emphasis on enterobacterial systems and early events involving the donor cell. This chapter surveys current understanding of these conjugation systems, with a focus on final stages leading to the establishment of the immigrant plasmid in the new host cell.

NATO ASI Series, Vol. H 103
Molecular Microbiology
Edited by Stephen J. W. Busby,
Christopher M. Thomas and Nigel L. Brown

Paradigms of conjugatively transmissible plasmids

Lederberg and Tatum's choice in the mid-1940s of the K-12 strain of *Escherichia coli* for their pioneering studies of 'sexual' reproduction and recombination in bacteria was remarkably fortuitous in that the strain was unusually fertile through the carriage of a transfer-derepressed plasmid, called sex factor F, which can integrate stably into the bacterial chomosome. The conjugative fertility conferred on *E. coli* K-12 by plasmid F led to the strain becoming the workhorse for seminal studies of bacterial genetics and, subsequently, of molecular biology. Not surprisingly, the transfer (Tra) system of plasmid F is exceptionally well documented (Frost *et al.*, 1994).

Another well studied set of conjugative elements comprise the IncPα group, typified by antibiotic-resistance plasmids isolated in the Birmingham Accident Hospital in 1969 and represented by the similar if not identical RP4 and RK2 plasmids. IncP plasmids stand out for their broad transfer and replication-maintenance ranges. For these reasons, considerable efforts have been invested in elucidating the molecular biology of their transfer with the result that the IncPα system now serves as a paradigm of the conjugation process (Pansegrau *et al.*, 1994).

The conjugation process mediated by IncF and IncPα plasmids can be divided operationally into three stages. The first, described as mating-pair formation, culminates in a specific surface contact between the donor and recipient cells. Effective pairing in turn triggers a series of DNA processing reactions which, in the systems examined to date, cause transfer of a unique plasmid strand to the recipient cell. The final stage is the establishment of the immigrant plasmid in the new host cell. As described here, establishment may be facilitated by plasmid proteins that are transferred from the donor cell or synthesised early in the infected recipient.

There exists a second class of conjugatively transmissible element - the mobilisable plasmid. These elements encode their own DNA processing apparatus but no mating pair-formation system. Hence, a mobilisable plasmid can only be transferred from a donor cell that harbours a coresident conjugative plasmid to determine the intercellular contact necessary for conjugation. Specificity is often observed between the ability of different conjugative plasmids to cause transfer of a mobilisable plasmid, which may reflect the need for a functional match between the mating and DNA processing systems encoded by the separate plasmids. Well characterized mobilisable plasmids are the colicinogenic element ColE1 and members of the IncQ group, exemplified by RSF1010 or R1162. An important feature of IncQ plasmids is their broad host range.

Mating-pair formation (Mpf)

Conjugative pili. The extracellular conjugative pilus is a key feature of enterobacterial mating systems. The F-encoded pilus and its biogenesis are particularly well documented (see review of Frost *et al.*, 1994). The organelle is viewed as a dynamic entity which extends by energy-dependent addition of pilin subunits at its base and retracts by depolymerization of subunits into the inner cell membrane. The structure is thought to establish the contact between the donor and recipient cell and, as a result of retraction, to bring the cells into surface contact. Cells brought into juxtaposition by F pilus retraction are subsequently stabilised in shear- and SDS-resistant aggregates by the products of two *tra* genes. One of these, TraN, is an outer membrane protein containing a surface-exposed domain which might interact with the lipopolysaccharide of the recipient cell (Anthony *et al.*, 1994).

While the extended F pilus consists of an array of subunits of a single polypeptide, namely the processed and N-acetylated product of *traA*, its assembly requires the products of 13 other known transfer genes. Each of these proteins has a specific location in the cell envelope. The F pilus has been observed to emerge from a zone of adhesion between the inner and outer membranes. Possibly the pilus assembly proteins interact at such an adhesion zone to form a complex that supports the secretion and uptake of pilin. In addition, the ensemble of proteins may have a structural role in DNA export. The extended pilus has a hollow core of 2-nm diameter: presence of a similar pore in the basal structure might provide the conduit for transfer of single-stranded DNA across the cell envelope.

The RP4-encoded mating system is likewise complex. Judged by requirements for self-transfer of RP4 between *E. coli* strains, the system consists of products of ten genes in the Tra2 core region, (*trbB, trbC, trbD, trbE, trbF, trbG, trbH, trbI, trbJ* and *trbL*) plus *traF* of the Tra1 region (Fig.1). Requirement of these proteins for pilus assembly, uptake of donor-specific phages and plasmid transfer itself provides circumstantial evidence that the mating system has a direct role in DNA transport. A number of Tra2 products contain potential transmembrane segments or are subject to signal peptide processing, indicating that the proteins are localised in the cell envelope. In confirmation, most of the Tra2 gene products have been detected in both membrane fractions, particularly in the outer membrane. A minority is found in the periplasm; these Tra2 proteins may interact with the membrane proteins to provide a specific transport channel (Haase *et al.*, 1995; J Haase & E Lanka, personal communication).

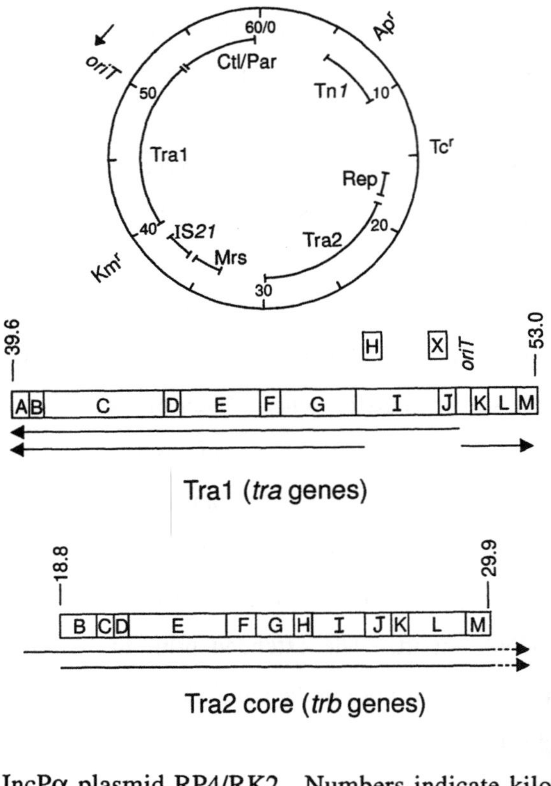

FIG. 1. Map of IncPα plasmid RP4/RK2. Numbers indicate kilobase coordinates. Abbreviations are: Ctl/Par, central control and partition; Mrs, multimer resolution system; Rep, replication; Ap^r, Km^r, Tc^r, resistance to ampicillin, kanamycin and tetracycline, respectively. Tn*1* and IS*21* are transposable elements. The arrow at *oriT* indicates the clockwise direction of DNA transfer. Horizonal lines with arrowheads under the maps of the Tra1 region and the Tra2 core indicate transcripts and the direction of their synthesis (for details, see Pansegrau *et al.*, 1994).

Environmentally important differences exist between the mating systems of F and RP4. The F system is described as 'universal' because it is able to cause bacteria in a liquid medium to conjugate with similar efficiencies to cells on a surface. In contrast, RP4 determines a 'surface-preferred' system that functions more effectively when cells are on a surface rather than in liquid. The pili of the two plasmids are morphologically different in that the F pilus is thick and flexible, whereas the RP4 pilus is thin and rigid.

RP4 pili are detectable on a minority of cells but appear to aggregate in bundles detached from cells. Possibly the extended structure is inessential for conjugation and a short stub suffices to make a mating contact between cells on a surface. This idea is consistent with electron microscopical observations that donor-specific PRD1 phages are in direct contact with the cell surface (Haase *et al.*, 1995).

While most conjugative plasmids examined in *E. coli* specify pilus-like structures, no such extracellular appendage has yet been associated with transfer systems of Gram-positive plasmids. Alternative methods for effecting the mating stage in conjugation have been described. One mechanism, observed for enterococcal pheromone-responding plasmids such as pAD1, involves a surface layer of proteinaceous microfibrils determined by the plasmid. This 'aggregation substance' functions as an adhesin to interact with a 'binding substance', thought to consist of lipotechoic acid, on the surface of the recipient cell (Clewell, 1993).

Interactions with the recipient cell envelope. A number of investigations have addressed the question of whether productive cell contact in *E. coli* conjugation requires a specific molecular structure in the envelope of the recipient. One approach has involved isolation of *E. coli* ConF$^-$ mutants, selected for a deficiency in receiving plasmid F during conjugation in a liquid environment. The mutants were found to be altered either in the lipopolysaccharide or the major outer membrane protein OmpA. However, the normal phenotype was restored when conjugation was allowed to occur on the surface of a solid medium, implying that surface mating bypasses one or more of the stages in liquid mating. The precise contributions of LPS and OmpA remain unclear but LPS might be involved in the initial interaction of the pilus and the recipient cell surface. ConF$^-$ *ompA* mutants show plasmid-specificity, in that they mate efficiently with donors of IncI1 and IncP plasmids. Likewise, LPS mutants show specificity, even for donors of different F-like plasmids. The specificities are not defined by the sequence of the pilin subunit. If the pilus tip is involved in recognising particular moieties in the LPS, there might be an unidentified adhesin at the tip which fulfils the function (Anthony *et al.*, 1994; Frost *et al.*, 1994). The nature of the cellular interactions for conjugation of cells on a solid surface remain to be resolved.

DNA processing systems

Relaxosomes. DNA processing reactions in conjugation have been reviewed in detail elsewhere (Lanka & Wilkins, 1995). Current molecular evidence indicates that transport of a specific single strand of DNA is a unifying feature of conjugation systems examined to date (Fig. 2). Transfer is initiated at a *cis*-acting origin of transfer (*oriT*) site, which may range in size from 38 bp (R1162) to ~250 bp (plasmid F). The transfer initiation complex includes *oriT* and an enzyme called relaxase. This protein catalyses specific cleavage and rejoining within *oriT* at a unique position known as the nick (*nic*) site. In the systems cited as paradigms, the first step in cleavage involves a nucleophilic attack on the phosphoester bond at the nick site by a tyrosine residue in the N-terminal portion of the relaxase. The cleavage product contains the relaxase molecule covalently linked to the 5' end of the opened strand. The energy of the phosphoester bond is conserved in the DNA-relaxase linkage and used in a strand joining activity that reverses the cleavage reaction. Such DNA cleavage-joining reactions are typical of a class of enzymes called DNA strand transferases. The class includes other enzymes with central roles in DNA metabolism, such as DNA topoisomerases.

Relaxosomes are believed to exist as an equilibrium of two negatively supercoiled forms of DNA, one being uncleaved and the other containing the specific cleavage at *nic*. The superhelicity of the second form is thought to be maintained by the molecule of relaxase, which is linked to the 5' end at the cleavage via the phosphotyrosyl bond and believed to hold the 3' end tightly via non-covalent bonds. The structure dissociates when plasmid DNA is purified by methods involving agents that denature proteins, releasing an open circular form of DNA containing the covalently-linked relaxase molecule. Such a plasmid DNA-protein complex that releases open circular DNA was historically called 'relaxation complex'.

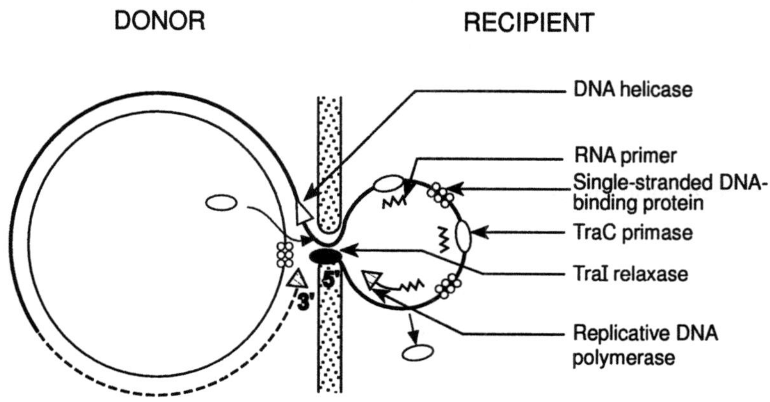

FIG. 2. Scheme of the processing of plasmid RP4 during conjugation. Reactions and proteins are discussed in the text.

Analysis of relaxosomes formed *in vivo* and in reconstitution experiments using purified proteins have shown a requirement for accessory proteins. The additional proteins in the F system are the site-specific DNA-binding proteins TraY and integration host factor (IHF): these form a protein-DNA complex that promotes the binding of relaxase, known as TraI protein. The accessory protein in the RP4 system is TraJ: this binds to negatively supercoiled *oriT* DNA as a prelude to the binding of the cognate relaxase, which is also known as TraI protein. RP4 TraH protein was found to stabilize the complex and TraK to increase the yield of cleaved *oriT* DNA. The relevant RP4 genes are clustered around *oriT* in the relaxase operon (Fig. 1), but on plasmid F, *traI* and *traY* lie at opposite ends of the long 33-kb transfer operon. Naming of the F and RP4 relaxases as TraI proteins is fortuitous and has no functional significance.

In addition to being a relaxase, F TraI protein has interrelated ssDNA-dependent ATPase and helicase (DNA unwinding) activities. In fact, the protein was characterized as DNA helicase I of *E. coli* before its role as a relaxase was defined. TraI helicase is highly processive and proceeds in the 5' to 3' direction on the bound DNA strand. It is often assumed that TraI is the enzyme responsible for unwinding the plasmid strand destined for transfer but this function has yet to be confirmed genetically. If the enzyme is a conjugative helicase, it can be inferred from its 5' to 3' polarity of translocation to move on the strand undergoing export. No IncP Tra protein has been found to have DNA-unwinding activity. Possibly conjugative unwinding of IncP plasmids is mediated by one or more host proteins.

Two of the most mysterious aspects of conjugation are the nature of the signal indicating formation of an effective mating contact and the mechanism that translocates the signal to the relaxosome to initiate the DNA transfer process. F TraD and RP4 TraG are members of a family of proteins which are thought to effect a functional match between the relaxosome and the mating apparatus, thereby facilitating the release of the single-stranded transfer intermediate and its export through the DNA transfer pore. TraG-like proteins contain two highly conserved nucleotide binding motifs. Such features might be expected for a DNA helicase.

Circularisation of the transferring strand. A central feature of the general model of conjugation is that DNA transfer is coupled to synthesis of a replacement strand in the donor cell, as in the rolling circle mode of replication (Fig. 2). In this model, the 5' terminus remains covalently attached to the relaxase molecule responsible for the initiating cleavage at *nic*, while the 3'OH terminus of the transferring strand is elongated continuously by the replicative DNA polymerase of the donor cell. The product is a transfer intermediate that is greater than unit plasmid size. Displacement of a unit length of DNA is envisaged to expose the regenerated *nic* region in single-stranded form, potentiating its recognition by the relaxase molecule linked to the 5' end. The enzyme is then thought to cleave the regenerated *nic* region and join the nascent 3'OH terminus to the 5' terminus to which the protein is linked.

Genetic experiments with small mobilisable plasmids containing two *oriT* sites as direct repeats have been instructive in testing predictions of the rolling circle mechanism. Constructs containing the 38-bp *oriT* of R1162, as developed by RJ Meyer and colleagues, have been particularly revealing.

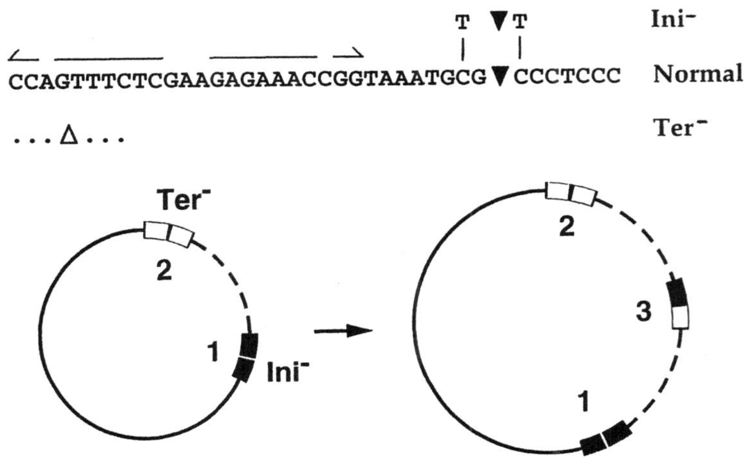

FIG. 3. Formation of greater-than-unit-length plasmids during conjugation. DNA transfer occurs in the clockwise direction. Boxes are *oriT* regions. The continuous line indicates the vector replicon. The broken line represents a spacer containing a reporter gene whose copy number can be measured by a colorimetric test. The nucleotide sequence is of R1162 *oriT*; the wedge (▼) indicates the nick site (adapted from Erickson & Meyer, 1993).

Figure 3 illustrates an example which demonstrates that the 3' terminus generated by the initiating cleavage is extended by replication. One *oriT* site (1; Ini⁻) was mutant with two C to T point mutations (Fig. 3): these prevented initiation of transfer by inhibiting specific cleavage. The second *oriT* (2; Ter⁻) contained a small deletion removing the outside arm of an inverted repeat necessary for terminating a round of transfer. Hence, transfer can only be initiated at a site 2 and terminated at a site 1. Another important feature of the construct is that one of the regions between the *oriT* repeats contained the replicon, which must be transferred to give a plasmid transconjugant, whereas the other region contained a genetically marked spacer.

The great majority of transconjugant colonies were found to harbour plasmids that were greater than unit length and contained three copies of *oriT* and two copies of the spacer. The novel *oriT* site (3) was recombinant, containing the normal sequence except for the C to T transition at position 30. Such a recombinant structure would be generated during rolling circle transfer by the joining of the 5' terminus of a specifically cleaved *oriT* 2 to the 3' terminus of a cleaved *oriT* 1. The requirement for transfer of the replicon sector will be satisfied when termination fails at site 1 in the first round of transfer to occur at a second copy of that site, generated by rolling circle extension of the 3'OH terminus formed by the initiating cleavage (Erickson & Meyer, 1993).

It is unknown whether the specific *oriT* recombination for circularising the transferred DNA occurs in the donor or recipient cell. Most models place the relaxase at the entrance to the DNA transport pore: the perceived advantage is that the protein can scan the transferring DNA for the regenerated *nic* site as a prelude to a concerted cleavage-joining reaction that terminates a round of transfer and generates a circular plasmid. The question of whether or not relaxases enter the recipient cell is discussed in a later section. Physical demonstration of relaxase transfer would require a method capable of detecting a single molecule of a protein in a cell.

Families of transfer systems

One interesting evolutionary relationship exists between the RP4 Tra system and the Vir system of the Ti plasmid. The latter is a native of the plant pathogen *Agrobacterium tumefaciens* and has been studied extensively because it causes tumours in plants. Tumourigenicity involves transfer of a plasmid sector of ~25 kb, called T-DNA, into the nucleus of the plant, where it is integrated into the plant genome. The T-DNA is flanked in the plasmid by two 25-bp direct imperfect repeats, called border sequences, and is transferred in single-stranded form commencing at the right border. This sequence is therefore classified as an *oriT* site.

Comparisons of predicted amino acid sequences and of gene arrangements indicate that the RP4 Tra2 core region is related to the Ti *virB* operon. The latter is thought to encode the mating structure for T-DNA transfer from agrobacteria to plant cells. Furthermore, the RP4 Tra2 and Ti *virB* operons are judged to be homologous to the Ptl operon of *Bordetella pertussis*, which determines export of the pertussis toxin

(Pansegrau *et al.*, 1994). These functionally different systems are inferred to comprise a family of macromolecular transport mechanisms.

Another set of homologies centres on the relaxosome. The border sequences of pTi are evidently homologous to the nick region of RP4 *oriT*, defined as the sector of ~20 nucleotides surrounding the nick site (Fig. 4). Furthermore, the enzyme cleaving the T-DNA borders, namely VirD2, is apparently homologous to the relaxase (TraI) of RP4: the two proteins possess similar enzymic activities and structural arrangements of conserved motifs in their N-terminal portions (Balzer *et al.*, 1994).

RP4 (IncPα)	t t c a c c T A T C C T G ▼C c c g g
R751 (IncPβ)	t t c a c a C A T C C T G ▼C c c g c
R64 (IncI1)	a t t g c a C A T C C T G ▼T c c c g
pTiC58 LB	a c a a t a T A T C C T G ▼C c a c c
pTiC58 RB	c c a a t a T A T C C T G ▼T c a a a
Consensus	Y A T C C T G ▼Y

FIG. 4. Nucleotide sequence alignment of nick regions of the RP4 type. The nick site is indicated by the wedge (▼). Further details are given in Lanka & Wilkins (1995).

Homologies also exist between components of the RP4 system and the transfer system of the IncI1 group of enterobacterial plasmids, typified by the ColIb-P9 and R64. I1 plasmids encode a particularly elaborate transfer system of ~50-kb which is apparently hybrid in specifying two types of conjugative pilus. Homologous loci on RP4 and IncI1 plasmids include the nick region (Fig. 4) and at least two genes specifying relaxosome proteins (Feruya & Komano, 1991; Balzer *et al.*, 1994). These genes are *nikB* of I1 plasmids and its counterpart RP4 *traI*, and *nikA* of I1 plasmids and its homologue RP4 *traJ*.

Microbial eukaryotes as recipients

E. coli donors of F and P plasmids can donate DNA to microbial eukaryotes, such as the yeast *Saccharomyces cerevisiae* (Heinemann & Sprague, 1989). Using an *E. coli*-

yeast shuttle vector based on the 2μm nuclear plasmid of yeast and containing the *oriT* of the cognate conjugative plasmid, we have found that RP4 is considerably more effective in causing trans-kingdom transfers than the systems encoded by sex factor F or the IncI1 plasmid ColIb-P9. Yeast transconjugants acquiring the shuttle vector by RP4-mediated conjugation were recovered at frequencies of up to 10^{-4} per recipient in a one-hour mating. Exponentially growing yeast cells were found to be more competent as recipients than stationary phase cells. Recipient proficiency was relatively haploid specific, reaching an optimum value early in the S phase of the cell cycle. The latter phenomenon might reflect changes in the cell wall at bud emergence (S Bates, BM Wilkins & AM Cashmore; in preparation).

The remarkable fertility of RP4-mediated conjugation of *E. coli* and yeast raises the question of whether or not IncP plasmids contain genes that facilitate transfer between distantly related organisms. Existence of such genes is suggested by the finding that RP4 transfer between *E. coli* strains requires only part of the plasmid's transfer system. Genetic requirements were identified using a binary system in which portions of the Tra1 and Tra2 regions were cloned in compatible vectors and individual genes were inactivated by insertion of a linker containing translational stop codons in all reading frames. This method established that *E. coli* conjugation requires the eleven Mpf proteins (ten from the Tra2 core plus TraF), TraG, and the four Tra1 core products involved in relaxosome function, namely TraH, I, J and K (see Haase *et al.*, 1995).

Using the approach described above, we have found that the essential requirements for *E. coli*-yeast conjugation are the same as those identified for conjugation of *E. coli* by the Lanka laboratory. In particular, only the core of the Tra1 region is needed. What are the roles of the other Tra1 loci? Possibly they are required to promote conjugation under different physiological conditions or to cause effective transfer between other bacterial species, as discussed in the next section.

The factor(s) allowing RP4 to transfer to *S. cerevisiae* with substantially greater frequencies than those obtained for F or ColIb remains to be established. One possibility is that the molecular interactions between donor and recipient cell envelopes are less stringent in the case of RP4-mediated mating. Another hypothesis is that establishment of the immigrant DNA in eukaryotic cells is facilitated by some feature of the RP4 DNA processing system.

Proteins transferred to the recipient to promote plasmid establishment

While there is no general mixing of the cytoplasmic contents of bacteria in conjugation, certain transfer gene products are transported from the donor to the recipient cell to promote establishment of the incoming DNA strand. Specific examples are plasmid DNA primases, and the single-stranded DNA-binding protein (VirE2) and relaxase (VirD2) encoded by the Ti plasmid.

Plasmid DNA primases. DNA primases play an important role in replication by synthesising the short oligoribonucleotides required for *de novo* initiation of DNA strand growth by a DNA polymerase. Some plasmids encode a primase as part of their conjugative transfer system. Classic examples of such enzymes are the larger (210-kDa) Sog polypeptide of ColIb-P9 and the 120-kDa TraC protein of RP4. Experiments with insertion mutations in RP4 *traC* indicate that RP4-encoded primase is only required in conjugation of some bacterial strains and species, where it promotes an event in the recipient cell. This requirement can be satisfied by enzyme made in the donor cell. The explanation is that plasmid primases provide a conjugation-specific priming mechanism that ensures efficient conversion of the transferred strand into duplex DNA in the newly infected cell. In some bacterial strains, host-encoded primer-generating systems can substitute, thereby alleviating the requirement for the plasmid enzyme (Merryweather *et al.*, 1986).

This model has been expanded by the molecular demonstration that plasmid primases are unusual among donor cell proteins in being transferred abundantly and unidirectionally to the cytoplasm of the recipient cell by a conjugation-dependent process. The *sog* and *traC* genes encode large sequence-related polypeptides as separate in-frame translation products. Each of the two Sog proteins (210- and 160-kDa) is transferred with an estimated stoichiometry of 250 molecules per plasmid strand. The enzymes purify as a single-stranded DNA-binding proteins and are probably transmitted to the recipient in a complex with the transferring DNA. Unlike IncP and IncI1 plasmids, most of the other groups of plasmid identified in *E. coli*, including F-type plasmids, lack a DNA primase gene and are apparently transferred devoid of associated proteins (Rees & Wilkins, 1990).

Single-stranded DNA-binding proteins. Proteins are also transferred in pTi-mediated conjugation of agrobacteria and plant cells. One such protein is VirE2, a single-stranded DNA-binding (SSB) protein encoded by the Ti plasmid. VirE2 is inessential for the production of the T-strand and its export from the donor bacterium, but is necessary for the establishment of the T-DNA in the plant cell. Previous models invoked transfer of VirE2 in a complex with the T-strand. However, recent genetic investigations indicate that the protein normally enters the plant cell independently of the T-DNA (Sundberg *et al.*, 1996). Irrespective of the mode of entry, VirE2 and T-DNA are thought to exist in a complex in the plant cell.

Diverse bacterial plasmids of the F and I complexes of incompatibility groups carry *ssb* genes of the type that are homologous to *E. coli ssb.* The similarity is such that the plasmid SSBs can substitute for *E. coli* SSB in bacterial DNA replication and repair. SSBs of this type appear to be non-transmissible, since the method used to demonstrate transfer of plasmid primases failed to detect transfer of a polypeptide of the size of *E. coli* SSB (Rees & Wilkins, 1990). Plasmid SSBs are thought to have a role in the establishment of the immigrant plasmid but the proteins are synthesised in the infected recipient cell, as discussed under leading region genes.

Relaxases. VirD2 of the Ti plasmid is tightly linked to the 5' end of the T-strand and is transferred from the donor bacterium into the nucleus of the plant cell. Here the protein promotes integration of the entrant DNA into a chromosome by a mode of illegitimate recombination that is initiated at a nick or break in host DNA. VirD2 transport raises the question of how the protein crosses the nuclear envelope. This structure consists of inner and outer membranes, apparently composed of conventional lipid bilayers, separated by a perinuclear space.

Nuclear protein transport occurs through pore complexes, which are large, proteinaceous, ring-like structures that traverse the nuclear envelope at sites where the membranes appear to be fused. Protein transport is an active process requiring a nuclear localization signal (NLS) within the polypeptide. A NLS may be located at different positions in a polypetide and typically comprises a short sequence containing a high proportion of positively charged amino acids (arginine and lysine). A number of proteins have been found to have a bipartite signal motif consisting of two basic amino acids, a spacer of ten residues followed by five amino acids of which three must be basic. Such a bipartite motif is found in many nuclear proteins but in few (<5%) non-nuclear proteins (Dingwall & Laskey, 1991).

The C-terminal region of VirD2 contains a bipartite motif, consisting of KR-eRKReR with a spacer of ten residues. The motif is active as a NLS and is thought to target the VirD2-DNA complex to the nuclear pore in a polar fashion. Nuclear uptake may be facititated by VirE2, which contains two bipartite motifs functional in nuclear localisation (Citovsky *et al.*, 1992; Howard *et al.*, 1992).

Such considerations raise the question of how the 2µm-based shuttle vector used in our *E. coli*-yeast conjugation experiments reaches the yeast cell nucleus, which is the normal location for the replication of the 2µm plasmid. Single-stranded circular DNA is transmitted efficiently into this compartment by transformation procedures (Simon & Moore, 1987), and the same may happen during conjugation. However, it is interesting to note that the central portion of RP4 TraI contains a putative bipartite NLS, composed of RRRR-nRIRR with a ten residue spacer. No motif suggestive of a bipartite NLS is apparent in the amino acid sequence of the relaxases of F and ColIb-P9, raising the possibility that the relative efficacy of RP4 in promoting conjugation of *E. coli* and yeast might be due to the capacity of the RP4 relaxase to traverse the nuclear cell envelope.

The barrier of DNA restriction and its evasion

One barrier to the establishment of the incoming plasmid is the restriction systems specified by the recipient cell. Classical restriction-modification (R-M) systems consist of a restriction endonuclease and a matching modification enzyme, or DNA methyltransferase, which recognise the same nucleotide sequence (Bickle & Krüger, 1993). Specific methylation of one or both DNA strands at the recognition sequence protects the DNA from cleavage by the cognate restriction enzyme.

R-M systems can be classified into three types based on enzyme composition, cofactor requirements, symmetry of recognition sequence and DNA cleavage position. Type I systems, which are historically associated with the Enterobacteriaceae and exemplified by *Eco*KI, are the most complex. The main enzyme is composed of three different subunits and catalyses both restriction and modification reactions. DNA cleavage occurs at variable and often considerable distances from the recognition sequences, which are asymmetric and hyphenated. Type II systems, typified by *Eco*RI, are simpler in that the methylase and endonuclease are separate enzymes. A type II endonuclease cleaves DNA within or close to the recognition sequence, which typically possesses two-fold rotational symmetry.

Allowing for different frequencies of restriction enzyme target sites, conjugative transfer of some plasmids is remarkably resistant to restriction compared to phage infections. For example, ColIb-P9 contains 20 *Eco*RI sites and ~7 *Eco*KI sites but transconjugant production by the unmodified plasmid is reduced by less than 80% in recipients specifying these enzymes. The resistance of ColIb is attributed to restriction-evasion mechanisms operating in conjugation.

In contrast to ColIb, transfer of IncP plasmids is remarkably sensitive to restriction, implying that IncP plasmids lack an active restriction-avoidance mechanism. Furthermore IncP plasmids are remarkably deficient of many of the 6-bp palindromes recognised by type II restriction enzymes. The probable explanation is that, in the absence of a restriction-avoidance mechanism, strong selection was imposed for the loss of restriction sites during the evolution of these plasmids (Wilkins *et al.*, 1996).

ColIb-P9 determines two types of restriction-evasion process (Read *et al.*, 1992). One is non-specific in acting against type I and type II restriction systems, and is manifest by a recovery of transconjugant production following an initial eclipse of several minutes. Recovery apparently requires the transfer of more than one copy of the plasmid to the recipient cell, as might occur in conjugation involving a rolling circle mode of DNA transfer. Recent evidence suggests that the recovery reflects a conjugation-induced breakdown of restriction in the mating cells, rather than the overwhelming of the restriction enzyme by substrate saturation (N Althorpe & BM Wilkins, unpublished data). ColIb-P9 encodes a second restriction-evasion mechanism, called ArdA, which operates specifically to alleviate restriction by type I enzymes (see next section).

Leading region genes promoting plasmid establishment

The leading region of a plasmid flanks the *oriT* site and is the first portion of the plasmid to enter the recipient cell during conjugation. Leading regions of diverse enterobacterial plasmids contain some highly conserved genes. Examples are the *ardA* antirestriction gene, and the plasmid *ssb* and *psiB* (plasmid SOS inhibition) loci which are located in a conserved module of ~3.5 kb (Chilley & Wilkins, 1995). While these genes are inessential for conjugation, they are thought to facilitate establishment of the incoming plasmid in certain genetic backgrounds. Another important feature of the genes is that they are expressed in a transient burst following entry into the recipient cell by a process called zygotic induction.

The *ssb* and *psiB* loci and alleviation of the SOS response. The bacterial SOS system is a regulatory network of DNA damage-responsive genes inducible by regions of single-stranded DNA. Plasmid *psiB* genes interfere with SOS induction as a function of the intracellular concentration of their product (Dutreix *et al.*, 1988). To monitor expression of ColIb *psiB* during conjugation, a *psiB*- *lacZ* operon fusion was constructed to allow the level of expression to be assessed through the accumulation of β-galactosidase. This approach showed that *psiB* is expressed at very low levels in vegetative cells but is subject to zygotic induction following entry into the recipient cell. The burst of expression was sufficient to prevent a weak SOS response observed in conjugation when the *psiB* gene was inactive. Thus, *psiB* apparently plays the role of preventing SOS induction by single-stranded DNA transfer without compromising induction of the stress response in cells containing damaged DNA (Jones *et al.*, 1992).

ColIb *ssb* was likewise found to be subject to zygotic induction. The physiological advantage conferred by the burst of plasmid SSB production is not clear. Bacterial SSB is not an abundant protein and the plasmid-determined homologue might alleviate SSB starvation during the processing of the immigrant plasmid (Jones *et al.*, 1992).

The *ardA* antirestriction gene. The ArdA system blocks the restriction function of a type I enzyme without commensurate inhibition of its modification activity. Genetic tests indicated that the system is active at low levels in vegetative cells harbouring ColIb but operates effectively when the plasmid is transferred by conjugation and not by transformation. The alleviation process requires expression of *ardA* in the newly infected recipient cell (Read *et al.*, 1992).

We have found that *ardA* is subject to zygotic induction, as assessed by the sensitive method of competitive reverse transcription-PCR. This procedure involved rapid isolation of RNA from conjugating cells, its conversion into DNA by reverse transcription (RT), followed by amplification of the DNA by PCR to give a 'test' product detectable by gel electrophoresis. The amount of test product was standardized against a constant amount of internal 'competitor'. The latter was derived by the RT-PCR procedure from RNA isolated from vegetative cells harbouring a recombinant plasmid with a synonymous mutation in *ardA*. The purpose of the mutation was to eliminate a restriction enzyme-cleavage site in *ardA*, thereby allowing the amount of test DNA to be distinguished from the uncleavable competitor DNA.

The results (Fig. 5) show that the amount of *ardA* mRNA increased rapidly during conjugation to reach a peak at 11 min. This corresponded to the time of maximum transfer activity in the population of cells (PM Chilley & BM Wilkins, in preparation). The possible advantage of zygotic induction is that it causes alleviation of restriction in the transconjugant cell while allowing activity of the restriction enzyme to recover in the resulting cell line. An important question is how does ArdA accumulate sufficiently rapidly to prevent destruction of the incoming plasmid. One possible answer centres on our finding that *ardA, psiB* and *ssb* are consistently orientated on natural plasmids such that the transcribed DNA strand is the transferred strand. We have found using inversion mutations that zygotic induction of *psiB* and *ssb* is orientation dependent. This finding raises the intriguing possibility that the leading portion of the transferring DNA strand is transcribed at optimal rate before the complementary strand is synthesised. The transferred strand itself will be insensitive to restriction because the substrate for restriction enzymes is duplex DNA (RA Roscoe & BM Wilkins, in preparation).

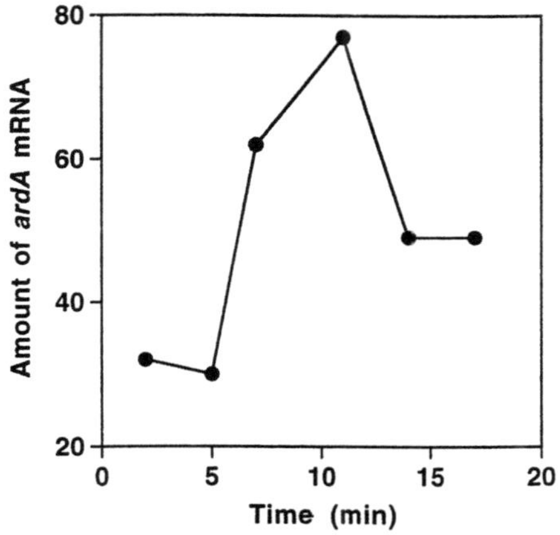

FIG. 5. Measurement by competitive RT-PCR of the relative abundance of *ardA* transcripts at various times in bacterial conjugation. Equal numbers of ColIb*drd-1* donors and recipient cells were mixed at t=0. Values on the ordinate refer to the amount of test RNA relative to a constant amount of competitor RNA, expressed as a percentage (data of PM Chilley).

Protection of the transconjugant cell from lethal zygosis

While most genes in a plasmid transfer region act to promote conjugation, one or two loci determine a system-specific barrier to transfer called entry or surface exclusion. Exclusion operates to reduce the transfer of a plasmid between cells containing the same or a closely related plasmid. Plasmid F carries two exclusion genes, *traS* and *traT*, which function additively. TraT is an abundant lipoprotein found in the outer cell membrane. The protein inhibits formation of mating aggregates, possibly by interacting with the putative adhesin at the tip of the pilus (Anthony *et al.*, 1994). TraS is an inner membrane protein which is inferred to prevent triggering of DNA transfer. Presumably TraS blocks some signal that is normally generated by the formation of a conjugative contact in the recipient and transduced to the relaxosome to initiate DNA export (Frost *et al.*, 1994).

One role of exclusion is inhibition of wasteful conjugation of cells containing the same plasmid. However, exclusion functions may be important in protecting the newly infected recipient cell from lethal zygosis. This killing phenomenon was observed when recipient cells were mixed with an excess of Hfr donors. Lethality was shown to be associated with a variety of physiological defects in the recipient, all of which might stem from alterations in the inner cell membrane caused by repeated acts of mating. Recipents mixed with an excess of donors carrying F in the autonomous state were immune to lethal zygosis, apparently due to acquisition of *traS* and/or *traT* (Skurray & Reeves, 1973; 1974; RA Skurray, personal communication). An explanation of the difference between the ability of F^+ and Hfr matings to induce killing is that all F^+ transconjugants receive the Tra region as an integral portion of the transferred plasmid. In Hfr-mediated conjugation, the Tra region is the last portion of the 4.7-mb cointegate to enter the recipient cell; in practice, the Tra region is transferred very rarely because ongoing cycles of transfer terminate spontaneously before the entire chromosome is transmitted. Hence, transfer of exclusion genes is rarely achieved in Hfr conjugations.

Acquistion of immunity to lethal zygosis may be particularly important when conjugating cells are in a surface-associated community. Imagine a plasmid sweeping through a microcolony of recipient cells in a wave of conjugation events. In the absence of a system for preventing repeated matings, newly infected cells might participate in repeated cycles of mating with neighbours, thereby accumulating membrane damage. Expression of the exclusion phenotype by transconjugant cells would confer protection.

Summary

Compared to mating-pair formation and the initiation of DNA transfer, understanding of events leading to the establishment of the immigrant plasmid is fragmentary. Most of the information has stemmed from the serendipitous discovery of plasmid genes, such as *sog, ssb, psiB* and *ardA*, which affect DNA metabolism in the newly infected cell. A more systematic study of the effects of conjugation on the physiology of the recipient cell may reveal important strategies for alleviating stress on the infected cell and for facilitating the installation of the incoming plasmid.

Acknowledgements

Unpublished work cited in this article was obtained with the support of MRC grant G9321196MB.

References

Anthony KG, Sherburne C, Sherburne R, Frost LS (1994) The role of the pilus in recipient cell recognition during bacterial conjugation mediated by F-like plasmids. Mol Microbiol 13:939-953

Balzer D, Pansegrau W, Lanka E (1994) Essential motifs of relaxase (TraI) and TraG proteins involved in conjugative transfer of plasmid RP4. J Bacteriol 176:4285-4295

Bickle TA, Krüger DH (1993) Biology of DNA restriction. Microbiol Rev 57:434-450

Chilley PM, Wilkins BM (1995) Distribution of the *ardA* family of antirestriction genes on conjugative plasmids. Microbiology 141:2157-2164

Citovsky V, Zupan J, Warnick D, Zambryski, P (1992) Nuclear localization of *Agrobacterium* VirE2 protein in plant cells. Science 256:1802-1805

Clewell DB (1993) Bacterial sex pheromone-induced transfer. Cell 73:9-12

Dingwall C, Laskey RA (1991) Nuclear targeting sequences - a consensus? Trends Biochem Sci 16:478-481

Dutreix M, Bäckman A, Célérier J, Bagdasarian MM, Sommer S, Bailone A, Devoret R, Bagdasarian M (1988) Identification of the *psiB* genes of plasmids F and R6-5. Molecular basis for *psiB* enhanced expression in plasmid R6-5. Nucl Acids Res 16:10669-10679

Erickson MJ, Meyer RJ (1993) The origin of greater-than-unit-length plasmids generated during bacterial conjugation. Mol Microbiol 7:289-298

Feruya N, Komano T (1991) Determination of the nick site at *oriT* of IncI1 plasmid R64: global similarity of *oriT* structures of IncI1 and IncP plasmids. J Bacteriol 173:6612-6617

Frost LS, Ippen-Ihler K, Skurray RA (1994) Analysis of the sequence and gene products of the transfer region of the F sex factor. Microbiol Rev 58:162-210

Haase J, Lurz R, Grahn AM, Bamford DH, Lanka E (1995) Bacterial conjugation mediated by plasmid RP4: RSF1010 mobilization, donor-specific phage propagation, and pilus production require the same Tra2 core components of a proposed DNA transport complex. J Bacteriol 177:4779-4791

Heinemann JA, Sprague GF (1989) Bacterial conjugative plasmids mobilize DNA transfer between bacteria and yeast. Nature 340:205-209

Howard EA, Zupan JR, Citovsky V, Zambryski PC (1992) The VirD2 protein of A. tumefaciens contains a C-terminal bipartite nuclear localization signal: implications for nuclear uptake of DNA in plant cells. Cell 68:109-118

Jones AL, Barth PT, Wilkins BM (1992) Zygotic induction of plasmid *ssb* and *psiB* genes following conjugative transfer of IncI1 plasmid ColIb-P9. Mol Microbiol 6:605-613

Lanka E, Wilkins BM (1995) DNA processing reactions in bacterial conjugation. Annu Rev Biochem 64:141-169

Merryweather A, Barth PT, Wilkins BM (1986) Role and specificity of plasmid RP4-encoded DNA primase in bacterial conjugation. J Bacteriol 167:12-17

Pansegrau W, Lanka E, Barth PT, Figurski DH, Guiney DG, Haas D, Helinski DR, Schwab H, Stanisich VA, Thomas CM (1994) Complete nucleotide sequence of Birmingham IncPα plasmids: compilation and comparative analysis of sequence data. J Mol Biol 239:623-663

Read TD, Thomas AT, Wilkins BM (1992) Evasion of type I and type II restriction systems by IncI1 plasmid ColIbP-9 during transfer by bacterial conjugation. Mol Microbiol 6:1933-1941

Rees CED, Wilkins BM (1990) Protein transfer into the recipient cell during bacterial conjugation: studies with F and RP4. Mol Microbiol 4:1199-1205

Simon, JR, Moore PD (1987) Homologous recombination between single-stranded DNA and chromosomal genes in *Saccharomyces cerevisiae*. Mol Cell Biol 7:2329-2334.

Skurray RA, Reeves P (1973) Physiology of *Escherichia coli* K-12 during conjugation: altered recipient cell functions associated with lethal zygosis. J Bacteriol 114:11-17

Skurray RA, Reeves P (1974) F factor-mediated immunity to lethal zygosis in *Escherichia coli* K-12. J Bacteriol 117:100-106

Sundberg C, Meek L, Carroll K, Das A, Ream W (1996) VirE1 protein mediates export of the single-stranded DNA binding protein VirE2 from *Agrobacterium tumefaciens* into plant cells. J Bacteriol 178:1207-1212

Wilkins BM, Chilley PM, Thomas AT, Pocklington MJ (1996) Distribution of restriction enzyme recognition sequences on broad host range plasmid RP4: molecular and evolutionary implications. J Mol Biol 258:447-456

Bacteriophage Mu

Martha M. Howe

Department of Microbiology and Immunology, The University of Tennessee-Memphis, Memphis, TN 38163, USA

1 Introduction

Bacteriophage Mu is a temperate phage of *Escherichia coli K-12* and several other enteric bacteria (for a comprehensive monograph and references prior to 1987 see Symonds et al., 1987; for reviews see Howe, 1987a; Harshey, 1988; Pato, 1989). As is typical for temperate phages, Mu infection can result in one of two alternative modes of development. If genes needed for Mu DNA replication and synthesis of phage morphogenetic and lytic functions are expressed, the phage will enter the lytic cycle, culminating with cell lysis in 45 to 60 minutes and release of ~100 phage particles per cell. On the other hand, if sufficient Mu repressor protein is made to prevent Mu replication and lytic gene expression, the Mu genome can enter the lysogenic cycle in which it integrates into the host chromosome and forms a repressed prophage which is passively replicated as part of the host chromosome. Cultures of a Mu lysogen contain a small number of spontaneously released phage particles; however, exposure to DNA damaging agents does not cause the prophage to enter the lytic cycle. In the laboratory, temperature-sensitive mutations in the Mu repressor gene (*c*ts) are used to confer heat-inducibility; upon transfer of the lysogenic culture to high temperature (usually 42°C), lytic development ensues.

At the time of its discovery in the early 1960's, Mu was recognized to possess an unusual property; about 1% of the lysogenic cells surviving Mu infection had new auxotrophic mutations. This ability of the phage to cause mutations in host genes upon lysogenization led to its name Mu, for *mu*tator (Taylor, 1963). These Mu-induced mutations arise by integration of the Mu genome into a gene, disrupting the continuity of the gene and preventing its function. It is the ability of Mu to integrate into essentially random sites in the host chromosome which leads to the broad spectrum of mutations caused by Mu and to the interest in understanding its integration mechanism.

The mature Mu phage particle consists of an icosahedral head, a short neck, a contractile tail with six tail fibers, and a noncontractile tail core (Fig. 1; for reviews and references see Howe, 1987; Harshey, 1988). The tail fibers confer the Mu host range specificity by interacting with the host cell wall lipopolysaccharide which serves as the receptor. Presumably, interaction of the tail fibers or base plate with the host cell wall triggers tail contraction, leading to injection of the tail core through the cell wall and release of the phage DNA into the cytoplasm.

NATO ASI Series, Vol. H 103
Molecular Microbiology
Edited by Stephen J. W. Busby,
Christopher M. Thomas and Nigel L. Brown

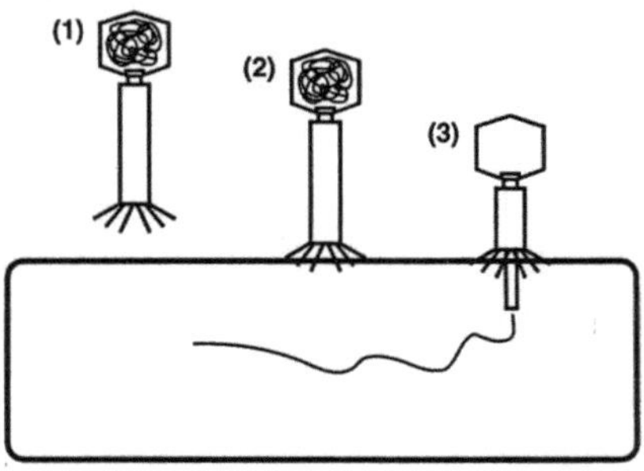

FIG. 1. Schematic representation of Mu adsorption and injection. (1) Diffusion of a Mu particle through the solution surrounding a sensitive cell; (2) interaction of the Mu tail fibers with the lipopolysaccharide layer of the cell wall; (3) contraction of the tail sheath and injection of Mu DNA into the cytoplasm.

2 Mu DNA

Mu DNA isolated from the mature phage particle is linear, double-stranded, and approximately 39 kb in length, with an estimated coding capacity of 35 to 45 genes (for references see Grundy and Howe, 1984; Howe, 1987a). Early studies of Mu DNA by electron microscopy of DNA heteroduplexes resulting from denaturation and reannealing revealed three regions of heterogeneity between DNA molecules isolated from different phage particles (Fig. 2). One region, called the invertible G segment, gives rise to single-stranded bubbles in heteroduplexes arising from annealing of DNA molecules containing the G segment in opposite orientations. Inversion of the G segment is catalyzed by the Mu Gin protein.

The other two heterogeneous regions are found at the ends of the mature phage DNA and consist of a random assortment of host DNA sequences which are packaged into the phage particle during packaging from a precursor in which the Mu genome is integrated into host DNA (Fig. 2). As will be discussed later, Mu DNA replicates by a process called replicative transposition, in which the Mu genome is always integrated within host DNA sequences and multiple independent copies of Mu DNA accumulate within the host chromosome. These copies are then packaged directly from this precursor by recognition of a packaging site (*pac*) near the left end of the Mu genome and packaging of a headful of DNA (Bukhari and Taylor, 1975; Harel et al., 1990). Cleavage of the precursor occurs within host DNA 50-150 bp to the left of the Mu genomic left end and 0.5-3.0 kb to the right of the genomic right end. The variable length of the right end host DNA segment is believed to result from the inherent imprecision in length determination of a headful packaging mechanism. Cutting within nearby host DNA sequences may serve to ensure that the Mu attachment sites (*attL* and *attR*), which are located at the ends of the Mu genome and needed for subsequent integration and replication, will be present in each phage particle. As shown in Figure 2, the host sequences present in the infecting Mu DNA are lost during the initial integration process.

Another unusual feature of Mu DNA is the presence of a novel modified base which protects Mu DNA from restriction when Mu particles made in one bacterial species then infect a different species (for reviews see Kahmann, 1984; Kahmann and Hattman, 1987). The Mu Mom protein catalyzes the modification of ~15% of the

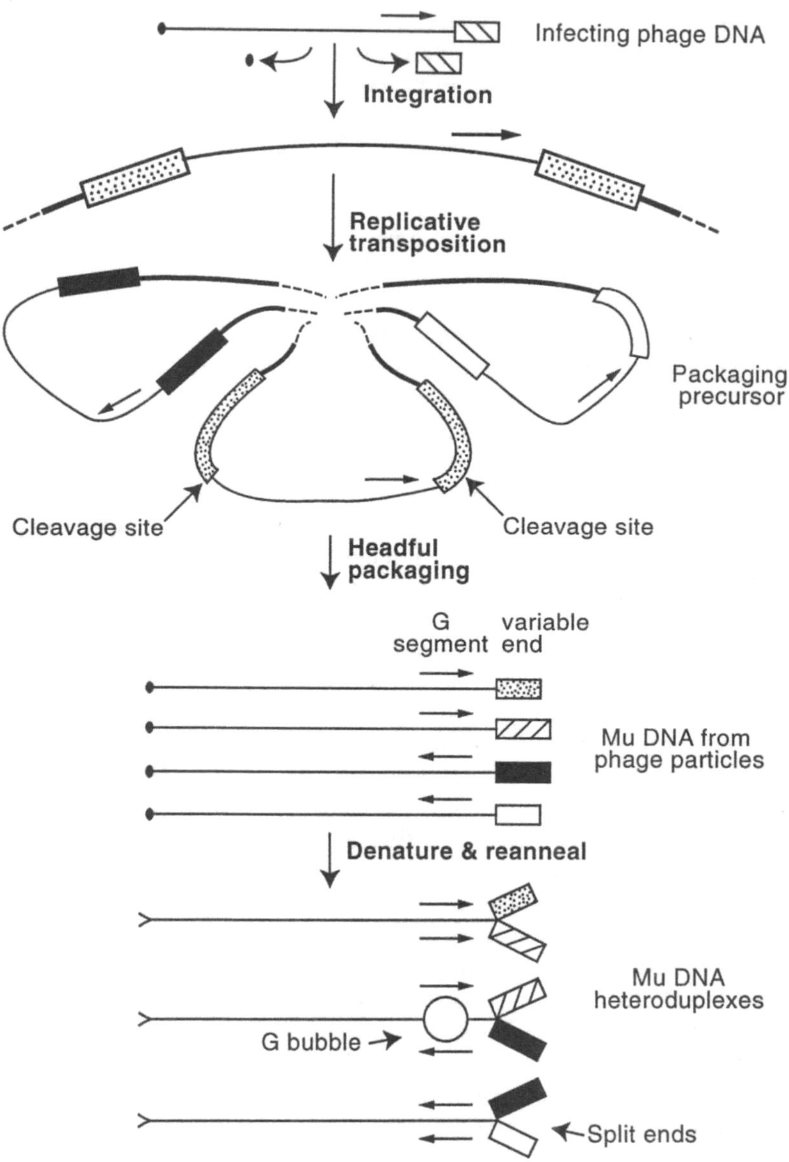

FIG. 2. Changes in Mu DNA structure arising during integration, replicative transposition, headful packaging, and following denaturation and reannealing of mature phage DNA. Boxes represent host DNA sequences flanking the Mu DNA; the short segment of host DNA at the left end of mature Mu DNA is indicated by a filled ellipse. These host DNA segments are found in single-stranded form in reannealed molecules. The easily visible single strands at the right end are referred to as split ends; due to their short length those at the left end are rarely seen. Horizontal arrows indicate the orientation of the invertible G segment, which gives rise to a single-stranded G bubble in DNA heteroduplexes generated by annealing of DNA strands containing G in opposite orientations.

adenine residues in the phage DNA to form acetamidoadenine. This widespread modification protects Mu DNA from cleavage by a number of restriction enzymes, both in nature and in the test tube. Inspection of the recognition sites for restriction enzymes which are affected by this modification led to the following proposed consensus sequence for Mom modification: C/G A G/C N Py.

3 Location and Function of Mu Genes

A genetic map of Mu showing the transcriptional organization of the genes is presented in Figure 3. The map shows two features which Mu shares with many viruses: (i) genes with related functions are located close to each other on the genome, resulting in a modular genome organization, and (ii) groups of viral genes are expressed sequentially, leading to their designation as early, middle, and late genes.

Genes *c* and *ner* at the left end encode repressors which function during lysogenic and lytic development, respectively (for reviews and references see Pato, 1989; Toussaint et al., 1994). The c protein maintains the dormant state of the Mu prophage in a lysogen and is responsible for the immunity of a lysogen to superinfection. The Ner protein functions during lytic development to reduce transcription from the promoters P_e and P_c. Ner function is critical to maintaining appropriate levels of the unstable A protein needed for replicative transposition, while at the same time reducing transcription of *c* and *attL* to levels which no longer inhibit transposition.

Genes A and B are required for integration and replicative transposition of Mu DNA; the A protein binds to the Mu *att* sites and catalyzes the strand transfer reaction; the B protein interacts with A and stimulates its activity. The B protein is also responsible for transposition immunity, the tendency of Mu to avoid transposing into its own DNA (Adzuma and Mizuuchi, 1989 and references therein). The roles of A and B in transposition will be discussed in detail in a later section.

The region between *B* and *C* (note that *C* and *c* are different genes) is often referred to as the semi-essential region since deletion and insertion mutations in this region significantly reduce but do not abolish phage growth (for a review see Paolozzi and Symonds, 1987). This region contains a number of open reading frames that were identified by DNA sequence analysis; several have been named based on the phenotypic effects of mutations. These include *kil*, involved in host cell killing; *gam*, inhibition of exonuclease digestion of Mu DNA; *sot*, stimulation of Mu DNA transfection efficiency; *arm*, amplified replication of Mu DNA; and *gem*, gene expression modulation (Ghelardini et al., 1989). The precise roles of many of these proteins in Mu growth are not yet understood.

The *mor* gene encodes an activator protein required for initiation of middle transcription from the middle promoter P_m (for references see Kahmeyer-Gabbe and Howe, 1996). Of the four open reading frames in the middle transcript, only the last, *C*, is essential for Mu growth; *C* encodes the activator for late transcription.

The *lys* gene is required for lysis of the host cell at the end of lytic development. The genes between *lys* and the G segment are involved in morphogenesis of the phage particle, encoding structural proteins of the phage head and tail and functions involved in their assembly (for reviews and references see Howe, 1987b; Harshey, 1988). Large

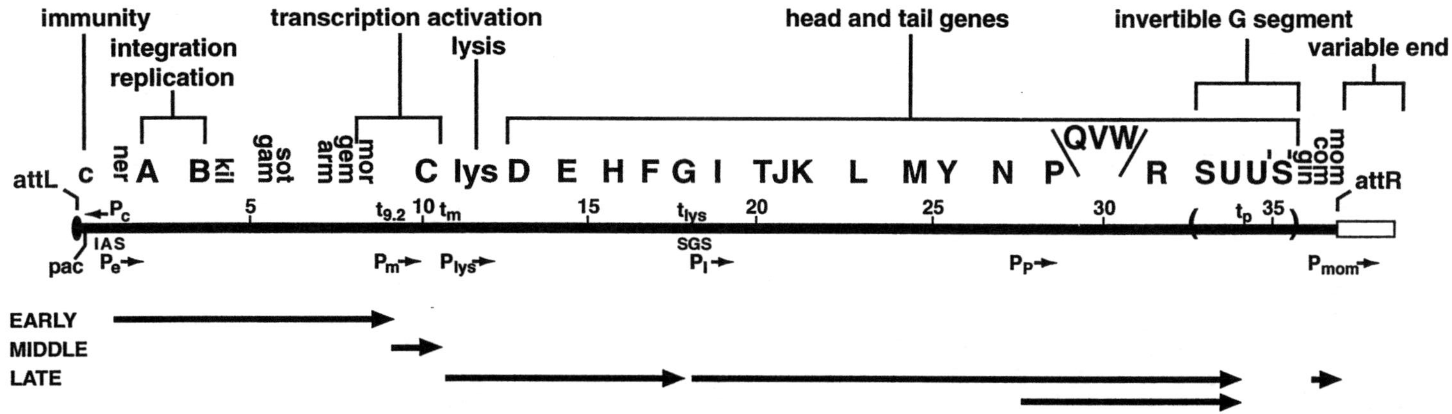

FIG. 3. Genetic map of bacteriophage Mu. The thick line with tick marks at 5-kb intervals represents the Mu genome; the filled ellipse and open box at the left and right ends, respectively, represent the attached host DNA sequences present in mature phage particle DNA. Mu genes and some of their functions are shown above the Mu genome. Promoters are designated P with an arrow showing the direction of transcription; terminators are designated t. Larger arrows below the map indicate the early, middle, and late transcripts made during lytic growth. For a more complete description of the functions of specific sites and genes, see the text.

portions of this region have not yet been sequenced, and our understanding of the function of specific gene products is still at a rather primitive stage. Electron microscopic analyses of lysates produced by phages defective in these genes suggest that Mu heads and tails assemble independently, with preheads being formed, filled with DNA, and then attached to tails. In the absence of tail fibers, tail fiberless but otherwise morphologically normal phage are made, suggesting that attachment of tail fibers to the assembled particles may be the final step in particle assembly.

The G segment contains two pairs of genes: S+U and S'+U' which encode two different sets of tail fibers that interact with different lipopolysaccharide receptors (Fig. 4; for reviews and references see Grundy and Howe, 1984; Koch et al., 1987). Since only the upstream pair of genes (adjacent to *R*) is expressed, the orientation of the G segment during lytic development determines the host range of the phage particles which are produced. Expression of S+U leads to particles able to adsorb to *E. coli K12*; whereas, expression of S'+U' confers an ability to infect different strains, such as *E. coli C, Enterobacter cloacae* and *Serratia marcescens.* Inversion of the G segment via site-specific recombination between the *gixL* and *gixR* sites at its ends is catalyzed by the Mu Gin protein and assisted by the host FIS protein. This G-*gin* system shares many features in common with other phage and bacterial inversion systems and has served as an excellent model for molecular dissection of the inversion process (Glasgow et al., 1989; van de Putte and Goosen, 1992).

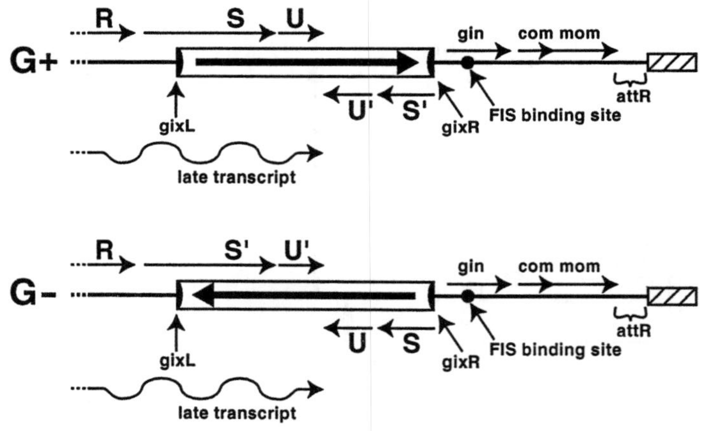

FIG. 4. Organization of genes and sites in the right end of Mu DNA containing the G segment in opposite (G+ and G-) orientations. Horizontal arrows represent open reading frames encoding the indicated proteins. The open box represents the invertible G segment with recombination sites *gixL* and *gixR* at its ends; the arrow inside the box indicates the orientation of the G segment. Note that only the C-terminal parts of S and S' are encoded within the invertible segment; the N-terminal amino acid sequences are encoded upstream of *gixL* and are identical in S and S'. The hatched box at the right end represents the Mu variable end host DNA.

The last two genes at the right end are *com* and *mom*, which are involved in regulation and synthesis, respectively, of the previously mentioned Mu-specific DNA modification.

There are also a number of sites at which specific Mu and host proteins act, including the previously mentioned *pac* site used in DNA packaging. The promoters and terminators which support initiation and termination of transcription, respectively, by the host RNA polymerase holoenzyme will be discussed later. The attachment sites *attL* and *attR* are roughly 0.2-kb regions which each begin at the respective terminal Mu nucleotide and include three binding sites for the Mu A protein. The IAS site (internal activating sequence), located 1 kb from the Mu left end, binds both Mu A and the host IHF (integration host factor) protein and is required for high levels of replicative transposition; deletion of the IAS reduces transposition to 1% of normal (for references see Baker, 1993). Lastly, the centrally-located strong DNA gyrase binding site, SGS, is required for normal levels of replicative transposition and may play a role in bringing the Mu attachment sites together for assembly of the transposition complex (Pato and Banerjee, 1996 and references therein).

4 Regulation: Lytic vs Lysogenic Development

Whether Mu infection results in lytic development or lysogeny depends on a number of complex regulatory interactions within the Mu early regulatory region (for references see Baker, 1993; Toussaint et al., 1994; van Ulsen et al., 1996). This region contains binding sites for the c and Ner repressors, the Mu A transposase, and the host IHF protein as well as two promoters recognized by the host RNA polymerase: P_c for c repressor biosynthesis and the early promoter P_e (Fig. 5). Binding of the Mu c repressor to multiple sites within the three operators (O1, O2, and O3) is cooperative and begins at low repressor concentrations with binding to O2; as the repressor level increases there is additional binding to O1 and then O3 to form a highly organized nucleoprotein complex (Krause and Higgins, 1986; Rousseau et al., 1996). Mu repressor bound at O2 inhibits transcription from P_e, thereby preventing early transcription in a lysogen. Repressor bound at O3 inhibits P_c, resulting in autoregulation of repressor levels in a lysogen. Binding of IHF to a site between O1 and O2, and the resulting DNA bend it introduces, stabilize repressor binding. In the absence of IHF, binding of the host H-NS protein occurs more broadly over the entire region to form a large nucleoprotein complex; this results in stabilization of the repressor-DNA complex and inhibition of transcription from both P_e and P_c.

In the absence of Mu repressor, eg., after heat induction of a Mu *c*ts lysogen, binding of H-NS to this region at least partially represses P_e (van Ulsen et al., 1996). Under these conditions, binding of IHF between O1 and O2 serves two functions i) IHF displaces H-NS, thereby relieving its repressive effect and ii) IHF directly interacts with host RNA polymerase bound at P_e, resulting in a stimulation of early transcription from P_e. Negative supercoiling also acts as a positive regulator of P_e and negative regulator of P_c (Krause and Higgins, 1986; Higgins et al., 1989). Once early transcription begins, several activities serve to reduce transcription from P_c:

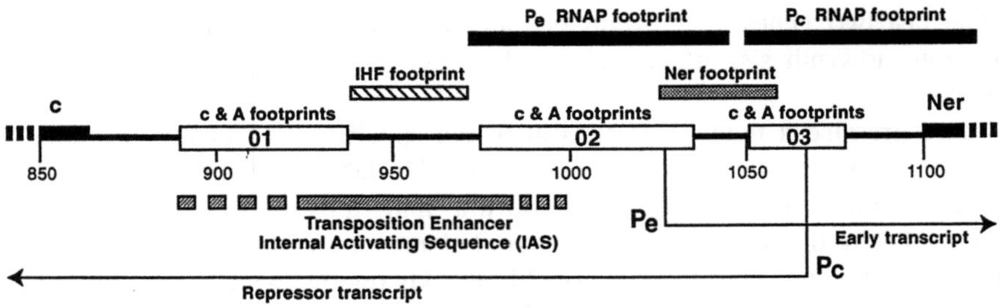

FIG. 5. The Mu early regulatory region. Boxes labelled footprint represent binding sites for the indicated proteins; O1, O2, and O3 are the operatorss for binding of multiple c repressor monomers during establishment and maintenance of the lysogenic state. P_e and P_c are the promoters for the early and repressor transcripts, respectively, with arrows indicating the direction of transcription. The filled dashed boxes at the ends of the diagram indicate the beginning of coding sequences for the *c* and *ner* genes. Numbers below the line are distances (in bp) from the left end of the Mu genome.

i) polymerases initiating at P_e either collide with those initiating at P_c or sterically occude binding of RNA polymerase to P_c, ii) the presence of negative supercoiling stimulates P_e and inhibits P_c, and iii) Ner protein produced from the early transcript binds between P_e and P_c to reduce transcription from both promoters. Binding of Ner is essential for normal phage development, possibly because transcription initiating at P_c may continue through *attL* and interfere with transposase binding to *attL* (Goosen and van de Putte, 1986).

5 Transcription during Lytic Development

In cells committed to lytic phage development, transcription of the 8.2-kb early region from P_e to the Rho-independent terminator $t_{9.2}$ reaches a peak at 4 to 8 minutes after heat induction and then declines due to the repressive action of Ner on P_e. Concomitantly, Mu DNA replication begins at approximately 8 minutes and increases rapidly until the end of the lytic cycle. Middle transcription from P_m also begins about 8 minutes after induction and requires the host RNA polymerase holoenzyme, the Mu-encoded activator protein Mor, and a replicating DNA template (for references see Kahmeyer-Gabbe and Howe, 1996). The 1.2 kb middle transcript terminates at the Rho-independent terminator t_m. Within minutes, C protein produced from the middle transcript binds to the four late promoters P_{lys}, P_I, P_P and P_{mom} and leads to activation of late transcription by the host RNA polymerase holoenzyme (for references see Zha et al., 1994).

Transcripts beginning at P_{lys} terminate ~7.6 kb downstream at t_{lys}, a Rho-independent terminator located just upstream of P_I (Zha et al., 1994). Transcripts initiating at P_I are extremely long and do not terminate before P_P which is located *within*, not *after*, the essential *N* gene. It appears that N must be translated from mRNA initiating at P_{lys} and continuing through P_P. Thus, P_P may function as an internal promoter to increase the level of mRNA for genes downstream of *N*. The ~6.3-kb P_P transcript terminates at t_P, which is a bidirectional terminator located between genes *U* and *U'* (Zha et al., 1994). Presumably those polymerases which initiated at P_I and continued transcription through P_P and beyond without falling off the DNA template also terminate at t_P, resulting in an unusually long ~15-kb transcript. At present, there are no other factors known to be involved in regulating transcription initiation from P_{lys}, P_I, or P_P; however, there has been little analysis of P_I or P_P, and none of the late promoters has been tested for possible dependence of transcription on DNA replication. In contrast, P_{mom} is subject to additional negative regulation by the host OxyR protein, which binds to P_{mom} when three GATC sites within P_{mom} are not modified by the host Dam methylase; thus, the absence of methylation leads to repression of transcription (Bolker et al., 1989 and references therein). Curiously, translation of Mom protein from the transcript is also negatively regulated; this occurs by formation of an RNA hairpin which sequesters the *mom* translation initiation region, reducing Mom translation. Com protein translated from the upstream region of the P_{mom} transcript binds to the RNA and prevents formation of the hairpin, allowing Mom translation to occur (Wulczyn and Kahmann, 1991 and references therein). It is argued that these regulatory mechanisms work together to prevent high-level expression of Mom until late in the lytic cycle since Mom expression is deleterious to the host cell.

6 Integration

Although integration and transposition of Mu DNA utilize many of the same sites and functions, there are significant differences in the processes by which they occur (for reviews and references see Mizuuchi and Craigie, 1986; Harshey, 1988; Pato, 1989; Haniford and Chaconas, 1992; Mizuuchi, 1992; Baker, 1993). The term "integration" is used for the process undergone by the mature phage DNA following its injection into the cell by an infecting phage particle. The injected DNA is linear and does not form a covalently closed circular molecule typical of many other infecting phage DNAs. Instead, in a subset of the molecules, the ends of Mu DNA appear to be held together by the Mu N protein, which is non-covalently bound at or near the DNA ends (Gloor and Chaconas, 1986 and references therein). This binding is sufficiently strong to allow stable supercoiling of the circular molecule, a likely prerequisite for integration. Whether this non-covalent circular form actually is an intermediate in the integration pathway is not yet known.

Analysis of the methylation state of the infecting phage DNA immediately after integration revealed that there is little or no replication during integration (Harshey, 1984). Thus, integration is referred to as a 'conservative', 'cut and paste', 'simple transposition' process. It requires the Mu A protein and both attachment sites and is stimulated by the Mu B protein. One possible model for events occurring during integration is shown in Figure 6. The Mu attachment sites are recognized by the Mu A transposase protein and single-strand nicks are introduced at the 3' ends of each Mu strand. At the same time staggered nicks are introduced into the target DNA 5 bp apart to generate 5-nt single-stranded 5' extensions. The strand transfer event involves joining of the 5' ends of target DNA to the 3' ends of Mu DNA. In the case of Mu

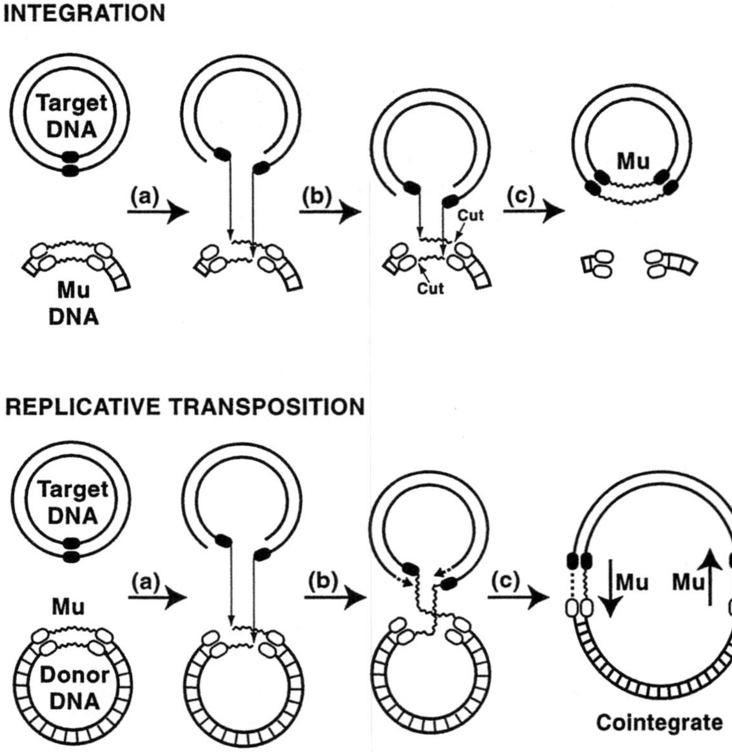

FIG. 6. Proposed models for events occurring during integration and replicative transposition of Mu DNA. These models contain many elements of the seminal models of Grindley and Sherratt (1979) and Shapiro (1979) with modifications suggested by many transposable element researchers. The two strands of Mu DNA are depicted by wavy lines, attached host DNA by solid cross-hatched lines, target DNA by solid lines, and newly synthesized strands by dashed lines. The open and filled ellipses represent the 5-bp sequence of target DNA which is duplicated during integration and transposition and found at the ends of the integrated Mu DNA. In step (a) in both models, single-strand nicks are introduced at the 3' ends of the Mu DNA strands; these 3' ends are then covalently joined to the protruding 5' ends produced by 5-bp staggered cleavage of the target DNA. For INTEGRATION, steps (b) and (c) involve cutting of the second strand at each end of Mu DNA, repair replication to fill in the 5-nt single-stranded gaps, and ligation to produce the covalently closed circular molecule containing the integrated Mu. The host DNA sequences originally attached to Mu DNA are released and lost. For REPLICATIVE TRANSPOSITION, steps (b) and (c) involve use of the 3' host DNA ends generated by the staggered cleavage of target DNA as primers for semi-conservative replication through the Mu DNA and ligation to the free 5' donor DNA ends to form the covalently closed cointegrate molecule.

integration (and simple transposition of other transposable elements) it is proposed that single-strand nicks are then introduced into the second strand at each Mu (or element) end to release the flanking DNA, which, in the case of Mu, would be the host DNA ends. Repair replication to fill in the 5-nt single-stranded gaps completes the process, generating the 5-bp direct repeats of target DNA found immediately flanking the integrated Mu. It is presumed that the host DNA sequences which are released are either degraded by nucleases or diluted away during growth of the cells. In "mini-Mu" phages in which large internal portions of the Mu DNA have been deleted, the very long variable end host DNA segments generated by headful packaging can participate in homologous recombination, leading to transduction of host markers at levels over and above that occurring by generalized transduction, which arises by rare mistaken packaging of host DNA instead of Mu DNA into phage particles (Faelen et al., 1979 and references therein).

7 Replicative Transposition

Once the initial integration event is completed, subsequent transposition events occur by a replicative transposition process in which cleavage of the second strand at the Mu DNA ends does not occur (see reviews listed in the previous section). Instead, after ligation of the 3' ends of Mu DNA to the protruding 5' ends of the target DNA, the remaining free 3' target DNA ends serve as primers for semi-conservative DNA replication through the Mu DNA and ending with ligation of the newly synthesized strand to the free donor DNA end remaining at the original Mu-host DNA junction. This process generates a cointegrate molecule in which the donor and target DNA molecules are connected by two copies of Mu DNA, one at each donor-target junction. In contrast to several other transposable elements which also generate cointegrates by replicative transposition, Mu does not possess an efficient site-specific recombination system for resolution of the cointegrate into donor and target molecules each containing one copy of Mu. Instead, it appears likely that the two new copies independently reinitiate the replicative transposition process, eventually leading to significant "scrambling" of the host chromosome and the formation of a wide variety of long linear and circular molecules containing one or more copies of the Mu DNA (Resibois et al., 1982). It is probably this DNA that serves as the precursor for headful packaging.

The ability of researchers to carry out transposition *in vitro* using mini-Mu plasmid DNA and purified proteins and to characterize many of the intermediates (for reviews and references see Mizuuchi and Craigie, 1986; Haniford and Chaconas, 1992; Baker, 1993) has led to a detailed molecular understanding of replicative transposition as a multi-stage process involving the sequential assembly of a nucleoprotein transposition complex, the strand transfer reaction, and then disassembly of the complex to produce a substrate which then recruits the host (*E. coli*) replisome to carry out the replication process.

Events begin with binding of six 75-kDa monomers of the Mu A transposase protein to the three binding sites located within each attachment site, *attL* (L1, L2, and L3) and *attR* (R1, R2, and R3), on a supercoiled DNA substrate (Fig. 7). The

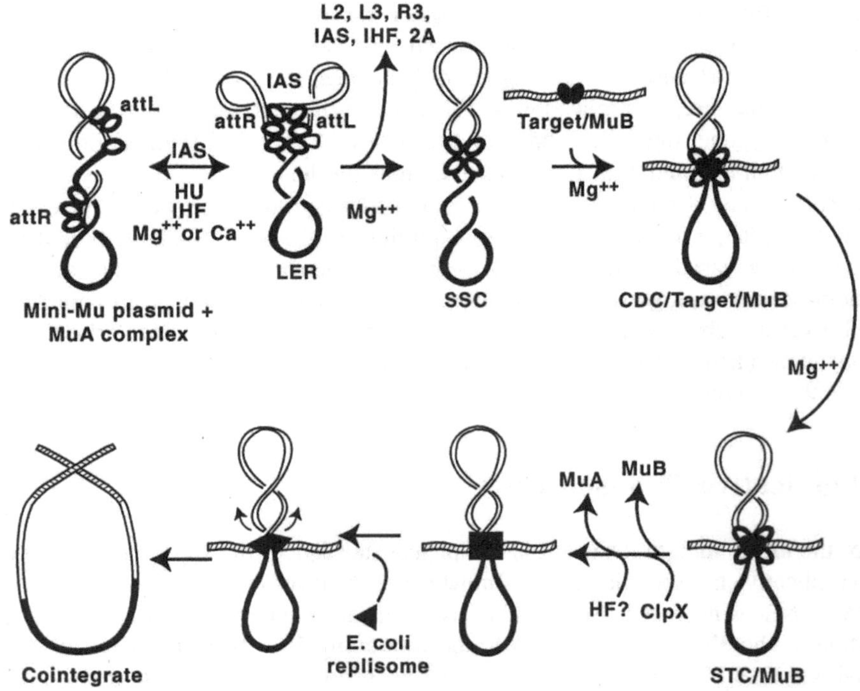

FIG. 7. Intermediates involved in the replicative transposition process. Open ribbons represent Mu DNA; filled ribbons represent vector DNA; hatched ribbons represent target DNA. Open and filled ellipses represent monomers of Mu A protein and B protein, respectively. The early stages of the process involve the following specific complexes: LER, left end-enhancer-right end complex; SSC, stable synaptic complex; CDC, cleaved donor complex; STC, strand transfer complex. Disassembly of the STC involves the action of ClpX and at least one as yet unidentified host factor (HF?) which produces an intermediate of unknown composition capable of recruiting the host replisome to carry out replication of the Mu DNA. For more information, see the text.

next step involves formation of a left end-enhancer-right end (LER) complex in which host proteins participate in the formation of a nucleoprotein structure in which *attL*, *attR*, and the Mu transposition enhancer (internal activating sequence; IAS) are brought into close proximity (Watson and Chaconas, 1996). The histone-like host HU protein binds between L1 and L2 and generates a strong bend which is believed to loop out the ~70-bp intervening DNA segment, thereby allowing interaction of A monomers bound at L1 and L2 (Lavoie and Chaconas, 1993). At the same time, the integration host factor protein (IHF) binds to the IAS in the early regulatory region and bends that DNA to allow two of the A monomers (probably those bound at L3 and R3) to utilize their second DNA binding regions to contact O1 and O2 within the enhancer. Negative supercoiling aids both in the formation of this complex and its stabilization. The LER complex then undergoes a major conformational change leading to the formation of a highly stable synaptic complex (SSC) containing an A tetramer bound to the "core sites" L1, R1, and R2. Formation of the SSC can be blocked by mutation of the terminal Mu nucleotide flanking the cleavage site, suggesting that there is specific recognition by A protein of the cleavage site in this SSC form (Watson and Chaconas, 1996). Following formation of the SSC, there is no longer a need for "accessory" binding sites L2, L3, and R3, the IAS or IHF protein, leading to the conclusion that their role is limited to assembly of the LER complex. When Mu B protein bound to target DNA interacts with the A tetramer in the SSC complex, it stimulates A to catalyze single-strand cleavage at the 3' ends of the Mu DNA, allowing relaxation of the supercoils in the vector DNA but maintaining supercoiling of the Mu DNA. The A tetramer in this cleaved donor complex (CDC) is stimulated by B to carry out transfer of the 3' Mu DNA ends to the 5' ends of the target DNA generated by 5-bp staggered cleavage, resulting in the STC or strand transfer complex. The strand transfer complex contains free 3' host DNA ends which could serve as primers for DNA replication; however, the presence of A prevents replication. Therefore, this nucleoprotein complex now requires disassembly in order to recruit the host replisome. The first step in disassembly is release of B protein from the complex. The second step involves recognition of the C-terminal portions of the A tetramer by the ClpX ATPase which serves as a molecular chaperone, resulting in an ATP-dependent transient conformational change which destabilizes A binding to DNA and also allows A to recruit host factors to the complex (Kruklitis et al., 1996; Levchenko et al., 1997 and references therein). The A protein which is released *in vitro* is not degraded and can function again for *att* site binding and transposition; the fate of released A *in vivo* is less clear since A is known to be unstable and subject to degradation by host proteases (Pato, 1989 and references therein). The host factors recruited to the complex have not yet been completely defined but they include the eight proteins which make up the replisome, namely DNA polymerase III holoenzyme, DnaB helicase, DnaC Protein, DnaG primase, DNA gyrase, DNA polymerase I, DNA ligase, and single-strand binding protein SSB.

At least two competitions may serve to regulate transposition. First, there is competition of Mu repressor and Mu A protein for binding to the transposition enhancer region; binding of repressor to O1 and O2 should prevent formation of the early LER complex, thereby directly inhibiting both transposition and production of the A and B proteins (Watson and Chaconas, 1996 and references therein). Second, there is competition between Mu B protein and ClpX for interaction with the C-terminal region of A protein; in this case B may confer resistance of the SSC, CDC, and STC complexes to disassembly by ClpX until strand transfer has been completed and Mu B protein released (Levchenko et al., 1997).

8 Epilogue

The preceding sections have presented a selected overview of Mu phage biology and molecular biology, leaving out or glossing over many interesting features, including the use of Mu and mini-Mu derivatives as *in vivo* genetic tools, the sophisticated molecular understanding of the G segment inversion process, and the molecular mechanism of activation of *E. coli* RNA polymerase at Mu promoters by different activator proteins. Nevertheless, it is hoped that the topics covered provide the reader with a useful and interesting overview of this novel phage and a convincing illustration of the usefulness of phages as model systems for studying molecular processes and their regulation.

9 Acknowledgements

Work from the author's laboratory was supported by the College of Medicine, University of Tennessee, Memphis; by National Science Foundation grants MCB-9305924 and MCB-9604653; and by a University of Tennessee Van Vleet Professorship.

10 References

Adzuma K, Mizuuchi K (1989) Interaction of proteins located at a distance along DNA: mechanism of target immunity in the Mu DNA strand-transfer reaction. Cell 57:41-47

Baker TA (1993) Protein-DNA assemblies controlling lytic development of bacteriophage Mu. Curr Opin Genet and Dev 1:708-712

Bolker MF, Wulczyn G, Kahmann R (1989) The *Escherichia coli* regulatory protein OxyR discriminates between methylated and unmethylated states of the phage Mu *mom* promoter. EMBO J 8:2403-2410

Bukhari AI, Taylor AL (1975) Influence of insertions on packaging of host sequences covalently linked to bacteriophage Mu DNA. Proc Natl Acad Sci USA 72:4399-4403

Faelen M, Toussaint A, Resibois A (1979) Mini-Muduction: a new mode of gene transfer mediated by mini-Mu. Mol Gen Genet 176:191-197

Ghelardini P, Liebart JC, Paolozzi L, Pedrini AM (1989) Suppression of the thermosensitive DNA ligase mutations in *Escherichia coli* K12 through modulation of gene expression induced by phage Mu. Mol Gen Genet 216:31-36

Glasgow AC, Hughes KT, Simon MI (1989) Bacterial DNA inversion systems. In: Mobile DNA (Berg DE, Howe MM eds), ASM Press, Washington DC, pp 637-659

Gloor G, Chaconas G (1986) The bacteriophage Mu N gene encodes the 64-kDa virion protein which is injected with, and circularizes, infecting Mu DNA. J Biol Chem 261:16682-16688

Goosen N, van de Putte P (1986) Role of Ner protein in bacteriophage Mu transposition. J Bacteriol 167:503-507

Grindley NDF, Sherratt DJ (1979) Sequence analysis at IS*1* insertion sites: models for transposition. Cold Spring Harbor Symp Quant Biol 43:1257-1261

Grundy FJ, Howe MM (1984) Involvement of the invertible G segment in bacteriophage Mu tail fiber biosynthesis. Virology 134:296-317

Haniford DB, Chaconas G (1992) Mechanistic aspects of DNA transposition. Curr Opin Genet Dev 2:698-704

Harel J, Duplessis L, Kahn JS, DuBow MS (1990) The cis-acting DNA sequences required for bacteriophage Mu transposition and packaging *in vivo*. Arch Microbiol 154:67-72

Harshey RM (1984) Transposition without duplication of infecting bacteriophage Mu DNA. Nature 311:580-581

Harshey RM (1988) Phage Mu. In The Bacteriophages, vol 1 (Calendar R, ed), Plenum Press, New York, pp 193-234

Higgins NP, Collier DA, Kilpatrick MW, Krause HM (1989) Supercoiling and integration host factor change the DNA conformation and alter the flow of convergent transcription in phage Mu. J Biol Chem 264:3035-3042

Howe MM (1987a) Phage Mu: an overview. In: Phage Mu (Symonds N, Toussaint A, van de Putte P, Howe MM eds), Cold Spring Harbor Laboratory Press, New York, pp 25-39

Howe MM (1987b) Late genes, particle morphogenesis, and DNA packaging. In: Phage Mu (Symonds N, Toussaint A, van de Putte P, Howe MM eds), Cold Spring Harbor Laboratory Press, New York, pp 63-74

Kahmann R (1984) The *mom* gene of bacteriophage Mu. Curr Top Microbiol Immunol 108:29-47

Kahmann R, Hattman S (1987) Regulation and expression of the mom gene. In: Phage Mu (Symonds N, Toussaint A, van de Putte P, Howe MM eds), Cold Spring Harbor Laboratory Press, New York, pp 93-109

Kahmeyer-Gabbe M, Howe MM (1996) Regulatory factors acting at the bacteriophage Mu middle promoter. J Bacteriol 178:1585-1592

Koch C, Mertens G, Rudt F, Kahmann R, Kanaar R, Plasterk R, van de Putte P, Sandulache R, Kamp D. (1987) The invertible G segment. In: Phage Mu (Symonds N, Toussaint A, van de Putte P, Howe MM eds), Cold Spring Harbor Laboratory Press, New York, pp 75-91

Krause HM, Higgins NP (1986) Positive and negative regulation of the Mu operator by Mu repressor and *Escherichia coli* integration host factor. J Biol Chem 261:3744-3752

Kruklitis R, Welty DJ, Nakai H (1996) ClpX protein of *Escherichia coli* activates bacteriophage Mu transposase in the strand transfer complex for initiation of Mu DNA synthesis. EMBO J 15:935-944

Lavoie BD, Chaconas G (1993) Site-specific HU binding in the Mu transpososome: conversion of a sequence-independent DNA-binding protein into a chemical nuclease. Genes & Dev 7:2510-2519

Levchenko I, Yamauchi M, Baker TA (1997) ClpX and MuB interact with overlapping regions of Mu transposase: implications for control of the transposition pathway. Genes & Dev 11:1561-1572

Mizuuchi K (1992) Transpositional recombination: mechanistic insights from studies of Mu and other elements. Ann Rev Biochem 61:1011-1051

Mizuuchi K, Craigie R (1986) Mechanism of bacteriophage Mu transposition. Ann Rev Genet 20:385-429

Paolozzi L, Symonds N (1987) The SE region. In: Phage Mu (Symonds N, Toussaint A, van de Putte P, Howe MM eds), Cold Spring Harbor Laboratory Press, New York, pp 53-62

Pato ML (1989) Bacteriophage Mu. In: Mobile DNA (Berg DE, Howe MM eds), ASM Press, Washington DC, pp 23-52

Pato ML, Banerjee M (1996) The Mu strong gyrase-binding site promotes efficient synapsis of the prophage termini. Mol Microbiol 22:283-292

Resibois A, Toussaint A, Colet M (1982) DNA structures induced by mini-Mu replication. Virology 117:329-340

Rousseau P, Betermeier M, Chandler M, Alazard R (1996) Interactions between the repressor and the early operator region of bacteriophage Mu. J Biol Chem 271:9739-9745

Shapiro JA (1979) Molecular model for the transposition and replication of bacteriophage Mu and other transposable elements. Proc Natl Acad Sci USA 76:1933-1937

Symonds N, Toussaint A, van de Putte P, Howe MM (1987) Phage Mu. Cold Spring Harbor Laboratory Press, New York

Taylor AL (1963) Bacteriophage-induced mutation in *Escherichia coli*. Proc Natl Acad Sci USA 50:1043-1051

Toussaint A, Gama M-J, Laachouch J, Maenhaut-Michel G, Mhammedi-Alaoui A (1994) Regulation of bacteriophage Mu transposition. Genetica 93:27-39

Van de Putte P, Goosen N (1992) DNA inversions in phages and bacteria. Trends Genet 8:457- 462

Van Ulsen P, Hillebrand M, Zulianello L, van de Putte P, Goosen N (1996) Integration host factor alleviates the H-NS-mediated repression of the early promoter of bacteriophage Mu. Mol Microbiol 21:567-578

Watson MA, Chaconas G (1996) Three-site synapsis during Mu DNA transposition: a critical intermediate preceding engagement of the active site. Cell:435-445

Wulczyn FG, Kahmann R (1991) Translational stimulation: RNA sequence and structure requirements for binding of Com protein. Cell 65: 259-269

Zha J, Zhao Z, Howe MM (1994) Identification and characterization of the terminators of the *lys* and *P* transcripts of bacteriophage Mu. J Bacteriol 176:1111-1120

Regulation of Bacteriophage λ Replication

Karol Taylor and Grzegorz Węgrzyn

Department of Molecular Biology, University of Gdańsk, Kładki 24, 80-822 Gdańsk, Poland

Abstract. Bacteriophage λ is both a model for basic biological studies on the molecular level and a commonly used tool in molecular cloning. In this review we summarise the control of bacteriophage λ development, placing particular attention to the mechanisms involved in the "lysis or lysogeny" decision and the regulation of replication of phage λ DNA, as well as of plasmids derived from this phage.

Keywords. Bacteriophage λ, control of virus development, regulation of gene expression, regulation of DNA replication, molecular cloning

1 Introduction

Bacteriophage λ is a virus which infects cells of *Escherichia coli* (for basic information, see Hendrix et al., 1983). This bacteriophage played a crucial role in the development of molecular biology including gene cloning (Thomas, 1993), and serves as a paradigm of molecular mechanisms of many general biological processes, for instance: repression and induction of transcription initiation, antitermination of transcription, site-specific recombination, a role for chaperone proteins in macromolecular assembly and DNA replication, and even the control of development.

The genome of bacteriophage λ consists of 48,502 base pairs of double-stranded DNA. In the λ virion, this genetic material is packaged in a linear form in the head of the phage capsid. The capsid is composed of the icosahedral head (54 nm diameter), tail (150 nm long, 15 nm diameter) and the tail fibre (25 nm long, 2 nm diameter). Upon adsorption on the surface of *E. coli* and penetration of the phage genome into the host cell, the linear DNA is converted immediately into the circular form due to single-stranded 5' extensions of 12 bases at both ends which are complementary to each other. Both ends (called *cos*) are ligated by the host DNA ligase and, following action of *E. coli* DNA gyrase, the phage genome becomes a negatively supercoiled structure.

On the linear map of the bacteriophage λ genome (Fig. 1) genes are clustered according to their role in λ development. Thus, a part of the λ genome encompassing the replication region may be cut out as a single DNA fragment and after circularisation it can replicate in *E. coli* cells as a plasmid (Fig. 1).

NATO ASI Series, Vol. H 103
Molecular Microbiology
Edited by Stephen J. W. Busby,
Christopher M. Thomas and Nigel L. Brown

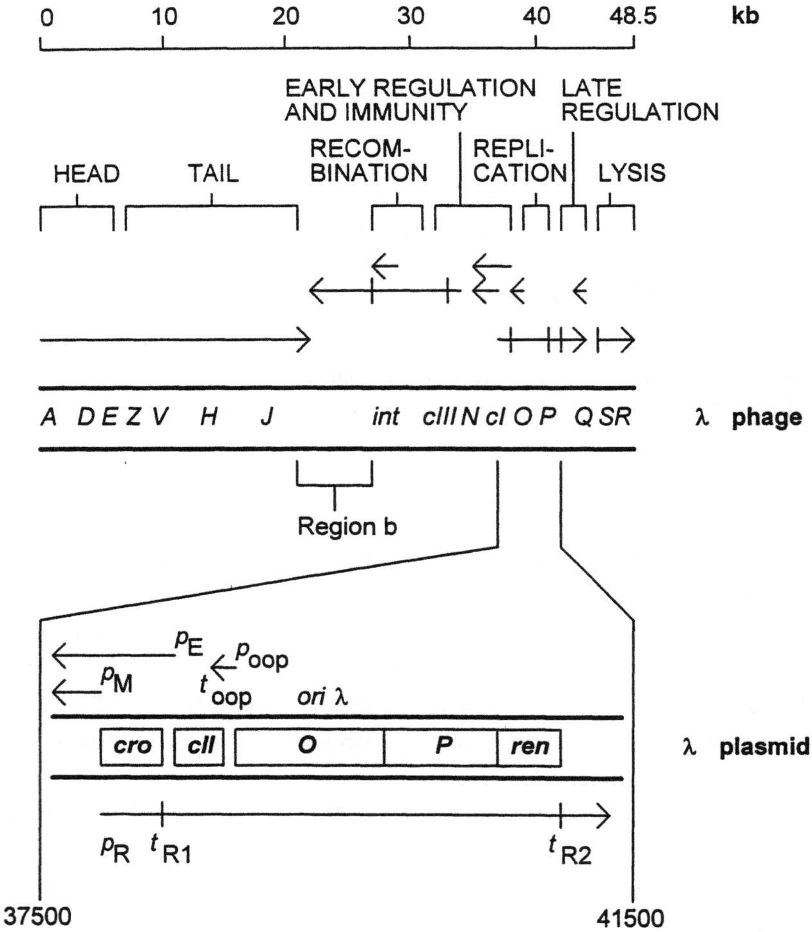

Figure 1. A genetic map of bacteriophage λ and λ plasmid. The scale at the top of the figure is given in kilobases (kb). Regions of the genome which contain genes coding for head proteins (HEAD), tail proteins (TAIL), recombination proteins (RECOMBINATION) controlling site-specific and general DNA recombination, regulators of expression of early genes (EARLY REGULATION AND IMMUNITY), replication proteins (REPLICATION), regulator of the expression of late genes (LATE REGULATION), and proteins responsible for host cell lysis (LYSIS) are indicated. Region *b*, encompassing a nonessential part of the λ genome, is also indicated. Positions of certain important genes are marked. Main leftward and rightward transcripts are marked by arrows; arrowheads indicate the direction of transcription. Main terminators are marked by short vertical bars crossing appropriate lines of transcripts. A fragment of the λ genome present in a typical λ plasmid is presented in the lower part of the figure. The p_R promoter and the *O* and *P* genes are essential genetic elements for replication of λ plasmid DNA, though typical plasmids derived from bacteriophage λ contains also *cro*, *cII* and *ren* genes. All promoters (p_R, p_M, p_E, p_{oop}) and terminators (t_{R1}, t_{R2}, t_{oop}) present in the replication region are indicated. The *oriλ* sequence (present in the middle of the *O* gene) is a region for initiation of λ DNA replication. Transcription from the p_R promoter produces mRNA for synthesis of the replication proteins (O and P) and at the same time activates *oriλ* (this process is called transcriptional activation of the *origin*). The t_{R1} site is a weak terminator and a significant proportion (15-50%) of transcribing RNA polymerase molecules are able to proceed beyond this region.

2 Two modes of bacteriophage λ development

There are two alternative pathways for bacteriophage λ development upon circularisation of its DNA inside the host cell (Fig. 2). The lytic cycle consists of phage DNA replication and expression of the phage late genes, i.e. genes coding for head and tail proteins and for proteins causing lysis of the host cell. This pathway leads to production of phage progeny (an average yield is about 100 virions per one infected cell), cell lysis and liberation of virions. The cycle is completed within about 30-40 min at 37°C in a bacterial host growing in a rich medium. Alternatively, the phage genome may be incorporated into a specific site of the host genome (between *gal* and *bio* loci) forming a stable prophage. Such a lysogenic cell can survive for many generations carrying an integrated phage genome which is replicated as a part of the bacterial chromosome. The most important phage gene active in the lysogenic cell is *cI* whose product is a strong repressor of the main lytic promoters, p_R and p_L, while at the same time activating its own promoter, p_M. When the lysogenic cell in endangered by unfavourable environmental conditions, the λ prophage can be induced. The induction results from CI repressor degradation mediated by the activated form of the host RecA protein (called RecA*). The activation of RecA occurs upon appearance of single-stranded DNA fragments with which RecA interacts. Therefore, induction of the wild-type λ prophage may be achieved upon treatment of lysogenic bacteria with DNA damaging agents such as UV light or inhibitors of DNA replication (like mitomycin C). After CI degradation, the p_R and p_L promoters become active, λ DNA is excised from the host chromosome and the lytic cycle begins; this leads to production of phage progeny and cell lysis as described above.

What are the conditions that influence which developmental mode should be chosen by the infecting λ phage? It is known that infection of one host cell by only one phage particle (i.e. low multiplicity of infection, m.o.i.), a high rate of bacterial growth and high temperature (e.g. 43°C) favour the lytic cycle. On the other hand, infection of one host cell by many phages (i.e. high m.o.i.), poor growth conditions and a low temperature (below 30°C) result in a higher efficiency of lysogenisation (Fig. 2). The "lysis or lysogeny" decision is a crucial step in the phage life cycle as the proper choice should ensure optimal phage propagation. The molecular mechanisms leading to establishment of one of two modes of λ development will be discussed below (for recent reviews and models, see Echols, 1986; Giladi et al., 1995; Szalewska-Pałasz et al., 1996; Obuchowski et al., 1997 and references therein).

3 The "lysis or lysogeny" decision

The decision whether to enter the lytic cycle or to lysogenise the host cell must reflect the current physiological state of the host cell being infected by λ phage; the intracellular conditions depend, in turn, on the environment. This "lysis or lysogeny" decision is governed by several regulators of transcription whose actions are presented schematically in Fig. 3.

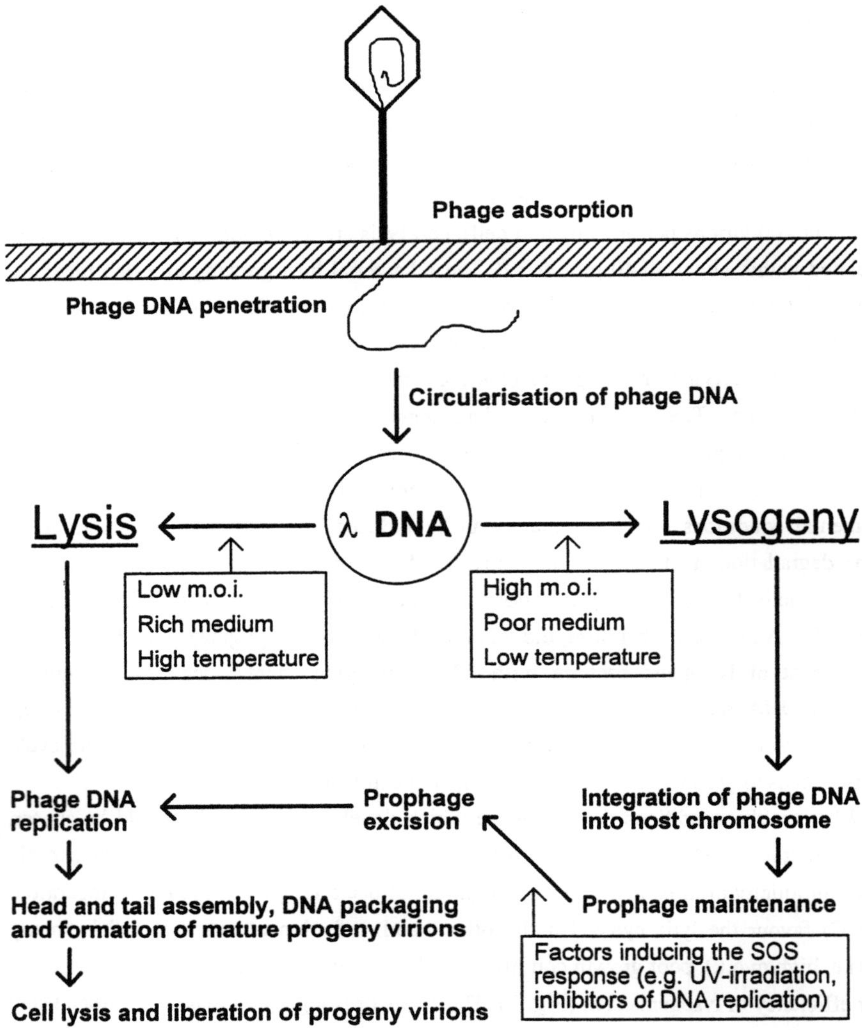

Figure 2. A scheme depicting two modes of phage λ development. The most important steps in the phage life cycle are marked. The two alternative developmental pathways are indicated as Lysis and Lysogeny (underlined) for the lytic and lysogenic pathway, respectively. Conditions and factors affecting the "lysis or lysogeny" decision of bacteriophage λ and those provoking prophage induction (i.e. excision of the prophage DNA and switch to the lytic pathway) are described in boxes; the box arrows indicate steps stimulated by appropriate conditions.

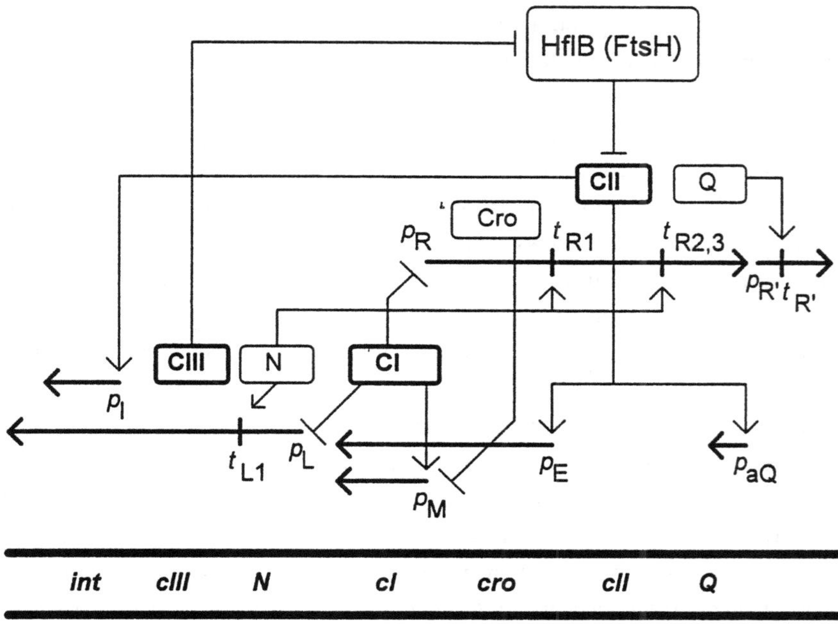

Figure 3. Bacteriophage λ regulatory network involved in the "lysis or lysogeny" decision. The phage genes playing the crucial role in the decision are marked in the lower part of the figure. Gene products are framed above mRNAs for their syntheses. Regulatory proteins (CI, CII and CIII) which are responsible for establishment and maintenance of the lysogenic pathway are marked in bold. Mutants carrying defect(s) in any of these genes form clear plaques on the lawn of wild-type host. in contrast to turbid plaques formed by wild-type λ phages. The most important promoters (p_R, p_L, p_I, p_M, p_E, p_{aQ}) are indicated; transcripts initiated at these promoters are marked as thick arrows (arrowheads indicate direction of transcription). The most important terminators (t_{R1}, t_{R2}, t_{R3}, $t_{R'}$, t_{L1}) are marked by short vertical lines crossing appropriate lines of transcripts. Positive regulation of gene expression (activation of promoters or transcriptional antitermination) by particular proteins is indicated by thin arrows. Negative regulation (repression of transcription initiation, degradation of the cII gene product by HflB (FtsH) protease, and inhibition of HflB (FtsH) activity by the CIII protein) is indicated by blunt-ended lines. The scheme is not drawn to scale.

For lytic development, the activity of the p_R and p_L promoters is necessary (the names are for Rightward and Leftward transcription, respectively). Initially, transcription from these early promoters provide mRNAs for synthesis of proteins Cro and N. Cro is a repressor of the p_M promoter, whose activity leads to production of the *c*I gene product (called also the lambda repressor) - a strong repressor of the p_R and p_L promoters. Repression of these promoters is crucial for the Maintenance of the prophage, hence the designation p_M. At higher concentrations, Cro is also a repressor of the p_R and p_L promoters. In the absence of other factors, transcription originating from p_R is largely (50 - 85%) terminated at the t_{R1} terminator while readthrough transcription stops at t_{R2}. Transcription initiated at p_L stops at the t_{L1} terminator. For efficient expression of more distant genes, the action of the *N* gene product, an antitermination protein, is necessary. Upon transcription of a DNA sequence placed between p_R and t_{R1} or p_L and t_{L1} (called *nutR* and *nutL*, respectively), the N protein interacts with a specific secondary structure in the transcript and mediates formation of a multiprotein complex (containing also bacterial proteins NusA, NusB, NusE and NusG) with RNA polymerase that prevents transcription termination at t_{R1}, t_{L1} and more distant terminators, including t_{R2} (Friedman and Court, 1995). Although t_{R1} is a weak ρ-dependent terminator and a significant proportion of transcribing RNA polymerase molecules are able to proceed beyond this region (which allows some expression of the replication genes *O* and *P*), the N-mediated transcriptional antitermination is absolutely necessary for both full expression of the replication genes and, perhaps more important, readthrough of t_{R2} and more distant terminators allowing expression of the *Q* gene. This gene codes for another antitermination protein that is indispensable for expression of the late (i.e. head, tail and lysis) genes. The late genes are under control of the $p_{R'}$ promoter, but also depend on Q-mediated antitermination at the $t_{R'}$ terminator for their expression (Fig. 3). Similarly to the above described rightward transcription, expression of genes situated beyond the t_{L1} terminator (e.g. *c*III and recombination genes) is dependent on N-mediated antitermination at this and more distant terminators. In summary, for lytic development, the activity of p_R and p_L promoters as well as transcription antitermination mediated by N and Q proteins are necessary. To achieve this, repression of these promoters by the CI protein must be prevented by Cro-mediated negative regulation of the p_M promoter.

There is a completely different scenario when the lysogenic pathway of λ development is chosen. In this case it is essential to switch off expression of the genes involved in the lytic mode. This is achieved by the action of the CI repressor which blocks transcription from p_R and p_L while stimulating its own synthesis by activation of the p_M promoter. However, shortly after infection the p_M promoter is inactive because (i) it is repressed by Cro as the *cro* gene is the first phage gene expressed after penetration of λ DNA into the *E. coli* cell, and (ii) it requires positive regulation by CI for maximal activity (*c*I is autoregulated). Therefore, there is another promoter, called p_E (i.e. promoter for Establishment of repressor synthesis and lysogeny), responsible for *c*I expression soon after infection. This promoter is positively regulated by the *c*II gene product. As *c*II is expressed from the p_R promoter it is clear that to enter the lysogenic pathway the phage must proceed a little way along the lytic pathway. The activity of p_E, as well as other CII-stimulated promoters: p_I (for Integrase gene expression) and p_{aQ} (for anti-*Q* transcript), is extremely low without positive regulation. The p_I promoter activity is necessary for expression of the *int* gene coding for Integrase, an enzyme which is responsible for integration of λ DNA into the host chromosome through a mechanism of site-specific recombination. Although the *int* gene may also be transcribed from the p_L promoter, its expression is inefficient in the presence of N activity due to the retroregulation mechanism. The *N* gene is transcribed from the p_L promoter and when p_L-initiated transcription proceeds beyond the t_I terminator due to N-mediated antitermination, RNA polymerase encounters the *sib* region whose transcription results in the appearance of an RNA structure recognised by RNase III and subsequent quick degradation of the *int* mRNA (note that p_I-initiated transcription does not produce mRNA for N synthesis, thus this transcription is terminated at t_I and the *int* mRNA is more stable). Expression of late genes is prevented by the activity of the p_{aQ} promoter which results in appearance of an RNA which is antisense to the mRNA for the Q protein synthesis, thus indirectly preventing antitermination at $t_{R'}$ through inhibition of Q production.

It is clear from this discussion that the antagonism between the Cro and CI repressors plays an important role in the choice between the lytic and lysogenic modes of λ development. If Cro wins, the expression of *cI* is switched off and all genes involved in the lytic pathway may be expressed. On the other hand, once CI is synthesised, it blocks the p_R and p_L promoters switching off synthesis of proteins engaged in the lytic development, including the Cro repressor. As shown in Fig. 4, the p_R and p_M promoters are oriented in opposite directions but they are placed very close one to another. Overlapping these two promoters are three operator sequences, called o_{R3}, o_{R2} and o_{R1}, recognised by both CI and Cro repressors. CI has the strongest affinity to the o_{R1} sequence. Thus even at relatively low concentrations the repressor is able to abolish the activity of the p_R promoter. In addition, when o_{R1} is occupied, CI can bind cooperatively to o_{R2}. This binding enhances the repression of p_R, but at the same time activates p_M. The Cro repressor, on the other hand, binds these operators in the reverse order. Thus it possesses the strongest affinity for o_{R3} (this results in effective repression of p_M), binds more weakly to o_{R2}, and has the weakest affinity for o_{R1} (this can lead to repression of p_R when Cro is present in high concentration). However, since early after λ DNA penetration the expression of *cro* is at the constant level and p_M is inactive, the most important factor in the "lysis or lysogeny" decision is the activity of CII as this dictates the level of CI. The *cII* gene product is very unstable *in vivo* with a half-life of about 1-2 min. It is degraded by the HflB/FtsH protease (Herman et al., 1993). This protease is partially inhibited by the *cIII* gene product.

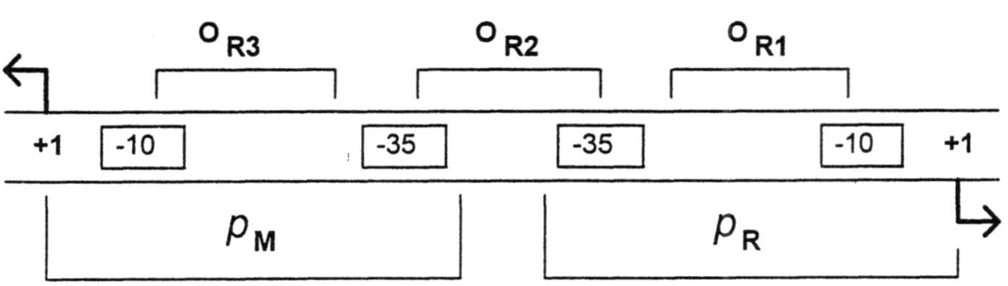

Figure 4. The promoter/operator region of the phage λ DNA fragment encompassing p_M and p_R. The sequences of TATA-boxes (-10), -35-boxes (-35) and transcription start sites (+1) of both promoters (p_M and p_R) are shown. The directions of transcription from p_M and p_R are marked by arrows. The o_{R3}, o_{R2} and o_{R1} sequences of the operator site, which are recognised by both CI and Cro repressors, are indicated.

As shown in Fig. 2, there are several factors influencing the "lysis or lysogeny" decision, each of them may directly or indirectly affect the very delicate balance between the level/activity of CII, CI and Cro. The efficiency of lysogenisation is higher when a host cell is infected by many phages rather than by few or just one. This phenomenon appears to be due to increased *c*III gene dosage. Since *c*III gene product is an inhibitor of the CII-digesting protease (HflB/FtsH), which is encoded by a single copy chromosomal gene, many copies of the *c*III gene mean an increased level of CIII protein and a more efficient inhibition of the protease. This results in a higher stability of CII that leads to establishment of the CI repressor synthesis and subsequent lysogenisation of the host cell. Concerning enhanced lysogenisation in starved bacteria, it was proposed that cyclic AMP, an alarmone which is produced in *E. coli* during its response to starvation conditions, negatively regulates HflB/FtsH activity (Herskowitz, 1985). Apart form the possible cAMP-mediated inhibition of CII degradation, a role for another alarmone, guanosine 5'-diphosphate-3'-diphosphate (ppGpp), in making the decision cannot be excluded. This alarmone is the main effector of the stringent response, a response of bacterial cells to amino acid starvation (though the level of ppGpp increases also in cells deprived of a carbon source). It is well known that this nucleotide affects transcription from many promoters (among others, the bacteriophage λ p_R promoter; Szalewska-Pałasz et al., 1994), thus its concentration may significantly affect levels of crucial players in the game of choosing one of the two developmental pathways. Finally, it appears that the stability of CII is higher at lower temperatures (either because of formation of a more stable CII conformation, or decreased activity of HflB/FtsH, or both) thus accounting for the efficient lysogenisation. The increased activity of the p_L promoter observed at low temperature (Giladi et al., 1995) may lead to higher expression level of the *c*III gene and, in turn, further stabilisation of CII.

4 Replication of λ DNA

At early times after infection, the replication of phage λ DNA occurs according to the *theta* (θ or circle-to-circle) mode, and is switched later to the *sigma* (σ or rolling-circle) mode. The plasmids derived from phage λ replicate according to the phage λ early replication mode, hence most of the recent information and all *in vitro* results concerning early λ DNA replication originate from studies on λ plasmids, traditionally called λ*dv*.

4.1 Early replication

The λ-coded replication proteins, λO and λP, are synthesised early after phage infection, since the *O-P* region is situated downstream of the p_R promoter and the weak t_{R1} terminator (Fig. 1). *In vivo*, λO is extremely unstable due to rapid hydrolysis by ClpXP protease (Gottesman et al., 1993). The λO protein is a prototype initiator protein, which binds to a specific DNA sequence, called the origin of replication, *ori*λ, and directs other viral and host replication proteins to this site in the process of replisome (or replication complex, RC) assembly. Genetic studies suggest that the N-terminal domain of λP binds to the C-terminal domain of λO, and formation of a λO-λP oligomer has been revealed *in vitro* (Żylicz et al., 1984). However, due to a strong interaction of λP with the host DnaB helicase, it seems that it is actually a λP-DnaB complex which binds to λO. The resulting complex, λO-λP-DnaB, is called the pre-primosome, since binding of DnaG primase to single-stranded DNA regions exposed by the action of the helicase occurs later in the RC assembly. In this pathway the *ori*λ-bound pre-primosome is the first intermediate in which λO is protected from the action of ClpXP protease (A.Węgrzyn et al., 1995a). In the λ pre-primosome λP inhibits the helicase action of DnaB which is indispensable for replication fork movement. Release of DnaB helicase from λP inhibition occurs due to the action of the hsp70 chaperone machine, composed of DnaK, DnaJ and GrpE proteins (Żylicz et al., 1989; Alfano and McMacken, 1989). It is worth remembering that in the λ plasmid replication system the activity of DnaK and DnaJ chaperones was demonstrated for the first time *in vitro* (Liberek et al., 1988). Loading of helicase molecules between the complementary λDNA strands at proper sites, in orientations that make possible further bi-directional replication, seems to be connected with the "transcriptional activation of *ori*λ", occurring due to p_R-initiated transcription, regulated by the host DnaA function (G.Węgrzyn et al., 1995a; G.Węgrzyn et al., 1996). Interaction with other host replication proteins, as Ssb, DnaG primase, DNA polymerase III holoenzyme and DNA gyrase results in the assembly of functional RCs, ready to initiate bi-directional replication, giving rise to two daughter λDNA circles at each replication round.

As opposed to the results obtained *in vitro* with polyclonal anti-λO antibodies (Alfano and McMacken, 1989), the studies *in vivo* clearly demonstrate that the λO initiator is an essential and intrinsic element of the RC. This structure (as a whole, or at least the part providing protection to λO from ClpXP protease) is inherited at each replication round by one of two daughter λ DNA circles (Szalewska-Pałasz et al., 1994). Therefore, RC assembly is restricted to the daughter circle which has not received the parental RC. The "old" RC-driven replication has been studied, after elimination of the *de novo* RC assembly, in two systems. In amino acid-starved *E.coli relA* bacteria the λO initiator cannot be synthesised (G.Węgrzyn and Taylor, 1992; G.Węgrzyn et al., 1992), and in *E.coli* wt cells growing in a complete medium the *Pts*1 mutation blocks RC assembly after a temperature upshift (A.Węgrzyn et al., 1996b) (Fig. 5). In both these systems λ DNA replication was found to be strongly λ*O*-dependent.

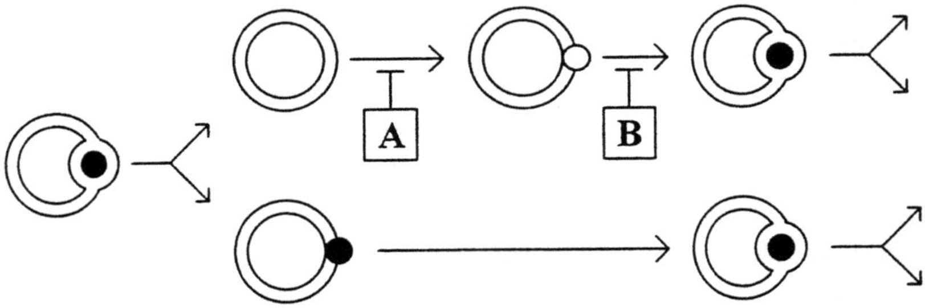

Figure 5. Two systems for studying λ plasmid replication carried out by the pre-assembled, inherited replication complex (RC, small filled circle). In both systems the assembly of a new RC has been blocked. (A) In amino acid-starved *E. coli relA* cells the λO initiator is not synthesised. (B) In *E. coli* wt growing in a complete medium the assembly of RC is blocked by *Pts*1 mutation at 43°C at the DnaA-dependent step, after formation of the pre-primosome (small open circle).

The old RC-driven replication, resistant to chloramphenicol, is dependent on both RNA polymerase and DnaA functions, suggesting that DnaA-regulated transcriptional activation of *ori*λ is here also required. Moreover, this replication is also dependent on the activity of the Hsp70 chaperone machine (A.Węgrzyn et al., 1995b; A.Węgrzyn et al., 1996b). It therefore seems that after termination of a replication round the parental RC, containing DnaB helicase blocked again by λP, remains bound to the annealed *ori*λ. Initiation of a next replication round would again require the DnaA-regulated transcriptional activation of *ori*λ coupled to the action of the Hsp70 chaperone machine. One may infer that Hsp70 chaperones do not cause dissociation of λP, but rearrange the DnaB helicase-containing structures (including the pre-primosome), releasing DnaB from λP inhibition. This model (Fig. 6) is in opposition to the results obtained *in vitro*, which demonstrate dissociation of a larger or smaller part of λP after chaperone action (Alfano and McMacken, 1989; Żylicz, 1993).

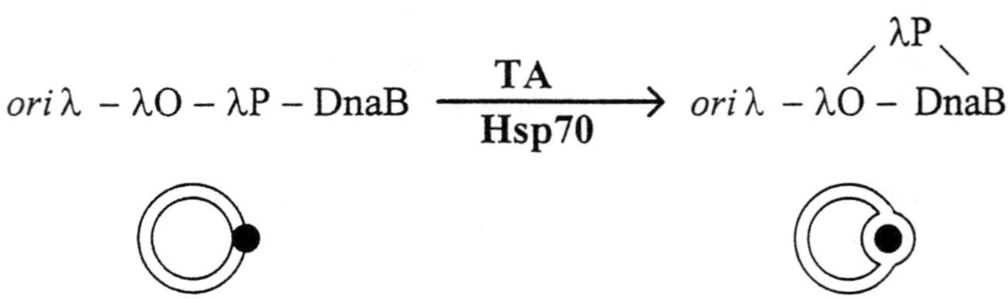

Figure 6. Postulated coupling of the DnaA-regulated transcriptional activation of *ori*λ (TA) with the release of DnaB helicase from λP inhibition due to DnaK, DnaJ and GrpE (Hsp70) chaperone action, causing loading of the DnaB helicase-containing structure (pre-primosome, or RC, small filled circle) between complementary λ DNA strands. Note that λP does not dissociate from the DnaB helicase-containing structures.

However, one may argue that the systems reconstituted from purified components did not reflect faithfully λ DNA replication *in vivo*. For instance, the *in vitro* replication systems used were not dependent on transcription, stringently required *in vivo*. Besides, *in vitro* efficiency of λ DNA replication is low, and initiation of the second replication round (when the DnaB-inhibitory function of λP is postulated to be important) does not take place.

4.2 Regulation of λ plasmid replication

The transient activity of the proteins that initiate a new round of replication seems to be a key element in theories on the control of this process. The λO initiator that is subject to rapid proteolysis by the ClpXP protease represents one of the best known examples. Rapid decay of an excess of λO, synthesised following a wave of p_R-initiated transcription, should make any uncontrolled re-initiation of replication impossible. Hence a wave of transcription would correspond to one, and only one, replication round, initiated by binding of λO to *ori*λ. However, stabilisation of λO caused by mutational inactivation of ClpXP protease did not change the transformation efficiency or copy number of λ plasmids, or the kinetics of λ phage growth (Szalewska et al., 1994).. Moreover, the replication pathway carried out by the inherited, λO-containing, RC cannot be explained by a model which assumes that the frequency of λO binding to *ori*λ determines the frequency of initiation events.

The above-mentioned wave of p_R-initiated transcription was assumed to be raised by a transient de-repression of the o_R operator, usually blocked by the dimeric λ Cro repressor (Matsubara, 1981). A decrease in the intracellular concentration of Cro as the cell volume increases during the growth cycle may cause this de-repression. The narrow time-window of the de-repressed state is closed by the increase in Cro level as a result of the wave of p_R activity. However, elimination of the autoregulatory loop of the λ Cro repressor (by exchange of the o_Rp_R-*cro* region for the operator-promoter region of the *lac* operon) had no deleterious effect on λ plasmid maintenance (Szalewska et al., 1994). Also the λ*cro*ts plasmid was well maintained, albeit at a 2-3 times higher copy number, after temperature upshift (A.Węgrzyn et al., 1996b).

It therefore seems, that the model of regulation of λ plasmid replication based on periodical events of Cro de-repression causing λ*O* gene expression and λO protein binding to *ori*λ is unlikely to be valid. At present, there remains only the DnaA-regulated transcriptional activation of *ori*λ as a possible element of this control. *In vitro* studies have been initiated in order to check if DnaA protein acts directly, *e.g.* through binding to DnaA boxes which are present downstream of p_R, but absent in the vicinity of *ori*λ. In contrast to the wild-type (cro^+) λ plasmid (Matsubara and Mukai, 1975), λcro^- plasmids replicate once and only once in the host cell cycle (A.Węgrzyn et al., 1996b). It seems, therefore, that in λ plasmid a Cro-mediated effect is superimposed on a basic regulatory mechanism that includes a re-

replication block during the cell cycle. The existence of a cell-cycle-specific regulatory mechanism in the replication of the phage λ genome cannot be required for λ lytic development. Is it required when the infecting phage DNA resides in a plasmid state for several cell generations, with suspended "lysis or lysogeny" decision? In this context it is worth remembering that phage DNA in λ*N*⁻ phage-infected cells replicates as a plasmid (Kleckner and Signer, 1977, and references therein).

4.3 Disassembly of the replication complex

Arguments presented in the preceding subsection as well as the construction of a viable λ*cro*-null plasmid clearly show that, contrary to previous claims (Matsubara, 1976; Berg, 1974), *cro* is not an essential plasmid gene. The previously presented studies on RC inheritance were performed in systems in which Cro-repression did not work (amino acid-starved *E. coli relA*), or the activity of Cro was mutationally eliminated. A temperature rise from 30 to 43°C did not cause RC disassembly in λ*cro*ts plasmid-harbouring bacteria growing in a rich medium, and inheritance of RC for more than 30 cell generations (and λ plasmid replication rounds) could be observed (A.Węgrzyn et al., 1996b). However, in these conditions, heat shock caused RC disassembly (monitored by the ClpXP-mediated decay of λO present in the RC) in λ*cro*⁺ plasmid-harbouring cells. The RC decay occurred due to heat shock induction of the *gro*E operon (A.Węgrzyn et al., 1996a), coding for GroEL and GroES chaperones, the elements of the Hsp60 machine, known of its role in λ phage head morphogenesis (Georgopoulos and Welch, 1993). This was the first demonstration of GroEL/S engagement in an *in vivo* disassembly of a highly organised protein structure under stress caused by temperature shift. This structure is dispensable for cell survival and it may be that its components are reutilised for the assembly of the host replication complex. It is possible that the observed phenomenon represents a general strategy of a cell endangered by unfavourable environmental conditions.

Merely increasing the cellular concentration of GroEL and GroES proteins was not sufficient to cause RC disassembly. The second requirement is a DNA gyrase-mediated negative supercoiling of λ plasmid DNA which counteracts DNA relaxation occurring immediately after the temperature upshift. We presume that the RC dissociates from *ori*λ during negative supercoiling, exposing the DNA-binding surface of λO (possibly hydrophobic) to the action of the GroE chaperone machine. It remains an open question as to why the RC disassembles after a temperature upshift of λ*cro*⁺ plasmid-harbouring bacteria, as opposed to RC stability in λ*cro*⁻ plasmid-bearing cells.

4.4 Switch to late replication

In a λ phage-infected *E. coli* cell only 2-3 of 50 circles produced by the *theta* mode of replication is switched to the *sigma* mode, and this explains the slow progress in the study of the mechanism of this process. According to the predominant hypothesis (Dodson et al., 1986), *sigma* may be preceded by one round of *theta* unidirectional replication initiated at *ori*λ, followed by displacement of the 5' end of the newly synthesised leading strand by its growing 3' end (Fig. 7).

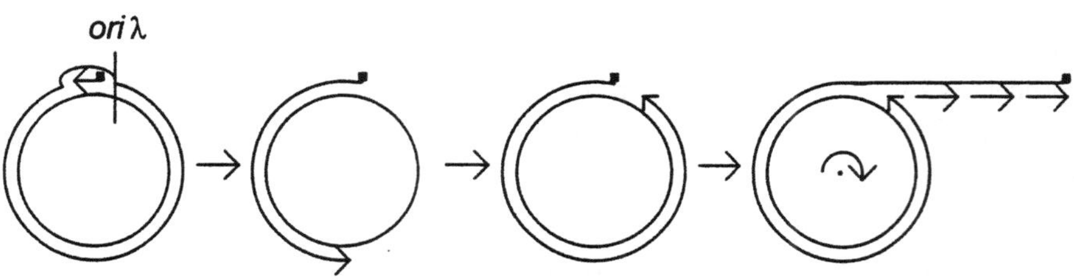

Figure 7. Hypothesis of the switch from early (*theta*) to late (*sigma*) replication mode of phage λ DNA. One round of unidirectional *theta* replication initiated at *ori*λ is followed by displacement of the 5' end (dot) of the newly synthesised leading strand by its growing 3' end (arrow). Only the fate of parental strand directing the synthesis of the leading strand is shown.

This hypothesis has been supported and supplemented by the results of experiments performed with λ-infected *E.coli dnaA*ts bacteria at 43°C (G.Węgrzyn et al., 1995b). Most of the parental phage DNA molecules, after one round of *theta* replication, switched to a replication with features of the *sigma* mode and produced progeny at high yield. These results have been interpreted according to the postulated role of DnaA-regulated transcription in the insertion of two λO-λP-DnaB pre-primosomes between the separated λ DNA strands in an orientation appropriate for subsequent bidirectional replication (see "Early replication"). In the absence of DnaA function only one RC would be installed, which would result in unidirectional replication. It is worthwhile mentioning that in an *in vitro* system of λ DNA replication reconstituted from purified proteins (DnaA was not included) replication was unidirectional, and there was a bias to bidirectionality when the system became transcription-dependent after supplementation with HU protein and RNA polymerase (Learn et al., 1993). In the *E. coli dnaA*ts strain used (a laboratory strain with a long history) the initiation of λ phage DNA replication at 43°C was dependent on the lambdoid prophage Rac (G.Węgrzyn et al., 1995b). However, this dependence was not confirmed when the *dnaA*ts mutation was present in a wild-type background. Therefore our corrected hypothesis is that during λ phage infection of *E. coli* wt the switch from bidirectional *theta* to unidirectional *theta*, and later to the *sigma* mode of replication is caused by consumption of the host *dnaA* function by the rapidly replicating phage DNA (Fig. 7). In the original hypothesis (Dodson et al., 1986) the crucial point is the switch from unidirectional *theta* to *sigma* replication mode, which occurrs due to a barrier in the growth of the leading strand represented by the *ori*λ-bound λO protein aggregate, observed under the electron microscope.

Since the pre-assembled RC is able to carry out a next *theta* replication round, it is quite probable that it may also be used in *sigma* replication. This prediction is supported by a study of bacteria infected with λ*O*ts or λ*P*ts mutants and shifted to the restrictive temperature at appropriate times (Klinkert and Klein, 1978). Late λ replication was found to be *O*-dependent corresponding to the presence of λO in the RC. Late λ replication was found to be *P*-independent. As the *P*ts1 mutation used blocks RC assembly this result is in accordance with the prediction that assembly of a new RC is not required during *sigma* mode of replication.

The role of the Cro repressor in directing the λ phage development towards the lytic pathway has been already discussed (Fig. 3). At moderate concentrations Cro blocks p_M-initiated transcription, decreasing the synthesis of the CI repressor (Fig. 4). Later on, at higher concentrations, Cro blocks p_L- and p_R-initiated early transcription, which become by then dispensable for phage progeny production. As discussed in the preceding paragraph, neither synthesis of λ replication proteins, nor transcriptional activation of *ori*λ should be required for late λ DNA replication carried out according to the *sigma* mode by the previously assembled RC. Therefore, the postulated consumption of DnaA in switching *theta* replication from a bi- to a uni-directional mode (and later to *sigma*) should occur before p_R-initiated transcription is completely blocked by Cro.

During the late phase of infection, *sigma* (rolling-circle, Fig. 7) DNA replication produces long concatemers of λ DNA, up to 10 genome equivalents in length. These long linear DNA products are specifically cut at the *cos* sites and serve as the substrate for the phage packaging system.

5 Bacteriophage λ as a tool in molecular cloning

Derivatives of bacteriophage λ and its genetic elements are widely used tools in molecular cloning. First, λ p_R and p_L promoters and the *c*I gene provide an efficient system for expression of a cloned gene in a stringently regulated manner. The CI protein is a strong repressor of p_R and p_L, thus the leakiness of these promoters is very low in the presence of CI activity. The *c*I857 allele codes for the thermosensitive repressor and increase of temperature (e.g. from 30 to 42°C) causes derepression of either promoter. As both p_R and p_L are relatively strong promoters, such a derepression results in effective expression of a cloned gene.

Region *b* of the bacteriophage λ genome, together with some genes positioned between this region and the *N* gene (Fig. 1), may be replaced by any DNA which then can be replicated and propagated together with the rest of the phage genome (Sambrook et al., 1989; Chauthaiwale et al., 1992). Such gene vectors are often used for construction of genomic libraries. A packaging system has been developed which allows construction of infective λ virions *in vitro* from separately produced phage heads and tails and recombinant DNA. Near complete genomic libraries of higher organisms can be prepared by this type of cloning. Vectors based on bacteriophage λ are also excellent tools for construction of single-copy gene fusions, suitable for investigation of gene expression in *E. coli*, as it is usually easy to obtain single-copy λ lysogenes. The maximum size of DNA that can be cloned in typical λ vectors is about 20-25 kb. Even longer fragments, up to 45 kb, can be cloned in so called cosmids which are modified plasmid vectors harbouring the *cos* sequence of phage λ genome. After cloning of a fragment of foreign DNA into such a plasmid, the recombinant DNA can be packaged *in vitro* into λ heads forming recombinant cosmid virions which are excellent carriers of the cloned DNA.

Recently, a series of general-purpose plasmid vectors based on the phage λ origin of replication has been constructed. These λ plasmid vectors are compatible with most other vectors in common use, and are useful for routine plasmid cloning as well as other application (Boyd and Sherratt, 1995). An amplification of plasmid vectors *in vivo* is generally very useful, and above described λ plasmid vectors can be efficiently amplified in *E. coli* strains by a simple method (Węgrzyn, 1995).

Finally, λ phage expression systems that produce foreign proteins fused to the surface of the virus particle have been described. Thus, λ can be used, apart from already known phage M13 systems, as a nice vector for display of proteins on the virion surface. For example, in the λ systems, a foreign protein may be fused to the C-terminus of the truncated phage main tail protein V by a peptide linker (Maruyama et al., 1994), or to the N-terminus of the phage head D protein (Sternberg and Hoess, 1995). Such systems provide many advantages, for example in protein purification or immunological examination of the investigated protein, and perhaps might by used even in gene therapy.

Acknowledgements

The authors are very grateful to Dr. Mark S. Thomas for discussions and for critical reading of the manuscript. The authors were supported by the Polish State Committee for Scientific Research (grants: 6 P04A 051 08 to K.T. and 6 P04A 059 09 to G.W.), USA-Poland Maria Skłodowska-Curie Joint Found II (grant MEN/HHS-96-225 to K.T.) and the University of Gdańsk (grant BW-0010-5-0256-6 to G.W.).

References

Alfano C, McMacken R (1989) Heat shock protein-mediated disassembly of nucleoprotein structures is required for initiation of bacteriophage λ DNA replication. J Biol Chem 264: 10709-10718

Berg D (1974) Genes of phage lambda essential for λdv plasmids. Virology 62: 224-233

Boyd AC, Sherratt DJ (1995) The CLIP plasmids: versatile cloning vectors based on the bacteriophage λ origin of replication. Gene 153: 57-62

Chauthaiwale VM, Therwath A, Deshpande VV (1992) Bacteriophage lambda as a cloning vector. Microbiol Rev 56: 577-591

Dodson M, Echols H, Wickner S, Alfano R. Mensa-Wilmot K, Gomes B, LeBowitz JH, Roberts JD, McMacken R (1986) Specialized nucleoprotein structures at the origin of replication of bacteriophage λ: localized unwinding of duplex DNA by six-protein reaction. Proc Natl Acad Sci USA 83: 7638-7642

Echols H (1986) Bacteriophage λ development: temporal switches and the choice of lysis or lysogeny. Trends Genet 2: 26-30

Friedman DI, Court DL (1995) Transcription antitermination: the λ paradigm updated. Mol Microbiol 18: 191-200

Georgopoulos C, Welch WJ (1993) Role of the major heat shock proteins as molecular chaperones. Annu Rev Cell Biol 9: 601-634

Giladi H, Goldenberg D, Koby S, Oppenheim AB (1995) Enhanced activity of the bacteriophage λ P_L promoter at low temperature. FEMS Microbiol Rev 17: 135-140

Gottesman S, Clark WP, de Crecy-Lagard V, Maurizi MR (1993) ClpX, an alternative subunit for the ATP-dependent Clp protease of *Escherichia coli*. J Biol Chem 268: 22618-22626

Hendrix RW, Roberts JW, Stahl FW, Weisberg RA, eds. (1983) Lambda II. Cold Spring Harbor Laboratory Press, Cold Spring Harbor, NY

Herman C, Ogura T, Tomoyasu T, Hiraga S, Akiyama Y, Ito K, Thomas R, D'Ari R, Bouloc P (1993) Cell growth and λ phage development controlled by the same essential *Escherichia coli* gene *ftsH/hflB*. Proc Natl Acad Sci USA 90: 10861-10865

Herskowitz I (1985) Master regulatory loci in yeast and lambda. Cold Spring Harbor Symp Quant Biol 50: 565-574

Kleckner N, Signer E (1977) Genetic characterization of plasmid formation of N^- mutants of bacteriophage λ. Virology 79: 160-173

Klinkert J, Klein A (1978) Roles of bacteriophage lambda gene products *O* and *P* during early and late phases of infection cycle. J Virol 25: 730-737

Learn B, Karzai AW, McMacken R (1993) Transcription stimulates the establishment of bidirectional λ DNA replication in vitro. C S H Symp Quant Biol 58: 389-402

Liberek K, Georgopoulos C, Żylicz M (1988) Role of *Escherichia coli* DnaK and DnaJ heat shock proteins in the initiation of bacteriophage λ DNA replication. Proc Natl Acad Sci USA 85: 6632-6636

Maruyama IN, Maruyama HI, Brenner S (1994) λfoo: a λ phage vector for the expression of foreign proteins. Proc Natl Acad Sci USA 91: 8273-8277

Matsubara K (1976) Genetic structure and regulation of a replicon of plasmid λ*dv*. J Mol Biol 102: 427-439

Matsubara K (1981) Replication control system in lambda dv. Plasmid 5: 32-52

Matsubara K, Mukai T (1975) Mode of replication of plasmid λdv. J Biochem 77: 373-382

Obuchowski M, Giladi H, Koby S, Szalewska-Pałasz A, Węgrzyn A, Oppenheim AB, Thomas MS, Węgrzyn G (1997) Impaired lysogenisation of the *Escherichia coli rpoA341* mutant by bacteriophage λ is due to the inability of CII to act as a transcriptional activator. Mol Gen Genet, in press

Sambrook J, Fritsch EF, Maniatis T (1989) Molecular Cloning: A Laboratory Manual. Cold Spring Harbor Laboratory Press, Cold Spring Harbor, NY

Sternberg N, Hoess RH (1995) Display of peptides and proteins on the surface of bacteriophage λ. Proc Natl Acad Sci USA 92: 1609-1613

Szalewska A, Węgrzyn G, Taylor K (1994) Neither absence nor excess of λ O initiator-digesting ClpXP protease affects λ plasmid or phage replication in *Escherichia coli.* 13: 469-474

Szalewska-Pałasz A, Węgrzyn A, Herman A, Węgrzyn G (1994) The mechanism of the stringent control of λ plasmid DNA replication. EMBO J 13: 5779-5785

Szalewska-Pałasz A, Węgrzyn A, Obuchowski M, Pawłowski R, Bielawski K, Thomas MS, Węgrzyn G (1996) Drastically decreased transcription from CII-activated promoters is responsible for impaired lysogenization of the *Escherichia coli rpoA341* mutant by bacteriophage λ. FEMS Microbiol Lett 144: 21-27

Thomas R (1993) Bacteriophage λ: transactivation, positive control and other odd findings. BioEssays 15: 285-289

Węgrzyn A, Węgrzyn G, Taylor K (1995a) Protection of coliphage λO initiator protein from proteolysis in the assembly of the replication complex *in vivo.* Virology 207: 179-184

Węgrzyn A, Węgrzyn G, Taylor K (1995b) Plasmid and host functions required for λ plasmid replication carried out by the inherited replication complex. Mol Gen Genet 247: 501-508

Węgrzyn A, Węgrzyn G, Taylor K (1996a) Disassembly of the coliphage λ replication complex due to heat shock induction of the *groE* operon. Virology 217: 594-597

Węgrzyn A, Węgrzyn G, Herman A, Taylor K (1996b) Protein inheritance: λ plasmid replication perpetuated by the heritable replication complex. Genes to Cells 1: in press

Węgrzyn G (1995) Amplification of λ plasmids in *Escherichia coli relA* mutants. J Biotechnol 43: 139-143

Węgrzyn G, Taylor K (1992) Inheritance of the replication complex by one of two daughter copies during λ plasmid replication in *Escherichia coli.* J Mol Biol 226: 681-688

Węgrzyn G, Pawłowicz A, Taylor K (1992) Stability of coliphage λ DNA replication initiator, the λO protein. J Mol Biol 226: 675-680

Węgrzyn G, Szalewska-Pałasz A, Węgrzyn A, Obuchowski M, Taylor K (1995a) Transcriptional activation of the origin of coliphage λ DNA replication is regulated by the host DnaA initiator function. Gene 154: 47-50

Węgrzyn G, Węgrzyn A, Konieczny I, Bielawski K, Konopa G, Obuchowski M, Helinski DR, Taylor K (1995b) Involvement of the host initiator function *dnaA* in the replication of coliphage λ. Genetics 139: 1469-1481

Węgrzyn G, Węgrzyn A, Pankiewicz A, Taylor K (1996) Allele specificity of the *Escherichia coli dnaA* gene function in the replication of plasmids derived from phage λ. Mol Gen Genet 252: 580-586

Zvlicz M (1993) The *Escherichia coli* chaperones involved in DNA replication. Phil Trans R Soc London B 339: 271-278

Zylicz M, Górska I, Taylor K, Georgopoulos C (1984) Bacteriophage λ replication proteins: formation of a mixed oligomer and binding to the origin of λ DNA. Mol Gen Genet 196: 401-406

Żylicz M, Ang D, Liberek K, Georgopoulos C (1989) Initiation of λ DNA replication with purified host- and bacteriophage-encoded proteins: role of the DnaK, DnaJ and GrpE heat shock proteins. EMBO J 8: 1601-1608

Replication and Maintenance of Bacterial Plasmids

Christopher M Thomas[1*], Grazyna Jagura-Burdzy[1,2], Kalliope Kostelidou[1], Peter Thorsted[1], and Malgorzata Zatyka[1,2]

[1]School of Biological Sciences, The University of Birmingham, Edgbaston, Birmingham B15 2TT, UK, And [2]Department of Microbial Physiology, Institute of Biochemistry and Biophysics, Polish Academy of Sciences, Warsaw, Poland.

1 Introduction

By definition, plasmids are non-essential extrachromosomal elements. However, most bacteria carry at least one plasmid which can vary from a few kb to hundreds of kb and the plasmid component of the DNA in a bacterial cell can be as high as 10% or more. A large proportion of naturally occurring plasmids (as opposed to ones that have been modified so that they cannot transfer) can transfer by one means or other - conjugative transfer, transduction or even transformation. Since a plasmid does not need to recombine with endogenous DNA to become established in a new strain it allows the genes carried on the plasmid to be accessible to many strains and species where the plasmid can replicate. In addition, if conjugative plasmids become integrated into the bacterial chromosome they can promote the exchange of chromosomal DNA. They thus facilitate sex between bacteria which results in the recombination of bacterial genes. In this way the plasmid component of the genome provides a resource of biodiversity on which bacterial communities can draw under conditions of selective pressure. Those individuals who possess the best combinations of chromosomal and plasmid-encoded genes for a specific physical or nutritional niche will dominate but other genes can persist as minor components within the community or in adjacent communities.

This article will focus on molecular aspects of the ways that plasmids have learnt to replicate efficiently and in a controlled way and the auxiliary mechanisms they have acquired to ensure that they are not lost from a bacterial population at high frequency. It will then focus on the IncP plasmids which are now the best understood self-transmissible plasmids.

2 Replication mechanisms and their control

The most fundamental property of a plasmid is that it is a unit of autonomous replication. Most plasmids are capable of out-replicating the chromosome, by existing as multiple copies or by replicating during the horizontal transfer process. This can be important since the phenotype conferred by a plasmid is often copy number-related. The ability of a DNA molecule to replicate autonomously may arise in a number of ways, but once it has acquired this ability it is unlikely to revert to a non-replicating state: such a mutant will be diluted out very rapidly since any phenotype it confers will be maintained by the plasmid molecules which are still

NATO ASI Series, Vol. H 103
Molecular Microbiology
Edited by Stephen J. W. Busby,
Christopher M. Thomas and Nigel L. Brown

capable of self replication. A number of ways of initiating replication are known (Figure 1) and this provides a means of grouping replicons.

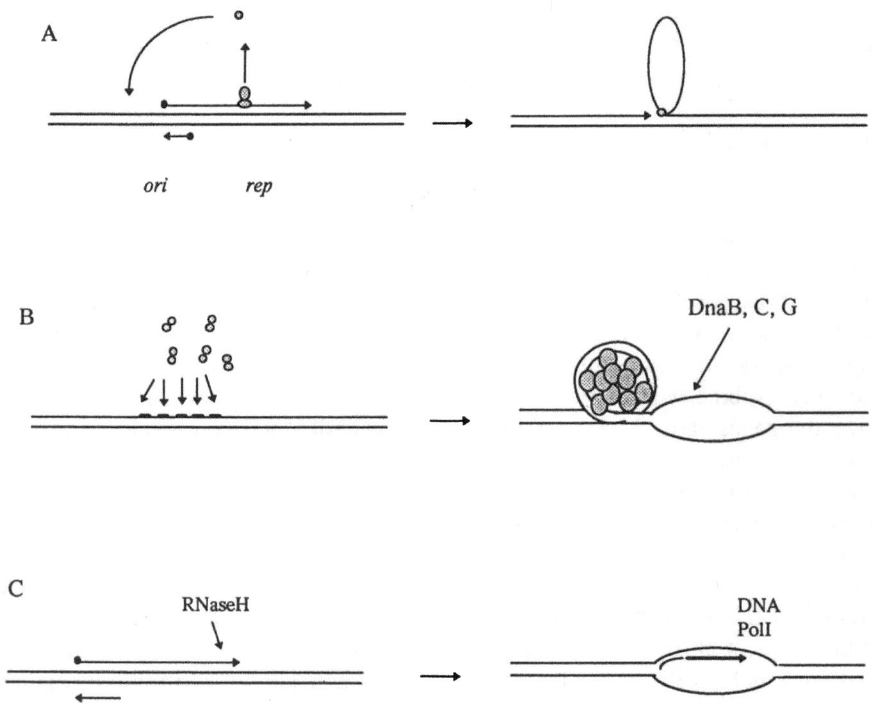

Figure 1. The three major modes of replication. A. Rolling circle replication in which initiation occurs by introduction of a nick at the origin by the Rep protein which then becomes covalently attached to the 5' end generated. DNA polymerase elongates from the exposed 3' end and displaces a single stranded copy which eventually is converted into double stranded form after initiation at the single strand origin. B. Iteron-activated theta-form replication in which binding of the Rep protein to a series of repeats termed iterons forms a complex which sequesters host DnaBC helicase to unwind the A+T rich region where leading strand synthesis is initiated. C. RNA transcript-initiated theta form replication as illustrated by ColE1. The RNAII transcript folds into a complex tertiary structure which remains associated with the replication origin where it is processed by RNaseH that has specificity for RNA-DNA hybrids. DNA Polymerase I uses this as a primer for initiation of leading strand synthesis. Control of initiation is via an antisense transcript, RNAI, which prevents RNAII from adopting the conformation it needs to associate with *ori* and be processed. Mutations in RNAI will have a complementary effect on the sequence of RNAII.

What controls replication of the bacterial chromosome is still not fully understood. However, it is known that initiation from all *oriC*s in the cell is co-ordinated at a particular time which is determined by cell mass. In contrast, plasmids do not replicate synchronously. They appear to replicate randomly with a frequency that increases as the cell grows, as recently confirmed even for plasmid F (Helmstetter et al., 1997). The molecule chosen for replication is selected randomly from the pool of available molecules in the cell. Despite this difference between control of chromosome and plasmid replication, the mechanisms by which plasmids regulate their replication and the way they maintain a certain average copy number during bacterial growth reveal important principles about bacterial molecular biology.

Pritchard, Barth and Collins (1969) proposed the inhibitor dilution model in which an inhibitor is synthesised in a burst after replication, effectively switching off further replication until the inhibitor is diluted out by cell growth. That replication can by switched off by a plasmid-specific inhibitor was indicated by the studies on ColE1-pSC101 hybrids performed in the laboratory of Stanley Cohen (Cabello *et al.*, 1976). Molecular studies have shown that numerous plasmids, both high copy number (ColE1 and pT181) and low copy number (the IncFII plasmids R1 and NR1), encode an antisense RNA molecule which can either block formation of primer for initiating leading strand synthesis, cause termination of transcription of the essential *rep* gene, or block translation of the *rep* mRNA (Wagner & Simons, 1994).

Antisense RNA as a controlling element has a number of advantages for a plasmid which may explain why it is so commonly found. First, it is easy to generate an antisense control circuit to regulate a replication event: all that is needed are mutations which generate a new constitutive promoter firing in the opposite direction to the replication signal. The product from such a promoter will accumulate in proportion to plasmid copy number and therefore inhibition should increase as copy number rises, providing a feedback loop which self-regulates copy number. Second, the half-life of most RNA species is quite short and therefore the level of the inhibitor should respond rapidly to changes in plasmid copy number. Third, the fact that the target and the inhibitor are encoded by the same piece of DNA allows the system to evolve rapidly - mutations which change the system will preserve the complementarity of the inhibitor and its target but can make it no longer susceptible to inhibition by the parental type.

An alternative control system is the handcuffing model which proposes that plasmids with iterons form complexes between their *oriV* regions mediated by their Rep protein. These complexes are proposed to block access to the origin by host replication functions and therefore stop replication until the cell has grown further. It is unclear whether these complexes should form straight after replication when the local concentration of *oriV* DNA is high or only when the average copy number rises. It is also not clear whether such complexes should break down passively as the cell grows or whether separation would depend on an active process as would occur during a partitioning cycle akin to the mitotic cycle. Although the handcuffing model for control currently has no rivals for plasmids that do not apparently depend on a diffusible RNA or protein repressor, there are many aspects of the model still to be tested.

2.1 Rolling circle replication

The most ubiquitous replicons use rolling circle replication (RCR; Gruss & Ehrlich 1992; Novick 1989). Such plasmids have been shown to depend on what is termed the double strand origin (*dso*), at which one strand is nicked to create a free 3' end to prime elongation by DNA polymerase III. This initiates copying of one strand and displacement of the other to create a free single strand, which becomes coated with single stranded DNA-binding protein. The *dso* is normally located just upstream of the gene, *rep*, encoding the protein that nicks the *ori*, or actually within the *rep* gene itself. There is also a requirement for a single strand origin which consists of an inverted repeat that allows RNA polymerase to generate a primer for lagging strand synthesis (Meijer *et al.* 1995). Control of RCR plasmid copy number and incompatibility can be mediated by protein repressor-operator interactions to regulate transcription as well as through the action of antisense RNA on transcription and translation of the gene.

2.2 Theta replication initiated by iterons

The second type of replication initiation is that in which the origin of theta replication is activated by binding of a Rep protein to a series of repeated sequences, termed iterons. This includes replicons such as the FIA replicon of F (Muraiso *et al.* 1987), the FIB replicon of P1 (Brendler *et al.* 1991), the origins of R6K (Filutowicz *et al.* 1986), RK2 (Perri *et al.* 1991), pSC101 (Manen & Caro, 1991) and others. In many cases DnaA protein and/or additional host proteins are needed for activation of the vegetative replication origin, *oriV*. The exact mechanism of activation is probably not identical in all plasmids of this type although it clearly involves induced melting of *oriV* sequences, normally within a short A+T-rich region (Mukhopadahyay & Chattoraj, 1993), followed by sequestration of DnaB/C helicase by the Rep protein to unwind the replication bubble. Although these specific interactions with host proteins might be expected to place constraints on plasmid host range, related plasmids are found in many species. Some of these plasmids that appear to have a restricted host range can extend their host range by simple point mutations (Fernandez-Tresguerres *et al.*, 1995). In the IncP plasmids, production of two related polypeptides may provide flexibility in the interaction with host protein (Durland & Helinski, 1987; Fang & Helinski, 1991; Shingler & Thomas, 1989). Copy number control and plasmid incompatibility appear to be due to Rep protein-iteron interaction and a handcuffing model has been proposed in which Rep binds together the replicated *oriV* regions so that access by replication functions is blocked and further initiation can only take place when the plasmids have been pulled apart (Kittel & Helinski, 1991; McEachern *et al.* 1989; Pal & Chattoraj, 1988).

2.3 Theta replication initiated by a processed RNA polymerase transcript

A third common replicon is that in which theta replication is primed by a processed form of a transcript produced by RNA polymerase (Itoh & Tomizawa, 1980). The best studied example of such a plasmid is ColE1 where a 500 nt preprimer transcript (RNAII) folds into a complex secondary and tertiary structure, allowing association with the replication origin, before RNaseH cleavage of the RNA in the RNA-DNA hybrid creates the 3' end from which DNA PolIII primes the leading strand synthesis. Control of replication occurs through RNA I, an antisense RNA which interacts with RNAII and sequesters it into an inactive form if attack occurs within a small window of time during the transcription and folding reaction (Tomizawa, 1986). Inactivation is irreversible. Interaction between RNAI and RNA II depends on kissing reactions between single stranded loops formed by the sequence-dependent folding. Mutations which alter these reactive loops change the rate of interaction and can significantly alter interactions between repressors from different plasmids (Lacatena & Cesareni, 1983). The plasmids like pAMβ1 from Gram positive bacteria may also use a similar mode (Bruand *et al.* 1993).

3 Auxiliary stable inheritance functions

A number of additional sets of genes are found on plasmids and contribute to ensuring that plasmid-free bacteria do not arise at high frequency (Figure 2). They are relevant to plasmid classification because they are often associated with particular replication systems. In some cases, otherwise-unrelated plasmids with the same auxiliary stability system may compete for stable inheritance, which in turn may lead to an incompatibility which is a commonly used method of plasmid classification.

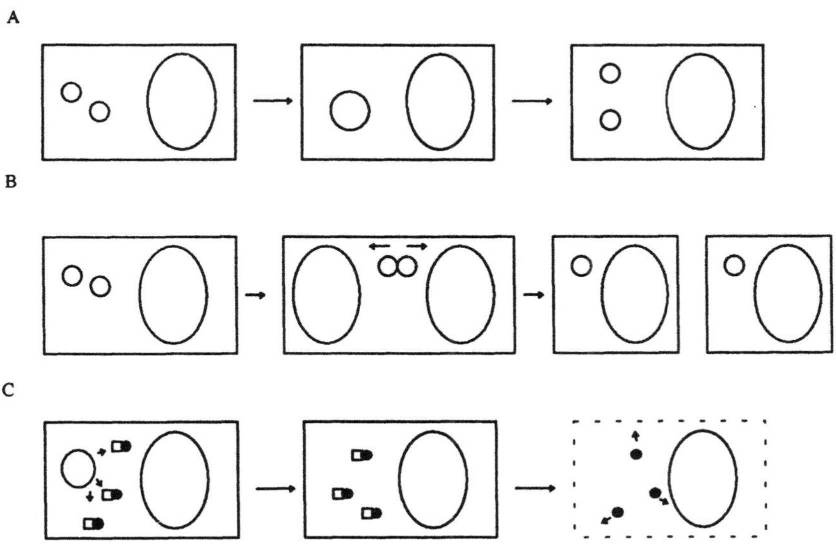

Figure 2. Principles of auxiliary stable inheritance functions. A. Multimer resolution. Homologous recombination leads to dimer formation while multimer resolution converts these back to monomers. B. Active partitioning. The currently favoured model is that an as yet poorly understood mechanism leads to pairing of plasmid molecules followed by active movement to either side of the cell division plane. C. Lethality to plasmid-free segregants. The plasmid produces molecules, illustrated here by the CcdA and CcdB proteins of F, which are left behind when the plasmid is lost. One of the molecules is responsible for host cell death while the other one prevents this effect. Differential decay of the two molecules unmasks the lethal effect.

3.1 Multimer resolution

Homologous recombination between identical plasmid molecules generates dimers or higher multimers. This can be a problem for both high and low copy number plasmids and consequently multimer resolution functions are found on all plasmids which replicate via a theta form. For low copy number plasmids, which replicate very few times in the cell division cycle, dimer formation can prevent active partitioning (see below). For high copy number plasmids, dimers have been proposed to be able to out-compete monomers because they have a greater chance of being picked for replication due to their possession of multiple origins of replication. For plasmids that control their replication by constitutive expression of a repressor in proportion to plasmid copy number, each replication of a dimer counts as two replication events and will appropriately reduce the chance of further replications. In addition the next round of replication has an even greater chance of picking a dimer because there are now two in the cell. Thus dimers tend to accumulate. This is called the dimer catastrophe hypothesis (Summers *et al.* 1993). The *cer* system of ColE1 is one of the best studied systems which promotes site-specific resolution of such dimers (Summers & Sherratt, 1984). There is circumstantial evidence that ColE1 dimers, that are in the process of resolution, synthesise an inhibitory RNA, termed Rcd, that blocks host cell division until after the dimer is resolved, thus further decreasing the chance of plasmid loss (Patient & Summers, 1993). Such a situation would fit with the idea that resolution of dimers is slow whereas the rate at which synapsis of the recombination site occurs is rapid.

Plasmids such as R6K which replicate bidirectionally to a terminus (Kolter & Helinski, 1978) may not respond to dimerisation in a simple way because for full replication of a dimer the replication fork must pass the terminus region. Also, in RCR plasmids the replication process will terminate when it reaches a second origin and this should generate monomer daughters even on a multimer template, so again this mode of replication may disfavour dimer accumulation (Gruss & Ehrlich 1992; Novick 1989).

3.2 Active partition

For low copy number plasmids, active partitioning is necessary to ensure that the plasmid is distributed in a non-random way between the cell compartments which define the next two daughter cells (Williams & Thomas, 1992). The rate of loss due to random partitioning depends on the number of independently segregating molecules present at septation. One problem with analysing partitioning functions is that we do not know whether plasmids behave as completely independently segregating molecules nor do we have very accurate ways of determining absolute plasmid copy numbers per bacterium. The distribution of plasmid copy numbers in individual bacteria may have a significant effect on plasmid stability because those bacteria at the low end of the copy number distribution may have a much greater chance of losing the plasmid than those at the middle and top end of the distribution. Thus the plasmid replication control circuits which prevent over-replication but stimulate replication when the copy number falls are important when considering plasmid loss rates.

Active partitioning is proposed to work via a plasmid pairing step (Austin, 1988; Williams & Thomas, 1992). Most simply this step could occur after replication has taken place and paired molecules could then be distributed to daughter cells by some sort of a mitotic process. The family represented by the *sop* and *par* genes of F and P1 are found on many plasmids (Motallebi-Vesharch *et al.*, 1990; Lobocka & Yarmolinsky, 1996) and relatives of these genes are also found on the chromosome of many bacterial species (Colin Bignell & CMT, unpublished) near the replication origin, *oriC*. The ParA protein is an ATPase which may provide the energy to drive the partition cycle. The ParB protein is normally a DNA-binding protein whose target is the *cis*-acting sequence which is needed for the cycle of partitioning and is often termed the centromere-like sequence by its analogy to the centromere in the mitotic cycle of eukaryotes. Different plasmids with the same *par* system may compete with each other but if the selection of pairs occurs preferentially between daughter molecules directly after replication, this may suppress the element of competition.

3.3 Lethality to plasmid-free segregants

A third class of stability system works by decreasing the chance of survival for plasmid-free segregants. One such mechanism is the production of plasmid-encoded bacteriocin to which plasmid-positive bacteria are immune. In this case the killing of plasmid-negative bacteria is from the outside. Stable plasmids which acquire an immunity system may have an advantage because their host will not die if it loses a competing plasmid carrying the bacteriocin gene. Of course bacteriocins are also an advantage for invading a community that has not encountered the bacteriocin before. Another mechanism to prevent survival of plasmid-free segregants is provided by genes encoding a lethal product whose effect is blocked when the plasmid is present but which becomes active when the plasmid is lost. The best studied examples of these are the *hok/sok* system of plasmid R1 (Gerdes *et al.* 1986) and the *ccd* system of F (Bernard & Couturier, 1992).

In the *hok/sok* system the *hok* (host killing) mRNA initially folds into a translationally inactive form. The way this appears to happen teaches us a great deal about mRNA structure and function. Translation of *hok* depends on coupling to the previous ORF, *mok*, which overlaps with *hok*. As the *mok/hok* region is transcribed the 5' end folds into what is termed a metastable loop which blocks translation of *mok* (Franch & Gerdes, 1996). However, when transcription is complete, the 3' end of the message loops back forming a favoured structure which displaces the original metastable structure and blocks translation of *mok* by sequestration with the 3' sequences. Degradation from the 3' end slowly removes these sequences and releases the mRNA into a translationally active form but one which is also sensitive to inactivation by the *sok* antisense RNA which is produced constitutively when the plasmid is present (Thisted *et al.*, 1995). When the plasmid is lost, the Sok RNA is degraded rapidly so that *hok* mRNA is translated and Hok kills the bacteria by causing membrane depolarisation (Gerdes *et al.* 1986). *hok/sok* are part of an extensive family whose phylogeny is very interesting (Gerdes *et al.* 1990). Some homologues are also encoded on the chromosome. Their function is not understood at present.

The *ccd* system differs in that killing by CcdB, which poisons DNA gyrase so that it introduces nicks into the bacterial DNA, is suppressed by an antidote protein, CcdA. On plasmid loss CcdA is inactivated and thus unmasks the killing effect of CcdB (Bernard & Couturier, 1990). With all of these systems, two different plasmids with the same stability mechanism and with the same regulatory or antidote specificity, will compete with each other. The normal consequence of plasmid loss will be suppressed when one of the two plasmids is lost: the plasmid that survives because it has a back-up system will have a considerable advantage and this may explain some of the advantage to the plasmid of acquiring a number of apparently redundant stability systems.

4 IncP plasmids

The IncP plasmids are important for bacterial genetics for a number of reasons (Thomas & Smith, 1987). They have a broad host range and are important agents for gene spread in complex bacterial communities. They have been found, not only found in clinical contexts (Ingram et al. 1993), but also in soil and aquatic environments (Hill *et al.*, 1992; Gotz *et al.* 1996). They carry a variety of phenotypic markers including antibiotic resistance, toxic ion resistance and ability to degrade xenobiotics such as chlorinated benzoate. Studies to date have shown the group to be quite diverse (Smith and Thomas, 1989). IncP plasmids include not only two major sub-groups (α and β) of single-replicon plasmids but also plasmids which seem to be dual replicons, or which have remnants of an IncP plasmid joined to a second plasmid. The recently isolated IncP plasmids seem to be a mixture: some are closely related to previously studied plasmids while others show considerable further divergence. They illustrate plasmid evolution within a single family. To help in this a variety of tools have been developed for identifying and classifying IncP plasmids (Gotz *et al.* 1996).

The best studied IncPα plasmids are RK2/RP4/RP1/R18 which are virtually identical to each other and probably represent different members of a clone that appeared and became widespread in 1969 (Ingram *et al.* 1973). The complete sequence of the 60099 bp genome has been compiled and the general properties reviewed (Pansegrau *et al.* 1994). These plasmids confer resistance to ampicillin, tetracycline and kanamycin, the latter two determinants being characteristic of this subgroup. Many derivatives of these plasmids with other transposons inserted have been isolated and probably arose in quite a narrow window of time (Villaroel *et al.* 1980).

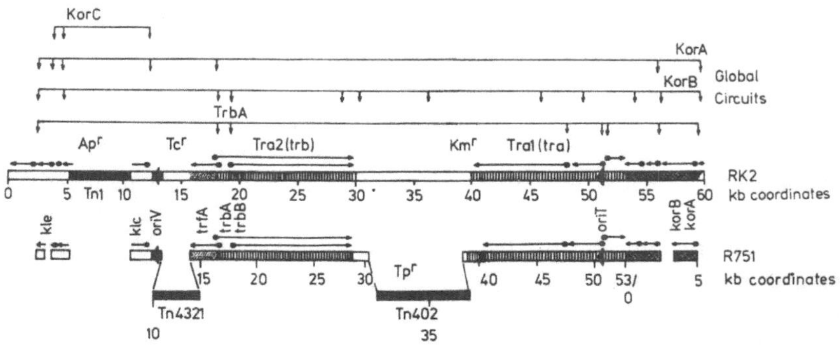

Figure 3. Alignment of the backbone of the RK2 and R751 showing the conserved replication (stippled regions), transfer (vertical hatching) and central control region (criss-cross hatching). Transposable elements are shown as solid blocks. Also shown are the regulatory circuits provided by KorA and KorB, encoded in the central control region, TrbA, encoded between the replication and mating pair formation operons, and KorC encoded in the *klc* operon.

The IncPβ plasmids are a much more diverse set of plasmids in the phenotypic characteristics and the transposable elements they carry. There seem to be two sub-sub-groups. R751 at present seems to be on its own while the cluster of R906, R772 and pJP4 all have an identically located insertion of a transposon which carries mercury resistance genes, indicating that they have a common ancestor which is different from R751 (Smith & Thomas, 1987). It is possible that there is a third group related to pBR60 (Burlage *et al.*, 1990) but further studies are required to substantiate this. Apart from R772 in which the *trfA*/*trb* region has been inverted, IncPα and IncPβ plasmids share a common organisation of replication and maintenance functions (Pansegrau & Lanka, 1987; Smith & Thomas, 1987). We have now completed the R751 sequence (Thorsted *et al.* unpublished; Accession number U67194) and this will help to define the IncP backbone.

IncP plasmids can be identified in a number of ways. Classical incompatibility tests have been used with both complete IncP plasmids RK2/RP4 (IncPα) or R751 (IncPβ) and derivatives such as the mini-*gal* derivatives of Davey *et al.* (1984). To our knowledge there is no good way using incompatibility to distinguish the α and β subgroups. However, reciprocal tests can identify those plasmids which appear to be hybrids or dual replicons which displace but which are not displaced by an IncP plasmid (Thomas & Smith 1987). IncP plasmids have also been identified by using the hybridisation probes of Couturier *et al.*, (1988) and this can be much more efficient for screening many newly isolated plasmids (Amuthan & Mahadevan, 1994). PCR probes have also been developed, for *oriV*, *trfA*, *oriT* and *korA*, allowing one to search for a source of IncP plasmids without having to isolate the particular strain carrying the plasmid (Gotz *et al.* 1996). Both the hybridisation and the PCR probes provide the potential to distinguish α and β plasmids since there are well defined RFLPs which differentiate between the two groups (Chikami *et al.*1985; Yakobsen & Guiney, 1983; Lanka *et al.* 1985). Degenerate PCR primers seem to provide a more sensitive way of identifying the plasmids. Thus the PCR products which represent true positive signals sometimes show too little sequence similarity to archetypal IncP plasmids to give a positive hybridisation signal. The complete sequence of both α and β plasmids provides a resource for defining primers for any additional region that may be of interest.

To determine the degree of similarity of a newly isolated IncP plasmid to the α and β archetypes a good method is restriction fragment length identity using restriction sites that occur frequently in the backbone combined with Southern blotting using probes for key regions of the IncP backbone. From the DNA sequence we recommend that *Sph*I and *Not*I are the most appropriate. The number of fragments produced for the major backbone sectors should allow an estimate of how similar the new plasmid is to one of the archetypes. In addition, if one observes conservation of a series of fragments which are contiguous in the archetype, then it should be possible to deduce that the whole of that portion of the backbone is conserved in the new plasmid, since it must be assumed that the restriction sites that define the ends of the fragments of identical mobility are the same sites. This can minimise the amount of work that needs to be done in studying a new plasmid.

Detailed molecular analysis has only been carried out on the IncPa plasmids RK2 and RP4, so the following sections deal mainly with work on these plasmids. However, sequence analysis of R751 and the limited functional analysis done on it so far, indicate that these features are conserved between both branches of the family.

4.1 The IncP replicon

An essential part of the plasmid core is the replication region which consists of two loci: the replication origin, *oriV* and the *trfA* gene which encodes the proteins that activate *oriV*. Replication proceeds unidirectionally from *oriV*. The minimal origin consists of five binding sites for TrfA (iterons), a DnaA binding site, an A+T-rich region where melting to initiate the replication bubble occurs, and an adjacent G+C-rich region which is proposed to aid the melting process. The DnaA binding site is essential in all species tested except *P.aeruginosa*. The *trfA* gene exhibits a structure which is quite common on the IncP plasmids: the possession of two translational starts so that the gene produces two related polypeptides, of 285 (TrfA2) and 382 (TrfA1) amino acids respectively. Deletions which inactivate the larger product do not affect replication in most species but do affect efficiency of replication in *P.aeruginosa*. Interestingly, the *trfA* gene of R751 shows a high degree of conservation in the coding region for TrfA2 but considerable divergence in the N-terminal extension which is unique to TrfA1. Tetracycline resistant derivatives of R751 which can be efficiently selected in *P.aeruginosa* show a reduced efficiency in establishment in this species which could be due to this divergence (Thorsted et al., 1996).

The complete *oriV* region has five more TrfA iterons. Upstream of the minimal *oriV* there are four, arranged as a single repeat, then an uneven group of three before the essential five. An extra copy in inverted orientation is present downstream. When *oriV* is present on a high copy number plasmid which would otherwise be compatible with an IncP plasmid, the plasmid will displace the IncP replicon (Thomas *et al.* 1981; Thomas *et al.* 1984). It is conceivable that this is the result of titration of TrfA, the IncP replication initiation protein. However, dissection of the *oriV* region shows that the inhibition observed is not directly proportional to the number of TrfA binding sites within the cloned *oriV* region (Shah *et al.* 1995). Since the TrfA binding sites are organised as a single repeat, an unevenly spaced group of three and an evenly spaced group of five it is possible that not all sites have the same "value". Indeed in experiments to study the ability of cloned regions of the plasmid to interfere with replication of a resident IncP plasmid, the single repeat and the group of three repeats have more inhibitory effect per TrfA binding site than the group of five (Shah *et al.* 1995). It has been proposed that plasmid replication may be controlled by what is termed "handcuffing", in which the Rep protein TrfA links together *oriV* regions after replication so that the origin is inaccessible for replication until the molecules have been pulled apart (Kittel & Helinski, 1991). The single repeat and the group of three repeats may be more potent in forming such complexes under the conditions tested (Shah *et al.* 1995).

Although handcuffing or a related phenomenon may be the reason why IncP plasmid replication does not increase exponentially when TrfA is provided, it is not clear that this is the only control. The *trfA* promoter is tightly regulated by products of the central control operon and also *trbA* (Jagura-Burdzy et al., 1992). Supply of excess TrfA in trans appears to cause a slight rise in copy number (Durland et al., 1990) as if the plasmid copy number is normally limited by the supply of TrfA but that if this limitation is removed then it hits a second barrier, which is proposed to be provided by handcuffing.

4.2 The IncP transfer system

The IncP transfer system is now one of the best understood because of the detailed analysis carried out by E. Lanka and coworkers in Berlin (reviewed in Pansegrau et al., 1994). A complete summary is beyond the scope of this article. The genes for the transfer are organised in two blocks, termed Tra1 and Tra2. The Tra1 region contains *oriT*, the transfer origin at which one strand is nicked to initiate rolling circle replication. This drives one strand of the plasmid into the recipient through the mating bridge. Divergent promoters flanking *oriT* direct the transcription of the genes *traKLM* in the clockwise direction and *traJIHGFEDCBA* in the anticlockwise direction. A third promoter is located upstream of *traG*. A nucleic acid-protein complex, termed the relaxasome, consisting of TraI, J, TraH and K assembles at *oriT* and is required for the nicking. These proteins also repress transcription from *traJp* and *traKp*, providing a way to limit their production when each *oriT* region is sequestered in a relaxasome (Zatyka *et al.*, 1994). The *oriT* region and the relaxasome proteins show similarities to functionally related genes in many Gram-negative and Gram-positive plasmids (Lanka & Wilkins, 1994). *traC* encodes a primase that can prime replication on the transferred strand. Host primase can substitute during the transfer to many species but in some crosses such as to

Salmonella and at lower temperature the fact that the IncP plasmids carry their own primase seems to increase the efficiency of formation of transconjugants.

The Tra2 region encodes most of the genes required to assemble the mating pair apparatus. This group of proteins is closely related not only to those of other *mpf* systems (for example, IncN and IncW) but also to the *vir* system of the Ti plasmid of *Agrobacterium tumefaciens* (Lessl *et al.* 1992; Motallebi-Veshareh *et al.* 1992). The *mpf* system is also related to the toxin export system of *Bordetella pertussis*. Individual genes, such as TrbB show similarities to even more widely spread functions involved in assembly of membrane-associated complexes.

4.3 Additional stable inheritance functions

The IncP backbone encodes a number of genes which are coregulated with the replication genes. However, apart from the partitioning functions associated with the central control region and described below none of these genes have been shown to be required for stable inheritance. By contrast, a group of five genes which are located between the *trb* and *tra* operons provide two types of auxiliary stability function. One of these encoded by *parABC* involves a site-specific recombinase in what is proposed to be a partitioning rather than a multimer resolution system (Eberl et al., 1994). Two other genes, *parDE*, provide a function which is lethal to plasmid-free segregants (Roberts *et al.*, 1994). These functions are not found in the IncPβ plasmid R751, so it seems likely that they were relatively recently acquired. If the ancestral plasmid had possessed these functions it is difficult to see how any plasmid that lost them would have survived the competition from its parent.

4.4 The central control region

The central control region includes the two global transcriptional repressors, KorA and KorB. This region also encodes an active partitioning mechanism, related to the F and P1 *sop* and *par* systems, which depends on both KorB and the putative ATPase IncC (Motallebi-Veshareh *et al.* 1990). We observed a number of years ago that the effect of *incC* in displacing IncP plasmids was dependent on the simultaneous presence of *korB*, on either the resident or the incoming plasmid (Thomas, 1986). Recent data suggests that the basis of this effect is that IncC increases the binding of KorB to at least some of its target operators (GJB & CMT, unpublished). One of these targets is the *trfA* promoter which is responsible for transcription of the essential replication gene *trfA*. The combined effect of *korA* and *korB* potentiated by *incC* appears to be sufficient to shut off *trfA* so tightly that replication can not occur efficiently. Since removal or mutation of the central control region has been demonstrated to reduce IncP incompatibility it may be that this is the basis of the major source of incompatibility between IncP plasmids.

It appears that the *incC korB* stability system in the IncPα plasmids is less effective than the equivalent function in the IncPβ plasmids (Macartney *et al.* 1997). This correlates with the fact that in IncPα plasmids there is a second partition/stability system (Gerlitz *et al.*, 1990) and this may have decreased the pressure on these plasmids to retain the efficiency of the *incC korB* system As a general point this illustrates the idea that when two systems with similar function are present on the same plasmid there may be more pressure to maintain the regulatory circuits of the individual systems than their original stabilising function.

Detailed analysis of the partitioning phenotype of the central control region shows that it depends only on the *incC* and *korB* genes and the KorB operator between the *upf54.4* and *upf54.8* genes, designated OB3. The region analysed includes two other OBs, OB1 which has a similar affinity for KorB as OB3, and OB2 which is about 10-fold weaker. Our test system consists of a shuttle plasmid that replicates from the high copy number pMB1 replicon in a $PolA^+$ strain but from the low copy number P7 replicon in a $PolA^-$ strain. The rate of loss per generation is about 4% in the absence of an active partitioning system. The central control region decreases the rate of loss to 1 to 2% per generation. Inactivation of *incC* or *korB* results in return to the unstable phenotype. However, inactivation of OB3 results in a less stable plasmid than the one without any partitioning system, That is, it is actively destabilised. Deletion of OB1 or inactivation of *incC* or *korB* removes this active destabilisation. Our current hypothesis is that destabilisation occurs because KorB can link two plasmids that have high affinity OBs, but that active partitioning can only be achieved by OB3. Therefore, when OB3 is inactivated the number of separate plasmids is reduced by KorB pairing and the plasmid is lost at a higher rate (Williams, Macartney & Thomas, unpublished). This provides good genetic evidence for the existence of a cycle of pairing and separation in plasmid partitioning.

It is of interest that homologues of *incC* and *korB* are found on a growing number of plasmids from *Agrobacterium*, *Rhizobium* and other species where they are located in the same operon as the essential *rep* gene and appear to play an auxiliary role in plasmid stability (Nishiguchi *et al.*, 1987; Tabata *et al.*, 1989) The basis for the incompatibility and replication control of these plasmids is of interest since in *Rhizobium* there seem to be many coexisting plasmids of this sort, derived from a common ancestor (Turner *et al.* 1996).

4.5 Coordination of core plasmid functions

The co-ordination of replication and transfer functions provided by the central control operon is unique to the IncP plasmids. KorA (101 amino acid subunits that form a dimer and possibly a tetramer) has seven binding sites on the IncPα genome, all near promoters, while KorB has twelve binding sites, some near promoters, others in inter- and intra-genic regions. There appears to be a hierarchy of binding strengths in both regulons. For the KorA regulon in particular we have shown that although the 10bp palindrome at the centre of the operator is essential for binding, it is not sufficient for full binding efficiency. An additional 2 bases to create a 12 bp palindrome are needed for class I operators along with three more bp at one end

(Jagura-Burdzy & Thomas, 1995). Mutational analysis of KorA has implicated the C-terminal region of the helix-turn-helix motif in recognising the outer ends of the operator. A highly basic C-terminal domain of the whole protein is implicated in co-operative interactions between KorA and KorB at promoters where the KorA operator overlaps the -10 region and KorB binds upstream of the -35 region (Kostelidou & Thomas, unpublished). One of the most interesting functions of KorA is to control a switch between the *trfAp* and the *trbAp* (Jagura-Burdzy & Thomas, 1994), the latter controlling production of TrbA, a third global regulator which provides cross-talk between the replication and transfer genes (Jagura-Burdzy & *et al.*, 1992).

KorB (364 amino acid subunit which forms both dimer and tetramer) binds at twelve sites on the backbone (Balzer *et al.*, 1992; Williams *et al.*, 1993). Some of these sites at close to promoters (*korAp, kfrAp, trfAp, klaAp*), or with 100 to 200 bp of a promoter (*trbBp* and *kleAp*). At all these sites it appears to repress transcription but how it does so when paced at a considerable distance is not known (Donia Macartney, GJB, KK, MZ & CMT, unpublished). Its other six sites are between or within genes and the function of KorB binding at these sites is not known. KorB contains a putative helix-turn-helix motif but the involvement of this in DNA recognition has not been confirmed (Theophilus & Thomas, 1987). Recent studies have shown that KorB can form higher order complexes on DNA containing a KorB operator and that the properties of these complexes are consistent with them representing paired molecules, held together by KorB (CMT & GJB, unpublished).

TrbA is encoded at the start of the *trb* operon (Jagura-Burdzy *et al.*, 1992). It is transcribed from a weak promoter that is face to face with the strong *trfA* promoter. When *trfAp* is unrepressed, *trbAp* is completely blocked but is switched on as the level of KorA in cell rises (Jagura-Burdzy & Thomas, 1994). TrbA has a monomeric size of 123 amino acids. There is an N-terminal DNA binding domain which is remarkably similar to phage and sporulation inhibitors of *Bacillus subtilis*. The C-terminal region shows a high degree of conservation with the C-terminus of KorA and we think that this indicates a common interaction with KorB (KK & CMT, unpublished). TrfB represses not only the Tra and *trb* promoters but most of the other backbone promoters as well (GJB & CMT, unpublished).

4.6 Evolution of the IncP backbone

The sequence comparison of the IncPβ IncPα plasmids reveals deletions and insertions which identify recently acquired genes as well as nonessential functions. It shows regions where there are strong functional constraints while other sectors seem to be much more prone to drift. For example, the *kfrA* gene, for which we have not yet been able to identify a function in plasmid maintenance or transfer, has a particularly high G+C content (70 to 80% as compared to the 60 to 65% for most of the rest of the backbone) and encodes a highly repetitive protein. The sequence comparison shows that many features of the promoter region and the polypeptide product in the two plasmids are conserved but that the iterated nature of the gene sequence with many repetitive G+C tracts has allowed a high degree of sequence drift, whose significance has yet to be established.

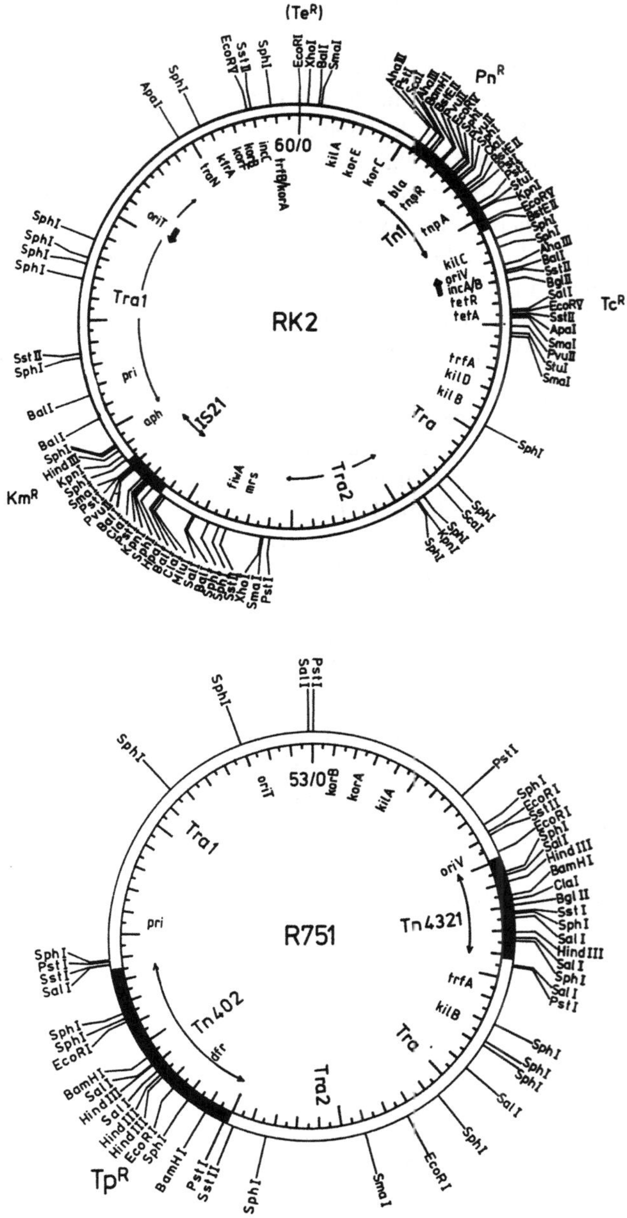

Figure 4. Physical and genetic maps of RK2 and R751 illustrating the clustering of restriction sites in the regions of the plasmid where there are phenotypic markers and transposable elements which are likely to have been acquired relatively recently.

A characteristic of DNA sequence of the backbones of the IncP plasmids is the high G+C content and the deficiencies in sites for many commonly used restriction enzymes. It was proposed many years ago that this deficiency is due to selection against sensitivity to restriction in a replicon which can replicate in many different hosts (Thomas *et al.*, 1980). The plasmid maps show clusters of restriction sites in the regions where transposable elements and phenotypic markers are found (Fig. 5; Pansegrau & Lanka, 1987; Smith & Thomas, 1987). The fact that these region are often of a lower G+C content fits with their being relatively recent acquisitions. It is proposed that the backbone has been subject to long-term selection against sensitivity to restriction barriers whereas the newly inserted DNA has not. This hypothesis has been given more credibility and specificity now as a result of studies by Wilkins *et al.* (1996) who showed that the IncPα backbone sequence is deficient in those hexanucleotide palindromes that are recognised by restriction enzymes that are found within Pseudomonas. On the other hand the backbone is abundant in GCCG and CGGC sequences which are characteristic of Pseudomonas genomes and this results in a surprisingly high frequency of restriction sites, for example *Not*I (5'GCGGCCGC3') which contain or are composed of these tetranucleotides. It is therefore suggested that plasmids may adapt with time to the genomic signatures of their host, which for IncP plasmids may have been the pseudomonads. The restriction site deficiency can be explained, not by transfer among all Gram-negative species but more simply by transfer among different Pseudomonas species. The reasons why bacterial genomes have different G+C contents is not known but the sequence characteristics of particular species may provide a more comprehensive basis for analysing plasmid-host relationships and deducing patterns of gene spread.

The IncP backbone sequence of the IncPβ plasmids has very similar sequence characteristics to the IncPα backbone. However, one unique feature stands out: a sequence motif that is found in the two regions of the plasmids which are almost invariably interrupted by insertion of transposable elements. When the *oriV-trfA* region of R751 and R906 was compared it was clear that although this region in both plasmids had transposon insertions, the sites of insertion were in different places so

oriV-trfA of R751				
		<<<<<<<<IVR >>>>>>>>		
R1	CTCGAG	CATCGCCACTTCCGGCGAGG	CAGTACCGCGC	*Xho*I-(*Acc*/*Sst*II)
R2	CTGCAG	CATCGTCACTTCCGGCGAGG	TGGCCA	*Pst*I-*Msc*I
R3	GTCGAC	CATCGCCACATCCAACGATG		*Sal*I
R4	GGGCCC	CATCGCCACTTCCGACGATG	ACGGCCG	*Apa*I-*Eag*I
R5	GGATCTC	CATCGCCACTTCCGACGATG	GAGATCC	(*Bam*HI)-(*Sac*I/*Bgl*II)
Tn*402* region in R751				
R6	CTCGAG	CATCGCCGCATCCGACGATG		*Xho*I
R7	CTGCAG	CATCGCCATTGCCGACGATG	GCACC	*Pst*I-(*Nar*I)
R8	GAGCTCG	CATCGCCACATCCGACGATG	CGCGCCGGC	*Sac*I-*Bss*HI/*Nae*I
R9	GTCGAC	CATCGCCAGGTCTGACGATG	GCGGCCGC	*Sal*I-*Not*I
R10	GGATCTCG	CATCGCCATTTCTGGCGATG	AGATCCACGG	(*Bam*HI)-(*Bgl*II/*Sac*I)
R11	TGGCCAG	CATCGTCAGAAATGGCGATG		*Msc*I
R12	GCCGGCCGGGCCTG	CATCGTCAGAGCCGACGATG		*Nae*I/*Bgl*II/*Eag*I
R13	CTGCAG	CACCGCCAGGAATGGCGATG	CCGGA	*Pst*I-(*Nae*I)

Figure 5. Repeated sequences which are associated with the apparent hot spot regions for insertion of transposable elements carrying phenotypic markers in the backbone of the IncPβ plasmid R751.

that it was possible to deduce what the uninterrupted backbone must have been like. In this backbone region there was a cluster of five repeated sequences, each repeat an additional eight copies of these repeats plus associated restriction sites in the region surrounding the insertion site for Tn*402* that encodes trimethoprim resistance. This region also encodes an open reading frame with similarity to adenine methylases (Thorsted *et al.*, unpublished). It prompts us to consider whether these repeats constitute hotspots for transposon acquisition which may in some way be related to plasmid encounter with restriction barriers.

5. Conclusions

The study of the core properties of bacterial plasmids have provided valuable insights into the fundamental molecular biology of bacteria. This not only applies to the dissection of isolated cassettes like those required for replication or stable maintenance. The IncP plasmids illustrate how consideration of a plasmid genome as a whole provides an additional level of understanding. One of the most striking features that emerged from studies on the IncPα plasmids was the way in which replication, transfer and stable inheritance (partitioning) functions are co-ordinately regulated (Fig. 4; Jagura-Burdzy *et al.* 1992; Jagura-Burdzy & Thomas, 1994; Zatyka *et al.*, 1995). This co-ordination is conserved in the IncPβ plasmids as well (Thorsted et al. 1996; Macartney *et al.* unpublished). Once a plasmid has acquired a useful regulatory system that allows the co-ordinate expression of its survival functions it is very unlikely to lose it. Evolution is of necessity unidirectional in the sense that the genes or genetic systems present at any one time can only be displaced by others that are better at propagating themselves, never by ones which are less able. One of the intriguing questions is what advantage the multiple levels of regulation confer. New approaches may be needed to answer this since simple genetic analysis of such global regulons is difficult to interpret and in some cases impossible because of the lethal effects of some mutations. We would maintain that the lessons so far suggest that completing the picture should be worthwhile.

Acknowledgements

GJB was supported by MRC project grants, while MZ was supported by project grants from The Wellcome Trust (034605). PT was supported by an EC Bridge Programme Fellowship and a BBSRC project grant. KK was supported by a Greek Government Scholarship.

References

Amuthan G, Mahadevan A (1994) Replicon typing of plasmids from phytopathogenic xanthomonads. Plasmid 32: 328-332

Austin S (1988) Plasmid partition. Plasmid 20: 1-9

Balzer D, Ziegelin G, Pansegrau W, Kruft V, Lanka E (1992) KorB protein of promiscuous plasmid RP4 recognises inverted sequence repetitions in regions essentail for comjugative transfer. Nucl Acids res 20: 1851-1858

Bernard P, Couturier M (1992) Cell killing by the F plasmid CcdB protein involves poisoning of DNA-topoisomerase II complexes. J Mol Biol 226: 735-745

Brendler T, Abeles A, Austin S (1991) Critical sequences in the core of the P1 plasmid replication origin. J Bacteriol 173: 3935-3942

Bruand C, Le Chatellier E, Ehrlich SD, Janniere L (1993) A fourth class of theta-replicating plasmids: the pAMβ1 family from Gram-positive bacteria. Proc Natl Acad Sci USA 90: 11668-11672

Burlage RS, Bemis LA, Layton AC, Sayler GS, Larimer F (1990) Comparative genetic organisation of incompatibility group P degradative plasmids. J Bacteriol 172: 6818-6825

Cabello P, Timmis K, Cohen SN (1976) Replication control in a composite plasmid constructed by *in vitro* linkage of two distinct replicons. Nature 259: 285-290

Chikami GK, Guiney DG, Schmidhauser TJ, Helinski DR (1985) Comparison of ten IncP plasmids: homology in the regions involved in plasmid replication. J Bacteriol 162: 656-660

Couturier M, Bex F, Bergquist PL, Mass WK (1988) Identification and classification of bacterial plasmids. Microbiol Rev 52: 375-395

Durland RH, Helinski DR (1987) The sequence of the 43-kilodalton *trfA* protein is required for efficient replication or maintenance of minimal RK2 replicons in *Pseudomonas aeruginosa*. Plasmid 18: 164-169

Durland RH, Helinski DR (1990) Replication of the broad host range plasmid RK2: direct measurement of intracellular concentrations of the essentail TrfA replication proteins and their effect on plasmid copy number. J Bacteriol 172: 3849-3858

Eberl L, Kristensen C, Givskov M, Grohmann E, Gerlitz M, Schwab H (1994) Analysis of the multimer resolution system encoded by the *parCBA* operon of the broad-host-range plasmid RP4. Mol Microbiol 6, 1969-1979

Fang F, Helinski DR (1991) Broad-host range properties of plasmid RK2: importance of overlapping genes encoding the plasmid replication protein TrfA. J. Bacteriol. 173: 5861-5868

Fernandez-Tresguerres ME, Martin M, Garcia de Viedma D, Giraldo R, Diaz-Orejas R (1995) Host growth temperature and a conservative substitution in the replication protein of pPS10 influence host range. J Bacteriol 177: 4377-4384

Filutowicz ME, Uhlenhopp O, Helinski DR (1986) Binding of purified wild-type and mutant π initiation proteins to a replication origin of plasmid R6K. J Mol Biol 187: 225-239

Franch T & Gerdes K (1996) Programmed cell death in bacteria - translational repression by messenger-RNA end-pairing. Mol Microbiol 21, 1049-1060

Gerdes K, Bech FW, Jorgensen ST, Loebner-Olsen A, Rasmussen PB, Atlung T, Boe L, Molin S, von Meyenberg K (1986) Mechanism of post-segregational killing by the *hok* gene product of the *parB* system of plasmid R1 and its homology with the *relF* gene product of the *E.coli relB* operon. EMBO J 5: 2023-2029

Gerdes K, Poulsen LK, Thisted T, Nielsen AK, Martinussen J, Andreasen PH (1990) The *hok* killer gene family in Gram-negative bacteria. New Biologist 2: 946-956

Gerlitz M, Hrabak O, Schwab H (1990) Partitioning of broad host range plasmid RP4 is a complex system involving site-specific recombination. J Bacteriol 172: 6194-6203

Gotz A, Pukall R, Smit E, Tietze E, Prager R, Tschape H, van Elsas JD, Smalla K (1996) Detection and characterisation of broad-host-range plasmids in environment bacteria by PCR. App Environ Microbiol 62: 2621-2628

Gruss A, Ehrlich SD (1989). The family of highly interrelated single-stranded deoyxribonucleic acid plasmids. Microbiol Rev 53: 231-241

Helmstetter CE, Thornton M, Zhou P, Bogan JA, Leonard AC, Grimwade JA (1997) Replication and segregation of a miniF plasmid during the division cycle of *Escherichia coli*. J Bacteriol 179: 1393-1399

Hill K.E, Weightman AJ, Fry JC (1992) Isolation and screening of plasmids from the epilithonwhich mobilise recombinant plasmid pD10. App Enviro Microbiol 58: 1292-1300

Ingram LC, Richmond MH, Sykes RB (1973) Molecular characterisation of the R-factors implicated in the carbenicillin resistance of of a sequence of *Pseudomonas aeruginosa* strains isolated from burns. Antimicrob. Ag. Chemother. 3: 279-288

Itoh T, Tomizawa J-I (1980) Formation of an RNA primer for initiation of replication of ColE1 DNA by ribonuclease H. Proc Natl Acad Sci USA 77: 2450-2454

Jagura-Burdzy G,Khanim F,Smith CA,Thomas CM (1992) Crosstalk between plasmid vegetative replication and conjugative transfer: repression of the *trfA* operon by *trbA* of broad host range plasmid RK2. Nucl Acids Res 20: 3939-3944

Jagura-Burdzy G, Thomas CM (1994) KorA protein of promiscuous plasmid RK2 controls a transcriptional switch between divergent operons for plasmid replication and conjugative transfer. Proc. Natl. Acad. Sci. USA 91: 10571-10575

Jagura-Burdzy G, Thomas CM (1995) Purification of KorA protein from broad host range plasmid RK2: definition of a hierarchy of KorA operators. J Mol Biol 253: 39-50

Kolter R, Helinski DR (1978) Activity of replication terminus of R6K in hybrid replicons in *Escherichia coli*. J Mol Biol 124: 425-441

Kittel B, Helinski DR (1991) Iteron inhibition of plasmid RK2 replication in vitro: evidence for intermolecular coupling of replication origins as a mechanism of replication control. Proc Natl Acad Sci USA 88: 1389-1393

Lacatena RM, Cesareni G (1983) Interaction between RNAI and the primer precursor in the regulation of ColE1 replication. J Mol Biol 170: 635-650

Lanka E, Furste PJ, Yakobso E, Guiney DG (1985) Conserved regions at the primase locus of the IncPα and IncPβ plasmids. Plasmid 14: 217-223

Lanka E, Wilkins BM (1995) DNA processing reactions in bacterial conjugation. Ann Rev Biochem 64: 141-169

Lessl M, Balzer D, Pansegrau W, Lanka E (1992) Sequence similarities between the RP4 Tra2 and Ti VirB region strongly support the conjugative model for T-DNA-transfer. J Biol Chem 267: 20471-20480

Lobocka M, Yarmolinsky M (1996) P1 plasmid partition - a mutational analysis of ParB. J Mol Biol 259: 366-382

Macartney D, Williams DR, Stafford T, Foster A, Thomas CM (1997) Divergence and conservation of the partitioning and global regulation functions in the central control region of the IncP plasmids RK2 and R751. Microbiol In press

Manen D, Caro L (1991) The replication of plasmid pSC101. Mol Microbiol 5: 233-237

McEachern MJ, Bott MA, Tooker PA, Helinski DR (1989) Negative control of plasmid R6K replication: possible role of intermolecular coupling of replication origins. Proc Natl Acad Sci USA 86: 7942-7946

Meijer WJJ, Venema G, Bron S (1995) Characterisation of single strand origins of cryptic rolling circle plasmids from *Bacillus subtilis*. Nucl Acids Res 23: 612-619

Miele L, Strack B, Kruft V, Lanka E (1991) Gene organisation and nucleotide sequence of the primase region of IncP plasmids RP4 and R751. DNA Seq 2: 145-162

Motallebi-Veshareh M, Rouch D, Thomas CM (1990) A family of ATPases involved in active partitioning of diverse bacterial plasmids. Mol Microbiol 4: 1445-1463

Motallebi-Veshareh M, Balzer D, Lanka E, Jagura-Burdzy G, Thomas CM (1992) Conjugative transfer functions of broad host range plasmid RK2 are coregulated with vegetative replication. Mol Microbiol 6: 907-920

Mukhopadhyay G, Chattoraj DK (1993) Conformation of the origin of P1 plasmid replication. Initiator protein-induced wrapping and intrinsic unstacking. J Mol Biol 231: 19-28

Muraiso K, Tokino T, Murotsu T, Matsubara K (1987) Replication of mini-F plasmid *in vitro* promoted by purified E protein. Mol Gen Genet 206: 519-521

Nishiguchi R, Takanami M, Oka A (1987) Characterisation and sequence determination of the hairy-root inducing plasmid pRiA4b. Mol Gen Genet 206: 1-8

Novick RP (1989) Staphylococcal plasmids and their replication. Ann Rev Microbiol 43: 537-565

Pal SK, Chattoraj D (1988) P1 plasmid replication: initiator sequestration is inadequate to explain control by initiator-binding sites. J Bacteriol 170: 3554-3560

Pansegrau W, Lanka E. (1987) Conservation of a common 'backbone' in the genetic organisation of the IncP plasmids RP4 and R751. Nucl Acids Res 15: 2385.

Pansegrau W, Lanka E, Barth PT, Figurski D, Guiney DG, Haas D, Helinski DR, Schwab H, Stanisich VA, Thomas CM (1994) Complete nucleotide sequence of Birmingham IncPα plasmids. Compilation and comparative analysis. J Mol Biol 239: 623-663.

Patient ME, Summers DK (1993) ColE1 multimer formation triggers inhibition of Escherichia coli cell division. Mol Microbiol 9: 1089-1095.

Perri S, Helinski DR, Toukdarian A (1991) Interaction ofplasmid encoded replication initiation proteins with the origin of DNA replication in the broad host range plasmid RK2. J Biol Chem 266: 12536-12543

Pritchard RH, Barth, BT, Collins J (1969) Control of DNA synthesis in bacteria. Symp Soc Gen Microbiol 19: 263-297

Roberts RC, Strom A, Helinski DR (1994) The parDE operon of broad-host-range plasmid RK2 specifies growth inhibition associated with plasmid loss. J Mol Biol 237: 35-51

Shah DS, Cross MA, Porter D, Thomas CM (1995) Dissection of the core and auxiliary sequences in the vegetative replication origin of promiscuous plasmid RK2. J Mol Biol. 254: 608-622

Shingler V, Thomas CM (1989). Analysis of nonpolar insertion mutations in the *trfA* gene of IncP plasmid RK2 which affect its broad-host-range property. Biochim Biophys Acta 1007: 301-308

Smith CA, Thomas CM (1987) Comparison of the organisation of the genomes of phenotypically diverse plasmids of incompatibility group P: members of the IncPß subgroup are closely related. Mol Gen Genet 206: 419-427

Smith CA, Thomas CM (1989). Relationships and evolution of IncP plasmids. In *Promiscuous Plasmids of Gram-negative Bacteria*, pp 57-77. Edited by C.M. Thomas. London: Academic Press.

Smith CA, Pinkney M, Guiney DG, Thomas CM (1993) The ancestral IncP replication system consisted of contiguous *oriV* and *trfA* segments as deduced from a comparison of the nucleotide sequences of diverse IncP plasmids. J Gen Microbiol 139: 1761-1766

Summers DK, Beton CWH, Withers HL (1993) Multicopy plasmid instability: the dimer catastrophe hypothesis. Mol Microbiol 8: 1031-1038

Summers DK, Sherratt DJ(1984) Multimerisation of high copy number plasmids causes instability: ColE1 encodes a determinant essential for plasmid monomerisation. Cell 36: 1097-1103

Tabata S, Hooykaas PJJ, Oka A (1989) Sequence determination and characterisation of the replicator region in the tumour-inducing plasmid pTiB6S3. J Bacteriol 171: 1665-1672

Thisted T, Sorensen NS, Gerdes K (1995) Mechanism of post-segregational killing - the secondary structure analysis of the entire *hok* messenger RNA from plasmid R1 suggests foldback structure that prevents translation and antisense binding. J Mol Biol 247: 859-873

Theophilus BDM, Thomas CM (1987) Nucleotide sequence of the transcriptional repressor gene KorB which plays a key role in regulation f the copy number of broad host rnage plasmid RK2. Nucl Acids Res 15: 7443-7450

Thomas CM (1986). Evidence for the involvement of the *incC* locus of broad host range plasmid RK2 in plasmid maintenance. Plasmid 16: 15-29

Thomas CM, Cross MA, Hussain AAK, Smith CA (1984) Analysis of the copy number control elements in the region of the vegetative replication origin of the broad host range plasmid RK2. EMBO J 3: 57-63

Thomas CM, Meyer R, Helinski DR (1980). Regions of the broad host-range plasmid RK2 which are essential for replication and maintenance. J Bacteriol 141: 213-222

Thomas CM, Smith CA (1987) Incompatibility group P plasmids: genetics, evolution and use in genetic manipulation. Ann Rev Microbiol 41: 77-101

Thomas CM, Stalker DM, Helinski DR (1981) Replication and incompatibility properties of segments of the origin region of the broad host range plasmid RK2. Mol Gen Genet 181: 1-7

Thorsted PB, Shah, DS, Macartney D, Kostelidou K, Thomas CM (1996) Conservation of the genetic switch of IncP plasmids but divergence of the replication functions which are major host range determinants. Plasmid 36: (in press)

Tomizawa J-I (1986) Control of ColE1 plasmid replication: binding of RNAI to

RNAII and inhibition of primer formation. Cell 47: 89-97

Turner SL, Rigottier-Gois L, Power RS, Armarger N, Young JPW (1996) Diversity of *repC* plasmid-replication sequences in *Rhizobium leguminosarum*. Microbiol 142: 1705-1713

Villaroel R, Hedges RW, Maenhaut R, Leemans J, Engler G, van Montagu MM, Schell J (1983). Heteroduplex analysis of P-plasmid-evolution: the role of insertion and deletion of transposable elements. Mol Gen Genet 189: 390-399

Wagner EGH, Simons RW (1994) Antisense RNA control in bacteria, phages and plasmids. Ann Rev Microbiol 48: 713-742

Wilkins BM, Chilley PM, Thomas AT, Pocklington M (1996) Distribution of restriction enzyme recognition sequences on broad host range plasmid RP4: molecular and evolutionary implications. J Mol Biol 258: 447-456

Williams DR, Thomas CM (1992) Active partitioning of bacterial plasmids. J Gen Microbiol 138: 1-16

Williams DR, Mottalebi-Veshareh M, Thomas CM (1993) Multifunctional repressor KorB can block transcription by preventing isomerisation of RNA polymerase-promoter complexes. Nucl Acids Res 21: 1141-1148

Yakobson EA, Guiney DG (1983) Homology of the transfer origins of broad host range IncP plasmids: definition of two subgroups of P plasmids. Mol Gen Genet 192: 436-438

Zatyka M, Jagura-Burdzy G, Thomas CM (1994). Regulation of the transfer genes of promiscuous IncPα plasmid RK2: repression of Tra1 region transcription by relaxasome proteins and by the Tra2 regulator TrbA. Microbiol 140: 2981-2990

Part 3

Expression

BACTERIAL GENE REGULATORY PROTEINS: ORGANISATION AND MECHANISM OF ACTION

Georgina Lloyd, Tamara Belyaeva, Virgil Rhodius, Nigel Savery and Stephen Busby

School of Biochemistry, University of Birmingham, Birmingham B15 2TT, UK

1. Introduction

The study of adaptive responses in bacteria led to the discovery of a panoply of proteins whose role is to modulate promoter activity, coupling the expression of specific genes to changes in the environment. These gene regulatory proteins are ubiquitous and are essential for all processes of adaptation and differentiation in all living cells. At first sight, this is a very complex topic, made incomprehensible by the sheer number of different factors and different modes of operation. The aim of this chapter is to tackle the central question of how gene regulatory proteins function, and to describe a small number of key examples from the microbial repertoire that reveal principles applicable to most systems.

It was a genetic study of the *Escherichia coli* lactose operon that led to the discovery of the first example of this class of regulatory protein (Muller-Hill, 1996). Monod and coworkers noted an abundant class of mutants in which ß-galactosidase expression was constitutive rather than inducible. The observation that inducibility could be restored by introducing genetic material *in trans* led directly to the idea that this material was coding for a repressor (and that the mutants had a defective repressor). This led to the suggestion that all regulation was due to repressors and that all genetic material was "tied up" with repressors that "lay awaiting" their cognate inducer. This view was subsequently challenged by experiments with the arabinose and maltose operons. Frequent mutations led to non-inducibility, but inducibility could be restored by introducing genetic material *in trans*. This led to the discovery of gene activator proteins: the mutations destroyed activator function but inducibility could be restored by genetic material encoding the activator.

Since the discovery of the genes encoding the *lac* repressor (*lacI*) and the arabinose (*araC*) and maltose (*malT*) activator proteins, dozens of genes encoding activators and regulators have been found and characterised. More than 10% of the *E. coli* genome codes for such gene regulatory proteins: some are specific regulators acting at one or two loci (e.g. the *lac* repressor), whilst others are global regulators controlling hundreds of genes in response to particular signals (e.g. the cyclic AMP receptor protein, CRP, which is triggered by the elevation of intracellular cAMP levels). Finally, some gene regulatory proteins appear to be "bystanders", whose main function is to organise the *E. coli* chromosome, but have been "co-opted" into playing a regulatory role at certain promoters (e.g. HNS and IHF proteins).

NATO ASI Series, Vol. H 103
Molecular Microbiology
Edited by Stephen J. W. Busby,
Christopher M. Thomas and Nigel L. Brown

2. Families of activator and repressor proteins

Gene manipulation techniques have simplified the characterisation of activators and repressors, and hundreds of sequences are now available. Simple sequence analysis shows that most activators and repressors belong to a relatively small number of families, members of which share family traits. The main families of activators and repressors are listed in Table 1.

In many cases, binding sites for activators and repressors have been identified by both genetics and biochemistry. The positions of base substitutions that interfere with activation (for activators) or repression (for repressors), together with the location of protected bases in footprinting experiments, have been sufficient to define factor-binding elements at many promoters. In most cases, activator binding sites are located further upstream of the transcript start site than repressor binding sites, although this is not a hard and fast rule (Collado-Vides *et al.*, 1991). The majority of repressor binding sites (operators) overlap the RNA polymerase (RNAP) binding site (defined from footprinting as bases between ~-45 and ~+20). At most activatable promoters, the activator binds between -30 and -100 (this applies to the most downstream activator at promoters regulated by multiple activators).

3. Triggering of transcription factors

Several different mechanisms control the activity of transcription factors. For example, many bacterial factors are controlled by the binding of small ligands and, in some cases, these ligands trigger binding of the factor to specific DNA targets at promoters. Thus the binding of CRP to the *lac* promoter is controlled by cyclic AMP: modulation of intracellular cAMP levels *via* adenyl cyclase activity couples CRP activity to the environment (Kolb *et al.*, 1993). When cAMP levels are low, CRP binds poorly and non-specifically to DNA, but when levels are higher, the affinity of CRP for specific sites is increased by several orders of magnitude. In other cases, ligands trigger the removal of factors from specific target sites (as, for example, with members of the Lac repressor family). As a general rule, for repressors, small ligands act as co-repressors for biosynthetic genes (e.g. for MetJ at the *met* biosynthetic genes) and as inducers for catabolic genes (e.g. for LacI at the *lac* operon). In contrast, with activators, small ligands are essential for binding to promoters controlling catabolic genes (e.g. CRP at the *lac* promoter), but destabilise binding to promoters controlling biosynthetic genes (e.g. NagC at genes for N-acetyl glucosamine biosynthesis). Although this type of regulation is found frequently, there are also many cases where both the triggered and non-triggered transcription factor bind specifically to the target promoter. In these cases, the trigger causes a reorganisation or conformational change of the factor whilst it is anchored to the DNA. A good example is MerR which is triggered by mercuric ions (Summers, 1992). Although both unliganded MerR and the MerR-mercuric ion complex occupy the same site at the MerT promoter, bound unliganded MerR does not distort the *merT* promoter spacer: distortion is triggered by binding of mercuric ions to the

Table 1: MAJOR FAMILIES OF BACTERIAL TRANSCRIPTION FACTORS

The AraC family (e.g. *E. coli* AraC, MelR, RhaS, RhaR, SoxS)
Domain Structure: N-terminal domain concerned with triggering by small ligand. C-terminal domain carries two helix-turn-helix motifs responsible for operator binding.
Main Properties: Transcription activators that overlap promoter -35 region. Bind to ~18 bp sequence in absence and presence of ligand.

The LysR family (e.g. *E. coli* LysR, OxyR, MetR, CysB)
Domain Structure: N-terminal domain carries helix-turn-helix motif responsible for operator binding. C-terminal domain concerned with triggering.
Main Properties: Co-inducer responsive transcription activators. Bind in absence and presence of ligand.

The CRP family (e.g. *E. coli* CRP and FNR)
Domain Structure: C-terminal DNA-binding domain carries helix-turn-helix motif. N-terminal domain concerned with triggering.
Main Properties: Transcription activators. Binding to target is ligand-dependent. Variety of promoter architectures.

The MerR family (e.g. *E. coli* SoxR and transposon-encoded MerR)
Domain Structure: No evidence for multiple domains. N-terminal region carries helix-turn-helix motif. C-terminal region concerned with triggering.
Main Properties: Transcription activators: target binding is ligand-independent.

The response-regulator family (e.g. *E. coli* NarL, NarP, UhpA, OmpR, PhoB)
Domain Structure: N-terminal domain (the response domain) triggered by phosphorylation. C-terminal domain carries helix-turn-helix motif that binds DNA.
Main Properties: Transcription activators that bind at a variety of positions in target promoters.

The Lac repressor family (e.g. *E. coli* LacI, GalR, PurR, CytR)
Domain Structure: N-terminal carries helix-turn-helix motif that binds DNA. C-terminal responsible for triggering.
Main Properties: Repressors. Bind as dimers but can form tetramer.

The MetJ repressor family (e.g. *E. coli* MetJ, 'Phage P22 Arc and Mnt repressors)
Domain Structure: Single domain.
Main Properties: Transcription repressors. Contact DNA via ß strand.

preformed MerR-*merT* promoter complex (see later). Similarly, the activity of members of the AraC family of transcription activators is triggered by an inducer ligand (e.g. AraC activity is triggered by arabinose) but, in each case, the protein binds to target promoters in both the presence and absence of inducer. In the case of the *araBAD* promoter, AraC protein in the absence of arabinose binds to two sites: *O2*, which is far upstream of the promoter, and *I2*, which is just upstream. Arabinose induces AraC to release *O2* and to occupy sites *I2* and *I1*, *I1* overlapping with the -35 region of the promoter (Lobell and Schleif, 1990). Transcription activation is due to occupation of site *I1* by AraC, likely because AraC interacts directly with Region 4 of the RNAP σ^{70} factor bound to the -35 region. Thus, in this case, the ligand triggers a rearrangement of the occupancy of available sites. It is possible to rationalise this type of mechanism in terms of economy: for a factor that interacts at no more than a few promoters in the cell, it is worthwhile "anchoring" the factor at target promoters rather than maintaining a pool of free-floating factor.

Apart from small ligand binding, two other mechanisms are used to control the activity of bacterial activators and repressors. First, the activity of a large number of factors (notably the response-regulator class) is controlled by covalent phosphorylation by membrane-bound kinases: this results in direct coupling of factor activity to events outside the cell. Second, the activity of many factors is controlled by their concentration in the cell, which, in turn, is controlled either by synthesis or turnover or both. The best examples of this are the bacteriophage lambda cI and cII proteins (Ptashne, 1986). Levels of cI in 'phage lambda lysogens are tightly autoregulated by a complex promoter, at which cI acts as both an activator and as a repressor. The initial burst of cI expression is due to induction of a second promoter which is totally dependent on the transient appearance of cII protein. cII protein is subject to rapid degradation by proteolysis, notably by the host *hfl* protease, and 'phage lambda-encoded cIII protein acts to counter this proteolysis. Although the case of cI and cII is an extreme example, it serves to make the point that regulation of the level of a transcription factor can be just as effective as regulation by ligand. Indeed many operons involved in the specification of virulence encode at least one regulatory protein that positively autoregulates that operon. This sets up a situation in which triggering of the regulatory gene product positively induces its own synthesis, thus creating a very effective switch.

4. Simple Activation

4.1 Models for activation

There are numerous instances where a single transcription activator is sufficient to stimulate transcription initiation by RNAP. Activation in many cases can be studied in simple *in vitro* systems consisting of promoter DNA, RNAP and the activator. Such studies show that transcription activators accelerate the formation of open complexes that are very similar to those forming at activator-independent promoters. This is a crucial point: activators do not create promoters, but make pre-existing poor promoters more efficient. Hence, most activator-dependent promoters are as susceptible to mutations in the -10 and -35 hexamer sequences as activator-independent promoters. This is vividly demonstrated at the *lac* promoter where CRP-dependent expression can be suppressed by substitutions in both the -10 and -35 hexamers and, conversely, the requirement for CRP can be simply short-circuited by improvement of the -10 hexamer. Moreover, CRP is essential to accelerate open complex formation (it does this by simply improving the initial binding of RNAP to the promoter), but it is not needed for transcription initiation once the open complex has formed (Tagami and Aiba, 1995). Because of fierce competition for RNAP in the cell, small differences in promoter efficiency will result in big differences in expression. To be effective as a switch, any activator need only contribute a few kilocalories to an open complex (one or two hydrogen bonds). Stated simply, activator-independent promoters contain sufficient DNA sequence information in their -10, -35 and UP elements for RNAP docking. Activator-dependent promoters carry one or more defective elements and thus RNAP needs help from activator proteins. Activators can function to accelerate closed complex formation, open complex formation or promoter clearance.

Two simple models have been advanced to explain how simple activators function. One model supposes that activators provide a direct contact point for RNAP, whilst the other postulates that there is no direct activator-RNAP contact, but that the activator alters the conformation of target DNA to make it more "attractive" to RNAP. To date, the only case where an activator has been clearly shown to function via an induced conformational change in the promoter DNA is MerR. In other cases, activators appear to function by making a direct contact with RNAP (Figure 1). Our current knowledge suggests that most activators can be categorised into one of two classes, depending on the location of their target contact site in RNAP (Busby and Ebright, 1994). Class I activators contact target sites in the C-terminal domain of the RNAP α subunit (αCTD), whilst Class II activators contact target sites in the C-terminal part (Region 4) of the RNAP σ subunit (here we are solely concerned with the major *E. coli* 70kDa σ factor).

4.2 The case of MerR: activation by conformational change

The MerR protein directly activates transcription of microbial mercury resistance genes expressed from the *merT* promoter. Unusually, bound MerR covers the DNA between the -35 and -10 regions, but, surprisingly, there is no evidence for direct interactions between MerR and RNAP. The key to the action of MerR lies in the observation that the distance of the spacer between the -10 and -35 elements at the *merT* promoter is greater than normal (19 base pairs instead of the optimal 17 base pairs). The MerR-mercuric ion complex causes a local underwinding of the *merT* promoter spacer DNA by 33°, causing realignment of the -10 and -35 elements (Ansari *et al.*, 1992; Summers, 1992). Thus, MerR activates transcription by provoking a change in the conformation of the target DNA that makes the *merT* promoter more attractive to RNAP (Figure 1A).

4.3 Class I activators interact with αCTD

Many transcription activators appear to function by making contact with αCTD (Ebright and Busby, 1995). The most studied example is the activation of the *E. coli lac* promoter by CRP binding to a site centred between base pairs -61 and -62 (position -61.5) upstream of the transcript start site (Ebright, 1993). Some evidence suggests that the primary effect of CRP might be to alter the promoter conformation: CRP induces a bend in the DNA target greater than 90° upon binding, and curved DNA sequences (introduced into the *lac* promoter by cloning) can supplant the requirement for CRP (Schultz *et al.*, 1991). However, the discovery that activation by CRP requires a 7 amino acid surface-exposed ß turn (residues 156-162: the CRP Activating Region) provides strong evidence of a crucial role for CRP-RNAP contacts. CRP carrying substitutions in this region can bind and bend *lac* promoter DNA normally, and yet is unable to activate transcription or to interact co-operatively with RNAP, showing that CRP-induced DNA bending alone is not sufficient for transcription activation. Several lines of evidence show that the Activating Region makes contact with αCTD and that this contact is essential for transcription activation (Busby and Ebright, 1994). First, photocrosslinking indicates that αCTD and the Activating Region of CRP are in close proximity in the ternary complex between CRP, RNAP and *lac* promoter DNA. Second, in footprinting experiments, wild-type RNAP, but not RNAP derivatives containing C-terminally truncated α, protects the DNA segment between the DNA site for CRP and the -35 element. Third, CRP-dependent transcription is reduced by the removal of αCTD or by some single amino acid substitutions in αCTD (these identify the CRP contact site). Finally, interactions between purified α and CRP can be detected and these interactions are disrupted by substitutions in the CRP Activating Region. The simplest model is that CRP activates transcription by recruiting αCTD to the DNA immediately upstream of the *lac* -35 element, thereby tightening the binding of RNAP to the promoter (Figure 1B). Stated simply, CRP compensates for defective promoter-RNAP interactions with protein-RNAP interactions.

Sequence analysis of the many promoters dependent on Class I activators shows that the binding sites for different activators are found at diverse locations, ranging from around -40 to upstream of -100, raising the question of how distally-bound activators can make direct contact with RNAP. Of course, one possibility is that there is no direct contact and that the effects of the activators are negotiated via conformational changes in the promoter DNA. However, there is little evidence for this and, in many cases where activators are distally bound, the intervening DNA can be bent such that direct contact is possible. In the case of activators that contact αCTD, diversity in binding site location is also made possible by the flexible linker between αCTD and αNTD, which permits αCTD to stretch to different locations (Blatter *et al.*, 1994). This is particularly apparent at different CRP-dependent promoters triggered by the Activating Region-αCTD interaction: such promoters are found with the DNA site for CRP centred near -61, -71, -81 or -91 and all are dependent on the same CRP-RNAP interaction. The flexible linker between αCTD and the remainder of RNAP, together with bending of the intervening DNA, permits the establishment of the same

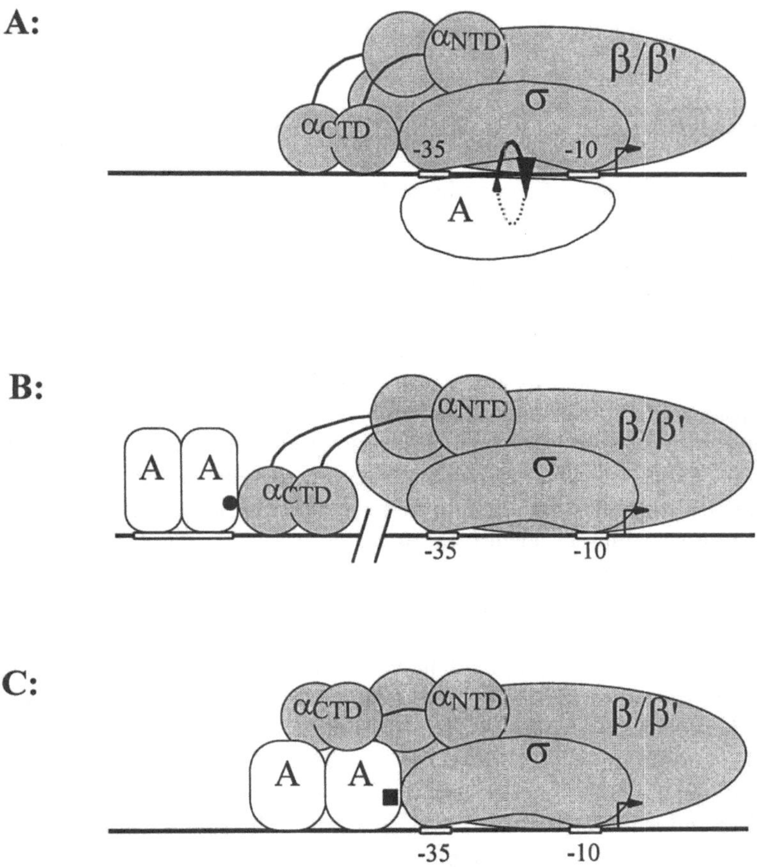

Figure 1: Simple transcription activation
The figure shows the likely organisation of RNAP and activator (A) subunits in transcriptionally competent complexes at different promoters. Although, for clarity, the DNA is drawn as a straight line, the path of the DNA is likely to be distorted, with the bends facilitating the different contacts.
A: The activator functions by altering the conformation of promoter DNA (see section 4.2): at *pmerT*, MerR twists the promoter DNA so that the -10 and -35 elements become aligned correctly to facilitate RNAP binding.
B: Class I activators bind upstream of RNAP and function by making contact with αCTD (see 4.3): at the *lac* promoter, the Activating Region in the downstream subunit of the CRP dimer (filled circle) contacts αCTD, recruiting it to the promoter.
C: Class II activators bind immediately adjacent to the -35 region and contact RNAP often via σ (see section 4.4): at the 'phage lambda P*rm* promoter, the Activating Region of cI (filled square) contacts Region 4 of σ^{70}, most likely facilitating contacts with the -35 element.

CRP-αCTD-DNA interaction at promoters with different architectures (Zhou *et al.*, 1994).

4.4 Class II activators interact with the RNAP σ subunit

Some transcription activators function by making contact with the RNAP σ subunit (Ishihama, 1993). These activators, known as Class II activators, bind to sites that overlap the -35 region of the target promoter. In contrast to the situation with Class I activators, there is very little flexibility in the location of Class II activators: this is because most Class II activators contact Region 4 of the RNAP σ subunit, and, unlike αCTD, the location of Region 4 during transcription initiation is fixed at the -35 region. The most studied example of a Class II activator is the bacteriophage lambda cI protein, which is essential for transcription initiation at the 'phage lambda P*rm* promoter. cI protein activates this promoter by binding to a site centred at -42 which overlaps the -35 element. Strong evidence for protein-protein contacts comes from the identification of single amino acid substitutions in cI that interfere with activation: these substitutions identify a 10 amino acid Activating Region, rich in negatively charged side chains, that is essential for transcription activation (Hochschild *et al.*, 1983). Two lines of evidence show that the target for this Activating Region is Region 4 of the RNAP σ subunit. First, the Activating Region in cI is located within the DNA-binding helix-turn-helix motif and is positioned adjacent to σ in ternary cI:RNAP:promoter complexes. Second, a single amino acid substitution in Region 4 of σ (R596H) is sufficient to suppress the effects of an activation-defective cI mutant (D38N) (Li *et al.,* 1994). Moreover, this suppression is allele specific (i.e. cI-D38N functions with σ-R596H but not wild type σ) suggesting that D38 of cI and R596 of σ are in close proximity. The simplest model is that cI activates transcription by helping Region 4 of σ bind to the -35 element, again tightening the binding of RNAP to the promoter (Figure 1C).

Transcription activation by many different activators (e.g. AraC, MalT and PhoB) is also affected by some single amino acid substitutions in Region 4 of the RNAP σ subunit, and it is probable that this region carries contact sites for most (though not all) transcription activators that bind at or near the -35 region (Ishihama, 1993). An important consequence of the binding of Class II activators at target promoters is that they block the binding of αCTD to its preferred target just upstream of the -35 region (Figure 1C). In some cases, this displaced αCTD is accommodated by binding to promoter DNA just upstream of the bound Class II activator. In other cases the displaced αCTD can also make a direct contact with the Class II activator and, in these cases, the Class II activator must be "ambidextrous" contacting both Region 4 of the RNAP σ subunit and αCTD (for an example, see Wing *et al.*, 1995). At more complex promoters, the αCTD that is displaced by the Class II activator is contacted by a second activator (usually a Class I activator) that binds further upstream (see Section 5.2.3).

5. Complex Activation

5.1 Types of complex activation

The regulation of most promoters is complex and expression is not simply dependent on one transcription activator. In many cases where two activators are involved, one activator is a "global" regulator, sensing a global metabolic signal and interacting at a large number of promoters, whilst the other is a "specific" regulator, triggering specific responses to a specific inducer at a very small number of promoters. Good examples of this are found at the *mal* and *mel* operons encoding genes for the catabolism of maltose and melibiose. Expression of these operons is co-dependent on CRP and the operon-specific regulators MalT and MelR. CRP is a global regulator that is triggered by elevations in cAMP levels (due to glucose starvation), whilst MalT and MelR are regulon-specific activators triggered by maltose and melibiose respectively.

Before considering the mechanisms responsible for co-dependence on two activators, it is important to understand that apparent co-dependence may not be due to both activators binding at the same promoter, but due to the synthesis of one activator being dependent on the activity of the other. For example, expression of the *melAB* operon (encoding an α-galactosidase and a melibiose transport protein) is strictly dependent on CRP and MelR, and yet the *melAB* promoter is activated by MelR alone (Webster *et al.*, 1988). However, the *melR* promoter is totally dependent on CRP, thus explaining the dependence of *melAB* expression on two activators: in fact, both the *melR* and *melAB* promoters are "simple" in that they are both controlled by just one activator. In contrast, although synthesis of the maltose-specific activator, MalT, is also dependent on CRP, expression from several promoters of the maltose regulon requires simultaneous binding of both CRP and MalT for transcription initiation (e.g. the divergent *malK-malE* promoters).

5.2 Mechanisms of complex activation

5.2.1 A possible role for nucleoprotein structures

One of the first complex regulatory regions to be dissected was the *malK-malE* intracistronic regulatory region carrying the divergent *malK* and *malE* promoters. Both promoters are dependent on the binding of both MalT and CRP: there are five DNA sites for MalT and three sites for CRP. An attractive suggestion (Raibaud, 1989) is that occupation of these sites leads to the formation of a nucleoprotein complex and it is this complex that triggers transcription activation (Figure 2A). Although this explanation is probably too simple for the *malE-malK* case (see below), it is an attractive model and may well be applicable at a host of promoters (many of which are so complex that no serious mechanistic studies have been attempted). The distinguishing feature of this model is that it supposes that the different factors bind co-operatively to form a structure.

5.2.2 Mechanisms involving repositioning

Transcription activation at the *E coli malK* promoter requires CRP binding to three sites and MalT binding to two sites upstream of the bound CRP (sites 1' and 2') and three sites downstream (sites 3', 4' and 5'). In the absence of CRP, MalT occupies three distinct downstream sites, 3, 4 and 5, that overlap with sites 3', 4' and 5'. Crucially, MalT can only activate transcription when it occupies sites 3', 4' and 5', probably due to direct interactions with σ. Thus the action of CRP (together with upstream-bound MalT) is to reposition downstream-bound MalT from an unproductive to a productive binding site (Figure 2B). This observation suggests a clear molecular basis for coactivation: the second activator is required to "nudge" the first (primary) activator from an abortive to a productive mode (Richet *et al.*, 1991). A similar scenario may be operating during AraC-dependent activation at the *araBAD* promoter, where activation is dependent on AraC and CRP: CRP appears to help AraC to break its association with site *O2* and thus to occupy *I1* (Lobell and Schleif, 1990).

Evidence to support repositioning models comes from two sources. First, mutations that improve MalT sites 3', 4' and 5' or AraC site *I1* decrease the CRP-dependence of the corresponding promoters. Second, at least in the case of the *malK* promoter, there is no absolute requirement for CRP to trigger the repositioning, and activation can be triggered by Integration Host Factor (IHF) that can also bend target sequences (Richet and Sogaard-Andersen, 1994). This is an interesting observation as it shows how an apparently unrelated protein can be recruited into an activation mechanism and, further, it suggests that the crucial feature in repositioning might be the protein-induced bend.

5.2.3 Mechanisms involving simultaneous touching

An alternative mechanism for complex activation is found at promoters where both activators make direct contact with RNAP. This situation is most common where a Class II activator is involved that binds overlapping the -35 region and makes contacts with Region 4 of the RNAP σ subunit. In such cases, promoter activity can be coupled to a Class I-type activator binding further upstream that contacts the displaced αCTD (Figure 2C). One example is expression from the *E. coli ansB* promoter (that controls the asparaginase gene) which is dependent on both FNR and CRP, binding near positions -41 and -91, respectively. Bound FNR overlaps the -35 region and makes direct contact with Region 4 of σ, whilst coactivation by CRP requires the CRP Activating Region which must make contact with αCTD (Scott *et al.*, 1995). The simplest model is to suppose that the FNR dimer displaces αCTD from its preferred location at the promoter, and αCTD thereby becomes available as a "target" for a second activator. A similar result has been found with a synthetic derivative of the 'phage lambda P*rm* promoter, which is dependent on cI protein bound at a site overlapping the -35 region and contacting σ. This promoter can be super-activated by CRP if a DNA site for CRP is introduced upstream: super-

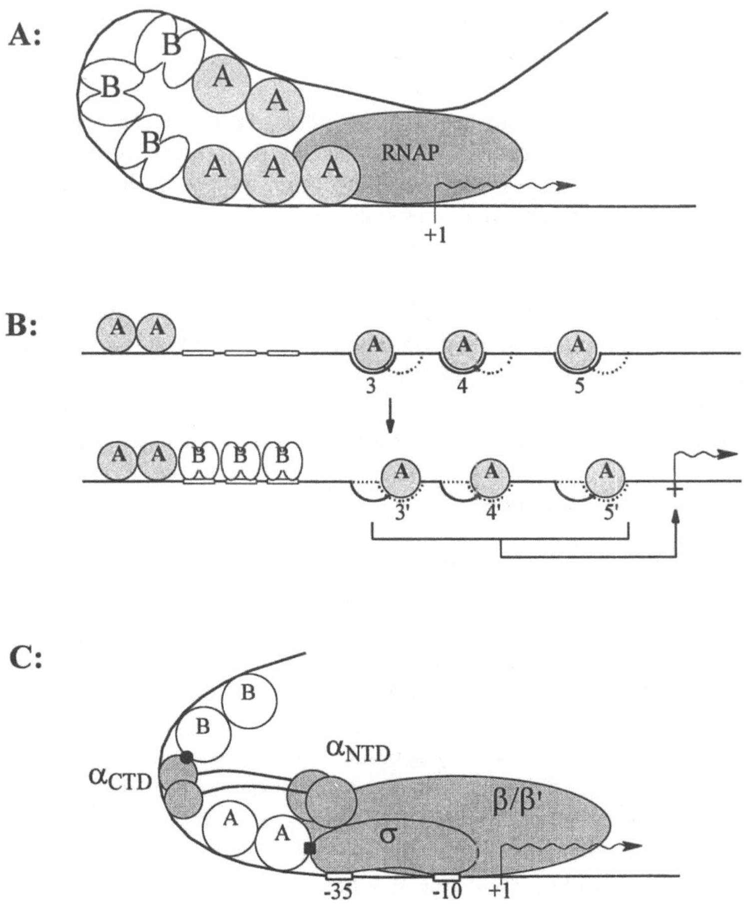

Figure 2: Complex transcription activation
The figure shows the likely organisation of RNAP and different activator subunits in transcriptionally competent complexes at complex promoters dependent on 2 activators, denoted A and B.
A: A nucleoprotein complex forms (see section 5.2.1): at the *E coli malB* locus containing the divergent *malE* and *malK* promoters, 3 CRP (B) dimers and 5 MalT (A) monomers form a complex.
B: One activator repositions another (see section 5.2.2): at the *E. coli malK* promoter, the binding of CRP (B) together with upstream-bound MalT (A), triggers the repositioning of downstream-bound MalT (A) to a position where it is competent for transcription activation.
C: Both activators make independent simultaneous contacts with RNAP (see section 5.2.3): at the *E. coli ansB* promoter, FNR and CRP make independent contacts with RNAP. CRP (B) functions as a Class I activator and contacts the flexible αCTD (filled circle), whilst FNR (A) functions as a Class II activator and contacts σ (filled square).

activation is dependent on the CRP Activating Region showing that CRP-αCTD interactions are essential (Joung *et al.,* 1994).

Although simultaneous touching provides an attractive model to account for co-dependent activation, it is necessary to explain why the binding of one activator is insufficient to trigger transcription initiation (this is not a problem with the repositioning model for co-dependent activation). This is especially crucial at FNR-dependent promoters such as the *E. coli ansB* promoter, which is totally inactive in the absence of upstream-bound CRP (Scott *et al.*, 1995). The simplest explanation is to suppose that the αCTD, which is displaced when FNR makes contact with Region 4 of σ, is somehow inhibitory to transcription activation, and that this inhibition must be overcome by contact with the upstream activator. Thus, the upstream activator behaves as an anti-inhibitor as well as an activator. This may be an important factor in explaining the activity of different naturally-occurring FNR-dependent promoters.

5.2.4 Recruitment of "bystanders"

All bacteria contain a number of very abundant small proteins that appear to play a role in maintaining DNA structure. Amongst these are HU and HNS, that were originally identified as "non-specific" DNA binding proteins present in sufficiently large amounts to be considered as proteins that could structure the bacterial chromosome. Similarly, IHF and Fis were discovered as host proteins essential for DNA rearrangements involving 'phage lambda and Mu. Although the activity of these proteins is not triggered by any specific signal, they have been recruited at a number of promoters where they contribute to activation and repression mechanisms. For example, in several cases, Fis and IHF behave as "simple" activators, binding upstream of promoter elements (Finkel and Johnson, 1992; Goosen and van de Putte, 1995). IHF also plays an important role at some "complex" promoters. For example, the *E. coli narG* promoter is coactivated by FNR and NarL, which bind around -41 and -200 respectively. IHF binding to target sites located between bound FNR and NarL is essential for activation (Schroder *et al.*, 1993). The simplest model is that IHF induces a bend that brings the upstream-bound NarL near to the FNR-RNAP complex, facilitating contacts with the distally-bound activator.

6. Simple and Complex Repression

6.1 The *lac* repressor

Simple repression occurs when a bound repressor covers essential promoter elements and thus blocks access by RNAP (Figure 3A). The *lac* repressor is usually taken as the paradigm for "simple" repression although, in fact, the *lac* repressor is not so simple because of the existence of multiple binding sites (see below). The *lac* repressor is a tetramer of identical subunits and belongs to a large family of repressors (Muller-Hill, 1996). The principal operator (*O1*) at the *lac* promoter is a pseudosymmetric sequence that overlaps the transcription startsite and accommodates two repressor subunits. This suggests a simple model in which the

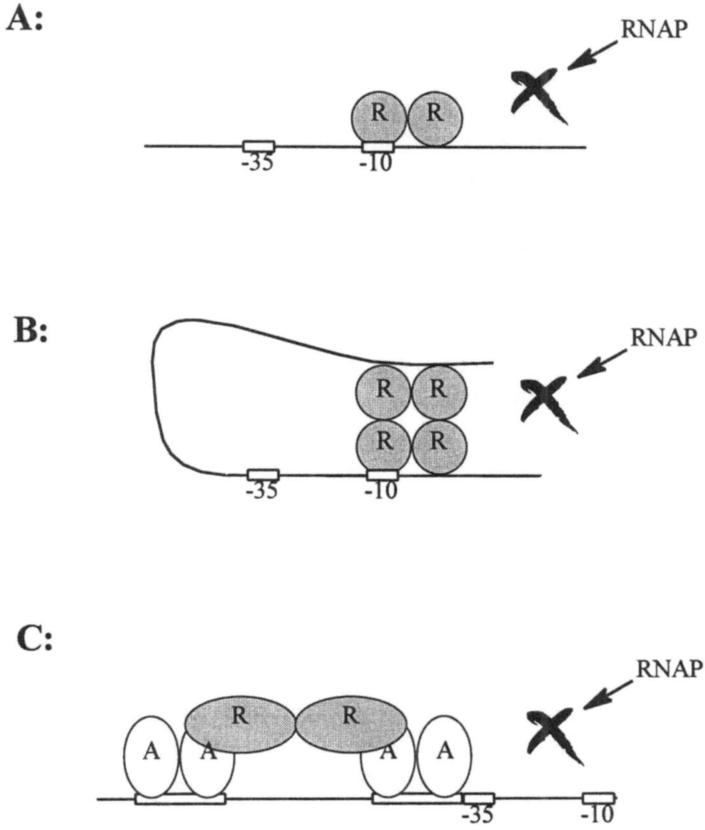

Figure 3: Mechanisms of Repression
A: Simple repression (see section 6.1): bound repressor directly overlaps a promoter element and occludes RNAP.
B: Interaction between distant repressors (see section 6.2): distally-bound repressors interact *via* a loop that enhances repression.
C: Repressors act by counteracting an activator (see section 6.3): promoters repressed by CytR depend upon CRP for activation. CytR binds to tandem-bound CRP at target promoters and suppresses CRP-dependent transcription activation.

access of RNAP to the *lac* promoter is sterically blocked by repressor, where the crucial factor in determining the efficiency of repression is the lifetime of the repressor-promoter complex. There will be competition between repressor and RNAP, and it is this competition that will determine the efficiency of repression. In fact, this competition is complex, depending on the exact juxtaposition of the RNAP-binding and repressor-binding sites and the kinetic characteristics of the promoter. As a general rule, repression becomes less efficient as the operator is distanced from the transcript start, implying that repressors are more efficient at interfering with transcription initiation than with transcript elongation. Promoters that are limited at the stage of the initial binding of RNAP are more sensitive to repression than promoters that are limited at the stage of isomerisation to the open complex or promoter clearance. RNAP will be a better competitor if the promoter is limited at a step after initial binding to the promoter.

6.2 Multiple repressor binding sites

At most repressible promoters, the DNA site for the repressor is located near to the transcript start, and in many cases two or more copies of the operator sequence are found. For example, promoters repressed by the MetJ or TrpR repressors can contain 2, 3, 4 or 5 copies of the operator sequence (Phillips and Stockley, 1996). In these cases, co-operative interactions between tandemly bound repressors help to exclude RNAP. In some instances, operators are duplicated but the second copy of the operator is remote from the transcript start. In these cases, repression due to repressor binding at the remote site requires DNA looping and repressor aggregation (Figure 3B). In the case of the *lac* operon, supplementary operator sites, *O2* and *O3*, are located 400 bp downstream and 93 bp upstream from *O1*: whilst *O1* alone is sufficient to ensure a good level of repression, optimal repression is contingent on the presence of both *O2* and *O3* (Muller-Hill, 1996). The simplest model to explain this is to suppose that tetrameric repressor simultaneously binds to *O1* and *O2* or *O3* and that this involves looping of the intervening DNA. In some cases, binding of multiple repressors to distant sites is essential for repression and not just an optional extra: for example, repression by the GalR repressor and the AraC protein (in the absence of arabinose it acts as a repressor) requires binding to distant sites via looping the intervening sequences. Interestingly, in the case of repression of the *gal* operon promoter by GalR, interaction between distally-bound GalR dimers requires HU binding to the intervening sequences (Aki *et al.*, 1996).

6.3 Complex repression and anti-activation

It is usually assumed that the crucial feature of repressors is their ability to bind DNA: thus occupation, whether by one or multiply-bound repressor molecules, is assumed to be both necessary and sufficient to shut out RNAP. However, there are a number of cases where the situation is not so simple and repression is "active", involving heterologous interactions between the repressor and other factors. A good example is the promoter of the *crp* gene, which is autoregulated by CRP binding to a single site downstream of the *crp* transcript start at around +40 (Hanamura and Aiba, 1991). In this case, repression is not due to bound CRP blocking RNAP elongation from the *crp* promoter, but rather due to the bound CRP activating a cryptic antisense promoter. Bound CRP recruits RNAP to this promoter, and it is the bound RNAP that blocks the *crp* promoter (note that repression is not due to antisense RNA made from this promoter). It is also likely that some repressors function actively by contacting RNAP and "jamming" it at a position where it is unable to initiate transcription (this is likely to be important for those repressors that can also function as activators).

Perhaps the best example of repression involving protein-protein interactions is the CytR repressor which represses the expression of genes scattered at ~10 different locations on the chromosome, involved in nucleoside catabolism and transport (Sogaard-Andersen *et al.*, 1991). Expression of all of these genes is dependent on CRP which, in each case, binds at tandem sites located around -40 and -93, although activation is due to binding at the -40 site. Although CytR is a member of the LacI family, alone it can bind only poorly to target promoters: for many years this was a puzzle as it was unclear how it could repress transcription. The key observation is that CytR only binds to target promoters when CRP is bound: rather than recognising a base sequence, CytR recognises the array of tandemly-bound CRP dimers separated by ~53 bp. Single amino acid substitutions in CRP that prevent CytR-dependent repression, whilst not interfering with transcription activation, have been isolated. These identify the CRP contact site for CytR as a surface exposed region that is distinct from the Activating Region and the DNA-binding motif. CytR-dependent repression thus involves repressor binding to an activator: the repressor acts as an anti-activator (Figure 3C).

6.4 Repressors come in many forms

Most repressors are proteins "dedicated" to repressing a small number of promoters and induced by a specific trigger. However, there are a number of cases where regulatory proteins fulfill a dual role. For example, AraC, MerR, TyrR and many more proteins can act as both repressors and activators. Clearly, if repression is due simply to interfering with the access of RNAP to a target promoter, then any activator also has the potential to behave as a repressor. It is instructive to consider CRP and FNR, which are both activators provided that they are correctly positioned with respect to promoter elements at a transcription start. Incorrect positioning of either protein will result in cAMP-induced or anaerobically-induced repression of overlapping promoters, and thus CRP and FNR can behave as both activators and repressors. This explains the observation that the synthesis of a small number of proteins is derepressed in *crp* and *fnr* mutants (Kolb *et al.*, 1993).

Finally, Fis, IHF and HNS have been recruited as repressors at a number of promoters (Finkel and Johnson, 1992; Goosen and van de Putte, 1995). For Fis and IHF, in some cases, repression is simple *via* binding to specific sites overlapping the

-10 or -35 elements at target promoters. In other more complex cases, IHF sites overlap sites for an essential activator, and the activator is displaced by IHF protein. Repression by HNS is more complex as this protein appears to recognise a DNA structure rather than a short operator sequence. Repression by HNS is due to ~100 bp blocks of sequence that are located either upstream or downstream from the target promoter. The simplest explanation is that these blocks act as nucleation sites for HNS which then silences neighbouring regions of the chromosome (Higgins *et al.*, 1990). Induction of promoters repressed by HNS can be triggered by changes in DNA structure (most likely supercoiling), which are linked to environmental factors such as osmolarity or temperature. It is also possible that the HNS "blockade" can be lifted by competition with transcription factors. For example, CRP-dependent activation of the *pap* operon may be partially due to displacement of HNS (induction becomes CRP-independent in a mutant *hns* background) (Goransson *et al.*, 1990).

7. Some perspectives

For any regulatory region we need to ask what are the factors that interact, why have they been selected, and how do they function? In most cases we now know which factors are operating. The question "why?" is more difficult, although it is reasonable to suppose that the interplay of activators and repressors at any target promoter ensures that the bacterium makes the right product at the right time. Thus, many promoters are under the dual control of a global and a specific regulator. In most cases, this is a repressor and an activator (as at the *lac* promoter) or two activators (as at the *mal* promoters). Thus expression is tied to a dual signal: both the global signal (cAMP in the case of the *lac* and *mal* promoters) and the specific signal (lactose or maltose) need to be present. Interestingly, there are few cases where regulation is effected by two repressors acting in tandem (the best examples are promoters repressed by both CytR and DeoR). Perhaps this reflects a preference for activators over repressors: repressors are costly, since they need to be maintained during periods of silence, whilst activators need only appear when the conditions are right. In most cases, we can now see the logic in why things are organised the way they are (or at least we can now profer explanations to satisfy our curiosity). Finally, the question "how?" continues to pose problems, mainly because of three reasons. First, at present, we lack structural information, second, we have only a sketchy understanding of the kinetics of the initiation process, and third, we do not really understand the importance of the bacterial folded chromosome structure in regulation. Fortunately, some of these gaps in our knowledge are being filled, and new structural and kinetic data will be the startpoint for understanding the making and breaking of the complex structures that regulate promoter activity.

Acknowledgements

Work from the authors' laboratory was supported by project grants from the U.K. Biotechnology and Biological Sciences Research Council and the Wellcome Trust.

References

Aki T, Choy H, Adhya S (1996) Histone-like protein HU as a specific transcriptional regulator: co-factor role in repression of *gal* transcription by GAL repressor. Genes to Cells 1:179-188

Ansari A, Chael M, O'Halloran T (1992) Allosteric underwinding of DNA is a critical step in positive control of transcription by Hg-MerR. Nature 355:87-89

Blatter E, Ross W, Tang H, Gourse R, Ebright, R (1994) Domain organisation of RNA polymerase α subunit: C-terminal 85 amino acids constitute a domain capable of dimerization and DNA binding. Cell 78:889-896

Busby S, Ebright R (1994) Promoter structure, promoter recognition, and transcription activation in procaryotes. Cell 79:743-746

Collado-Vides J, Magasanik B, Gralla, J (1991) Control Site Location and Transcriptional Regulation in *Escherichia coli*. Microbiological Reviews 55:371-394

Ebright R (1993) Transcription Activation at Class I CAP-dependent promoters. Mol Microbiol 8:797-802

Ebright R, Busby S (1995) *Escherichia coli* RNA polymerase α subunit: structure and function. Current Opinion in Genetics and Development 5:197-203

Finkel S, Johnson R (1992) The Fis protein: it's not just for DNA inversion anymore. Mol Microbiol 6:3257-3265

Goosen N, van de Putte P (1995) The regulation of transcription initiation by integration host factor. Mol Microbiol 16:1-7

Goransson M, Sonden B, Nilsson P, Dagberg B, Forsman K, Emanuelsson K, Uhlin B (1990) Transcriptional silencing and thermoregulation of gene expression in *Escherichia coli*. Nature 344:682-685

Hanamura A, Aiba H (1991) Molecular mechanism of negative autoregulation of *Escherichia coli crp* gene. Nucleic Acids Res 19:4413-4419

Higgins C, Hinton J, Hulton C, Owen-Hughes T, Pavitt G, Seirafi A (1990) Protein H1: a role for chromatin structure in the regulation of bacterial gene expression and virulence? Mol Microbiol 4:2007-2012

Hochschild A, Irwin N, Ptashne M (1983) Repressor structure and the mechanism of positive control. Cell 32:319-325

Ishihama A (1993) Protein-protein communication within the transcription apparatus. J Bacteriol 175:2483-2489

Joung J, Koepp D, Hochschild A (1994) Synergistic activation of transcription by bacteriophage lambda cI protein and cyclic AMP receptor protein. Science 265: 1863-1866

Kolb A, Busby S, Buc H, Garges S, Adhya S (1993) Transcriptional regulation by cAMP and its receptor protein. Ann Rev Biochem 62:749-795

Li M, Moyle H, Susskind M (1994) Target of the transcriptional activation function of phage lambda cI protein. Science 263:75-77

Lobell R, Schleif R (1990) DNA looping and unlooping by AraC protein. Science 250:528-532

Muller-Hill B (1996) *The lactose operon*. Walter de Gruyter, Berlin

Phillips S, Stockley P (1996) Structure and function of the *Escherichia coli met* repressor: similarities and contrasts with the *trp* repressor. Phil Trans Roy Soc Lond B 351:527-535

Ptashne, M (1986) *A Genetic Switch: Gene Control and Phage lambda*. Blackwell Scientific Publications & Cell Press

Raibaud O (1989) Nucleoprotein structures at positively regulated bacterial promoters: homology with replication origins and some hypotheses on the quaternary structure of the activator proteins in these complexes. Mol Microbiol 3:455-458

Richet E, Vidal-Ingigliardi D, Raibaud O (1991) A new mechanism for coactivation of transcription: repositioning of an activator triggered by the binding of a second activator. Cell 66:1185-1195

Richet E, Sogaard-Andersen L (1994) CRP induces the repositioning of MalT at the *Escherichia coli malKp* promoter primarily through DNA bending. EMBO J 13: 4558-4567

Schultz S, Shields S, Steitz T (1991) Crystal structure of a CAP-DNA complex: the DNA is bent by 90°. Science 253:1001-1007

Schroder I, Darie S, Gunsalus R (1993) Activation of the *Escherichia coli* nitrate reductase (*narGHJI*) operon by NarL and FNR requires Integration Host Factor. J Biol Chem 268:771-774

Scott S, Busby S, Beacham I (1995) Transcriptional coactivation at the *ansB* promoters: involvement of the Activating Regions of CRP and FNR when bound in tandem. Mol Microbiol 18:521-531

Sogaard-Andersen L, Pedersen H, Holst B, Valentin-Hansen P (1991) A novel function of the cAMP-CRP complex in *Escherichia coli*: cAMP-CRP functions as an adaptor for the CytR repressor in the *deo* operon. Mol Microbiol 5:969-975

Summers A (1992) Untwist and shout- a heavy metal-responsive transcription regulator. J Bacteriol 174:3097-3101

Tagami H, Aiba H (1995) Role of CRP in transcription activation at the *Escherichia coli lac* promoter: CRP is dispensible after the formation of open complex. Nucleic Acids Res 23:599-605

Webster C, Gaston K, Busby S (1988) Transcription from the *Escherichia coli melR* promoter is dependent on the cyclic AMP receptor protein. Gene 68:297-305

Wing H, Williams S, Busby S (1995) Spacing requirements for transcription activation by *Escherichia coli* FNR protein. J Bacteriol 177:6704-6710

Zhou Y, Merkel T, Ebright R (1994) Characterization of the Activating Region of *Escherichia coli* catabolite gene activator protein. Role at Class I and Class II CAP-dependent promoters. J Mol Biol 243:603-610

Bacterial Two-Component Regulatory Systems

Valley Stewart

Section of Microbiology, Cornell University, Ithaca, NY 14853, USA

Abstract

Two-component regulation involves a stimulus-sensitive sensor (histidine protein kinase) that controls the phosphorylation state, and therefore the activity, of its cognate response regulator. Specific two-component systems regulate bacterial response to a wide range of environmental conditions, and many variations on the basic theme have been identified. Specific sensor-regulator examples considered here are the NtrB-NtrC system controlling nitrogen assimilation enzyme synthesis, the EnvZ-OmpR system controlling outer membrane protein synthesis, and the NarX-NarL / NarQ-NarP system controlling anaerobic respiratory enzyme synthesis.

Keywords

Gene expression, signal transduction, two-component regulation

1 Introduction

Bacteria are remarkable for their capacity to sense and respond rapidly to changes in the chemical and physical composition of their milieu. A limited number of core mechanisms for effecting these responses are known, but many variations of these cores have evolved. This chapter introduces one widely-distributed core mechanism, so-called two-component signal transduction. The interested reader may consult many other sources for more information; I particularly recommend the volume edited by Hoch and Silhavy (1995) as a source of authoritative recent reviews on a wide range of two-component systems. Other recent reviews that I have found to be especially informative include those by Appleby, Parkinson and Bourret (1996), Ninfa (1996), Parkinson and Kofoid (1992), and Parkinson (1993). The classic review by Stock, Ninfa and Stock (1989) is well worth perusal.

Both protein components of these regulatory pairs are defined ultimately by primary amino acid sequence (Fig. 1). The *sensor* component contains a carboxyl-terminal domain, comprising approximately 240 residues, that is termed the histidine protein kinase or *transmitter* domain. The *response regulator* component contains an amino-terminal domain, comprising approximately 120 residues, that is termed the *receiver* domain (Parkinson and Kofoid, 1992). The transmitter domain is usually connected to a separate input domain, such as a ligand-binding domain, and the receiver domain is usually connected to an output domain, such as a DNA-binding domain. However, there are many known variations on this relatively simple scheme.

NATO ASI Series, Vol. H 103
Molecular Microbiology
Edited by Stephen J. W. Busby,
Christopher M. Thomas and Nigel L. Brown

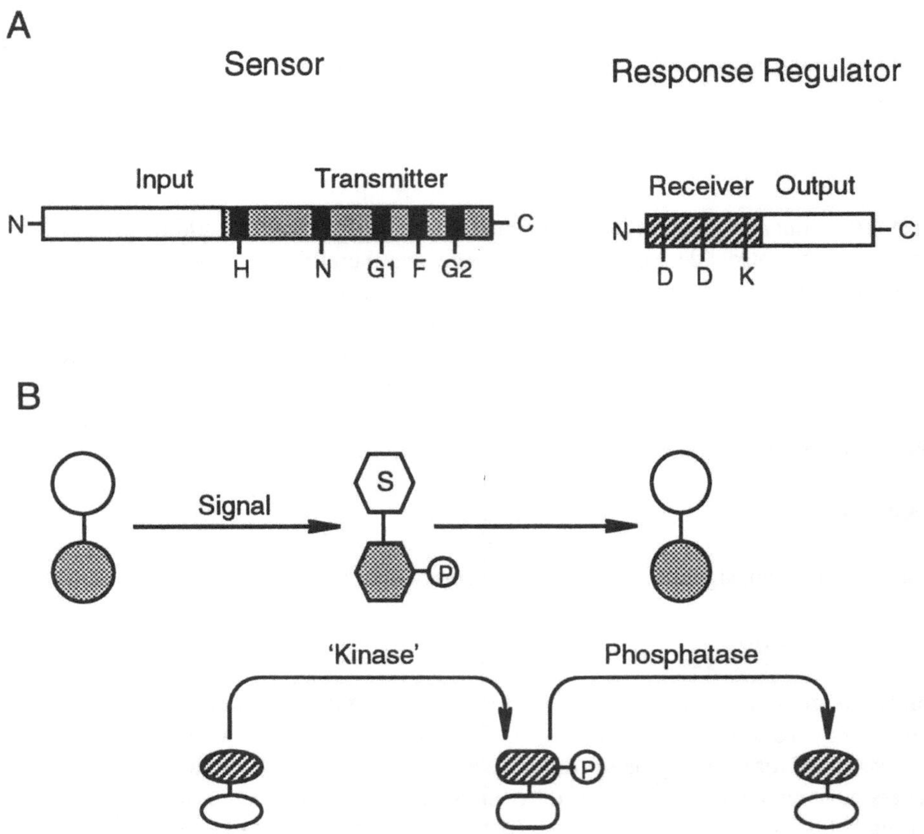

FIG. 1. The two-component paradigm. A. Schematic view of a generic sensor and response regulator. Amino- and carboxyl-termini are denoted N and C, respectively. Blocks H, N, G1, F and G2 in the transmitter domain, and the three residues Asp, Asp and Lys in the receiver domain, are conserved among most known representatives. Input and output domains vary greatly among different proteins. B. Transphosphorylation reactions. In response to a signal, the sensor autophosphorylates and serves as a substrate for response regulator phosphorylation. In the absence of signal, the sensor acts to dephosphorylate the phospho-response regulator.

The term two-component regulatory system has come to imply that the components communicate through a specific mechanism, transphosphorylation from His to Asp residues, involving these specific domains. Therefore, both partners must contain these well-defined domains in order to qualify as a true two-component regulatory system. For example, the the nitrogen regulatory protein NtrC contains an amino-terminal receiver domain, a central ATPase domain, and a carboxyl-terminal DNA-binding domain. The *nif* gene regulatory protien NifA shares similar ATPase and DNA-binding domains, and therefore is sometimes incorrectly classed as a two-component regulator, even though the amino-terminal domain of NifA shares no similarity with receiver domains. Likewise, although it has no receiver domain, the ToxR protein from *Vibrio cholerae* shares DNA-binding domain similarity with the response regulator OmpR. Describing NifA and ToxR as two-component-type regulators is not only potentially very confusing, it also serves to obscure the fact that the activities of these proteins are controlled in different, interesting ways.

Most two-component systems directly control gene expression, although the Che system controls behavior (chemotaxis). There are as many known variations on the basic two-component theme as there are known systems, and I make no attempt to describe them all. Rather, I will present the basic two-component model as currently understood, and then discuss a limited number of well-studied systems. Unless noted otherwise, these specific examples are of systems found in *Escherichia coli* and/or its close relative *Salmonella typhimurium*.

2 Transmitter and Receiver Domains

The transmitter domain, the defining domain of sensor proteins, is determined by a series of conserved sequence motifs, residues within which contribute essential functional groups for the transphosphorylation reactions. Details and references are readily accessible (Ninfa, 1996; Parkinson, 1993; Parkinson and Kofoid, 1992; Stock et al., 1995). The critical residue is an invariant His, which serves as the site for autophosphorylation. The sequence motif that contains this His residue is termed the H box (Fig. 1). Other conserved motifs include the N, G1, F and G2 boxes (Fig. 1), which form the catalytic center. (These motifs are designated by the one-letter symbol for a highly conserved residue within each, namely His, Asn, Gly and Phe, respectively.) The G1 and G2 motifs are likely involved in binding ATP. These conserved motifs are relatively short (about ten residues each), and the sequences between the motifs are generally not conserved among different transmitters. At this writing, there is no report of a three-dimensional structure for any transmitter domain.

Some transmitters lack one of the four boxes N, G1, F or G2. However, in evaluating a newly-determined sequence, the experimentalist is advised to ensure that at least three of these boxes, along with the critical H box, are clearly apparent before denoting the deduced polypeptide as a transmitter domain. Most claims to have discovered an 'unorthodox' transmitter are generally met with skepticism until supported by direct experimental evidence for biochemical function.

Transmitter domains are invariably connected to input domains, which are generally unrelated to each other (Fig. 1). Many (not all) sensor proteins are integral membrane proteins, with two or more transmembrane helices delimiting a periplasmic domain. In many cases, it is known or presumed that the periplasmic domain binds the signal ligand or other stimulus, leading to conformational changes that result in transmembrane signalling. Two well-studied soluble, cytoplasmic sensors are CheA and NtrB, involved in chemotaxis and nitrogen regulation, respectively. CheA activity is controlled by transmembrane methyl-accepting chemotaxis proteins (MCPs; Amsler and Matsumura, 1995), whereas NtrB activity is controlled by a series of cytoplasmic proteins that form a signalling cascade (Ninfa et al., 1995).

The receiver domain, the defining domain of response regulator proteins, is determined by overall sequence conservation, again with critical residues which contribute essential functional groups for the transphosphorylation reactions. Details and references are readily accessible (Parkinson and Kofoid, 1992; Parkinson, 1993; Volz, 1993; Volz, 1995). The critical residue is an invariant Asp, corresponding to position 57 in CheY (Asp-57), which serves as the site for phosphorylation (Fig. 1). Other highly-conserved residues include those corresponding to CheY residues Asp-12, Asp-13, Tyr-87 and Lys-109 (Fig. 1).

The three-dimensional structure of CheY has been extensively analyzed (Volz, 1993; Volz, 1995; Zhu et al., 1996). The polypeptide chain is arranged as an α/β barrel, with five α-helices and five β-sheets. Residues Asp-12, Asp-13, and Asp-57 form an acidic pocket, and residues Tyr-87 and Lys-109 contribute critical hydrogen bonds. A magnesium ion, essential for phosphorylation, is coordinated by Asp-12 and Asp-13. At this writing, the structure of phospho-CheY has not been reported, so the resulting conformational changes remain a matter of speculation. The receiver domain of NarL, a response regulator of nitrate-responsive gene expression, shares extensive structural similarity with CheY (Baikalov et al., 1996).

Receiver domains are generally connected to DNA-binding domains (Fig. 1). The sequences of these output domains themselves fall into related groups, so response regulators are often classified into families based on the nature of their DNA-binding domains (e. g., OmpR family, FixJ family, etc.). As noted above, NtrC and related response regulators contain a central ATPase domain, which is essential for activation of the σ^N- (σ^{54}-) dependent form of RNA polymerase (Porter, North and Kustu, 1995). Some receivers, such as CheY and Spo0F, a component of the sporulation phosphorelay in *Bacillus subtilis* (Appleby, Parkinson and Bourret, 1996), are not connected to additional output domains. Finally, receiver domains are sometimes found associated with transmitter domains, as in the sensors ArcB, VirA of *Agrobacterium tumefaciens*, and BvgS of *Bordetella pertussis* (Parkinson and Kofoid, 1992).

3. Transphosphorylation Reactions

Covalent modification in two-component signal transduction was discovered by Alexander Ninfa and Boris Magasanik, who were working to reconstitute nitrogen-regulated gene expression in vitro (Ninfa and Magasanik, 1986). A minimal transcription system containing DNA template, σ^N-RNA polymerase, nucleotide triphosphates, and purified NtrC (GlnG) and NtrB (GlnL) exhibits efficient transcription initiation. The NtrB protein is necessary only in catalytic amounts, and preincubation with nucleotides greatly stimulates overall activity. Pursuit of these initial observations revealed that NtrB transfers the γ-phosphoryl group from ATP to NtrC (Ninfa and Magasanik, 1986). At the same time, several groups had noted sequence similarities between seemingly unrelated proteins involved in bacterial chemotaxis and gene expression. Appreciation and integration of these observations under the rubric of 'two-component regulatory systems' (Nixon, Ronson and Ausubel, 1986), and the identification of NtrB-NtrC as an example of such a system, opened the way for interpretation of disparate observations from a variety of regulatory systems (Stock, Ninfa and Stock, 1989). Further work with select model systems — notably NtrB-NtrC, CheA-CheY, and EnvZ-OmpR — led to the identification of the His and Asp residues that serve as phosphorylation sites (Parkinson and Kofoid, 1992; Parkinson, 1993).

The first reaction is autophosphorylation of the conserved His residue in the transmitter domain (Fig. 1). This reaction is controlled in some systems in response to stimulus, and in other systems by the absence of stimulus. For example, EnvZ autokinase activity appears to be stimulated in the presence of stimulus (Pratt and Silhavy, 1995). By contrast, NtrB autokinase activity is inhibited by GlnB (P_{II}) protein in its non-uridylylated form, the signal of intracellular nitrogen sufficiency (Ninfa and Magasanik, 1986; Ninfa et al., 1995). Studies with mutant forms of various sensors indicate that transmitters are active as dimers, and that transphosphorylation can occur between subunits (Amsler and Matsumura, 1995; Ninfa et al., 1995).

The second reaction is phosphoryl transfer to the cognate receiver domain (Fig. 1). Small molecular weight compounds such as acetyl phosphate can serve as relatively efficient phosphodonors in this reaction, which demonstrates that the receiver domain is also an autokinase (Lukat et al., 1992). However, the cognate transmitter domain serves as the most efficient substrate for receiver phosphorylation, and it is convenient therefore to refer to transmitter 'kinase' activity as a shorthand designation for this activity. Presumably, unique sequences within the transmitter domain form contact surfaces that impart selectivity and specificity upon transmitter-receiver interactions.

The final reaction is dephosphorylation of the receiver domain, which in general is also catalyzed by the cognate transmitter (Fig. 1). In some cases, CheY for example, the phosphoryl linkage is inherently unstable, and the phosphorylated form of the protein has a half-life measured in seconds. (This fact has greatly complicated the structural determination of phospho-CheY). However, even in this case, a specific phosphatase (CheB) acts to dephosphorylate phospho-CheY upon appropriate stimulus (Amsler and Matsumura, 1995). Phospho-NtrC has a longer half-life, and dephosphorylation is stimulated by GlnB-activated NtrB (Ninfa et al., 1995). Finally, the phosphorylated forms of many response regulators, such as OmpR, are quite stable, yet their cognate sensors (e. g., EnvZ) catalyze efficient dephosphorylation (Fig. 1). Although the H box region of the transmitter is important, this phosphatase activity is not simply a reversal of the 'kinase' activity, because certain substitutions for the invariant His residue result in proteins that retain phosphatase activity even though kinase activity is abolished (Kamberov et al., 1994; Skarpohl, Waukau and Forst, 1997).

Thus, three related biochemical reactions control response to a given stimulus: sensor autophosphorylation, response regulator phosphorylation, and response regulator dephosphorylation. In principle, any of these phosphorylation reactions might be regulated, although most discussions center around the questions of sensor autophosphorylation and response regulator dephosphorylation. It is often presumed that response regulator phosphorylation is limited only by the availability of its substrate (phosphorylated sensor), and is therefore not directly regulated.

4 NtrB-NtrC: Control of Nitrogen Assimilation

Nitrogen is often the growth rate-limiting nutrient in natural environments, and enterobacteria can assimilate nitrogen from a variety of organic and inorganic sources. Specific peripheral pathways convert the nitrogen from these sources into ammonium and/or glutamate. Assimilation of this nitrogen into central metabolism requires the enzyme glutamine synthetase (encoded by *glnA*). Thus, bacteria respond to internal nitrogen limitation (Ikeda, Shauger and Kustu, 1996) by increasing the synthesis and activity of glutamine synthetase, and also by synthesizing the enzymes for peripheral pathways. The physiology of these processes has been well-reviewed (Magasanik, 1996; Merrick and Edwards, 1995).

Nitrogen-regulated gene expression is controlled by the response regulator NtrC, the phosphorylated form of which functions as a transcriptional activator (Fig. 2). NtrC contains a conventional amino-terminal receiver domain, a central ATPase domain, and a carboxyl-terminal DNA binding domain. Phosphorylation stimulates both NtrC multimerization, resulting in increased DNA binding affinity, as well as ATPase activity, which is required for transcription activation. Phospho-NtrC binds to enhancer sites upstream of the relevant promoters, and makes direct contact with the σ^N form of RNA polymerase through formation of a DNA loop. This protein-protein contact uses the energy of ATP hydrolysis to stimulate transcription initiation (Porter, North and Kustu, 1995).

The phosphorylation state of NtrC is controlled by the sensor NtrB, a soluble cytoplasmic protein (Fig. 2). NtrB contains a conventional carboxyl-terminal transmitter domain and a unique amino-terminal domain. In the absence of stimulus, NtrB autophosphorylates and serves as a substrate for NtrC phosphorylation ('kinase' activity). This default state corresponds to conditions of internal glutamine limitation.

The intracellular glutamine pool is monitored by GlnD (uridylyltransferase), which controls the covalent modification of GlnB (P_{II}). In response to glutamine limitation, GlnB-UMP is produced, whereas in response to glutamine sufficiency, unmodified GlnB predominates. Unmodified GlnB stimulates NtrB phosphoprotein phosphatase activity. This results in a decreased concentration of phospho-NtrC, the transcriptional activator, and both glutamine synthetase and peripheral nitrogen pathways are synthesized at low levels. GlnB, in both its modified and unmodified forms, also controls the activity of GlnE (adenylyltransferase), a regulator of glutamine synthetase enzyme activity (Ninfa et al., 1995).

Thus, control of phospho-NtrC levels is directly linked to the internal availability of nitrogen. The NtrB-NtrC two-component system couples the direct measurement of metabolic pool levels (via GlnD) to transcriptional control (NtrC) through a cascade of covalent modification reactions. Operation of this cascade in vivo evidently results in sensitive and rapid adjustments that provide very fine control of enzyme synthesis.

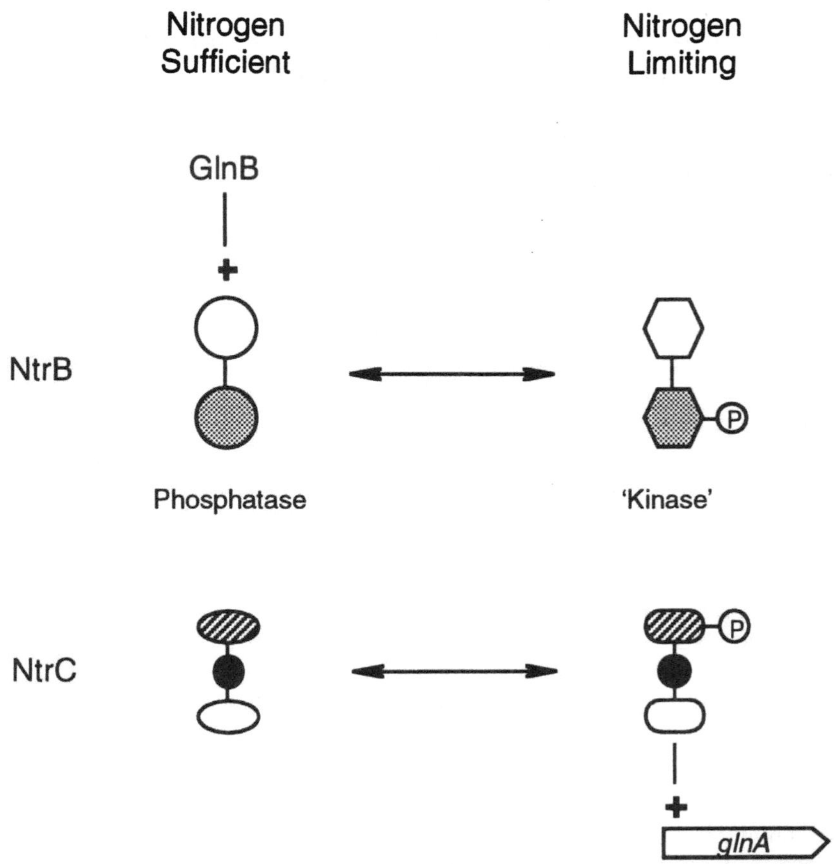

FIG. 2. NtrB-NtrC. During nitrogen-sufficient conditions, unmodified GlnB (P_{II}) shifts NtrB into the phospho-NtrC phosphatase conformation. During nitrogen-limiting conditions, NtrB autophosphorylates and acts as a substrate for NtrC phosphorylation. Phospho-NtrC activates transcription of *glnA* and other nitrogen-regulated genes. The ATPase domain of NtrC is depicted as a filled circle.

5 EnvZ-OmpR: Control of Outer Membrane Permeability

The Gram-negative outer membrane provides an effective barrier to both hydrophobic and hydrophilic compounds. However, cells must acquire small hydrophilic solutes for growth and nutrition. Specific outer membrane proteins, termed porins, form hydrophilic channels that allow solutes the size of glucose or smaller to pass into the periplasm, whereupon they are transported into the cytoplasm. Controlling the outer membrane permeability barrier is therefore of critical importance.

E. coli synthesizes two porins, OmpF and OmpC, with different permeability properties. Individual polypeptides form homotrimers, as well as heterotrimers with hybrid properties. Thus, very fine adjustments in outer membrane permeability are made possible through control of the relative amounts of OmpF and OmpC. Indeed, a variety of physiological parameters influence their relative rates of synthesis (Pratt et al., 1996). The most intensively-studied of these parameters is medium osmolarity.

Porin gene expression is controlled by the response regulator OmpR, the phosphorylated form of which functions as a transcriptional activator (Fig. 3). OmpR contains a conventional amino-terminal receiver domain and a carboxyl-terminal DNA binding domain. Phospho-OmpR binds to sites upstream of the relevant promoters, and makes direct contact with the α subunit of σ^{70}-containing RNA polymerase (Slauch, Russo and Silhavy, 1991). This protein-protein contact stimulates transcription initiation (Pratt and Silhavy, 1995).

The phosporylation state of OmpR is controlled by the sensor EnvZ, a cytoplasmic membrane protein (Fig. 3). EnvZ contains a conventional carboxyl-terminal transmitter domain, located in the cytoplasm, and a unique amino-terminal domain, located in the periplasm. The periplasmic and cytoplasmic domains are separated by a pair of transmembrane helices. Under conditions of high osmolarity, efficient EnvZ autophosphorylation results in the accumulation of high levels of phospho-OmpR. Under conditions of low osmolarity, EnvZ acts predominately as a phospho-OmpR phosphatase, which results in relatively low levels of phospho-OmpR. Unfortunately, the means by which EnvZ monitors osmolarity is unknown (Pratt and Silhavy, 1995).

Constant levels of porin are maintained irrespective of the growth condition, but the relative ratio of the two polypeptides exhibits considerable adjustment. Thus, the two target genes exhibit reciprocal regulation: in low osmolarity, *ompF* is predominately

---

FIG. 3. (Adjacent page) EnvZ-OmpR. A. Growth in low osmolarity medium. The equilibrium between the EnvZ phosphatase and 'kinase' conformations is shifted toward the phosphatase. The resulting low level of phospho-OmpR is sufficient to fill high-affinity activation sites in the *ompF* control region, resulting in high-level *ompF* expression. B. Growth in high osmolarity medium. The equilibrium between the EnvZ phosphatase and 'kinase' conformations is shifted toward the 'kinase'. The resulting high level of phospho-OmpR is sufficient to fill low-affinity activation sites in the *ompC* control region, resulting in high-level *ompC* expression, and also to fill low-affinity repression sites in the *ompF* control region, resulting in low-level *ompF* expression.

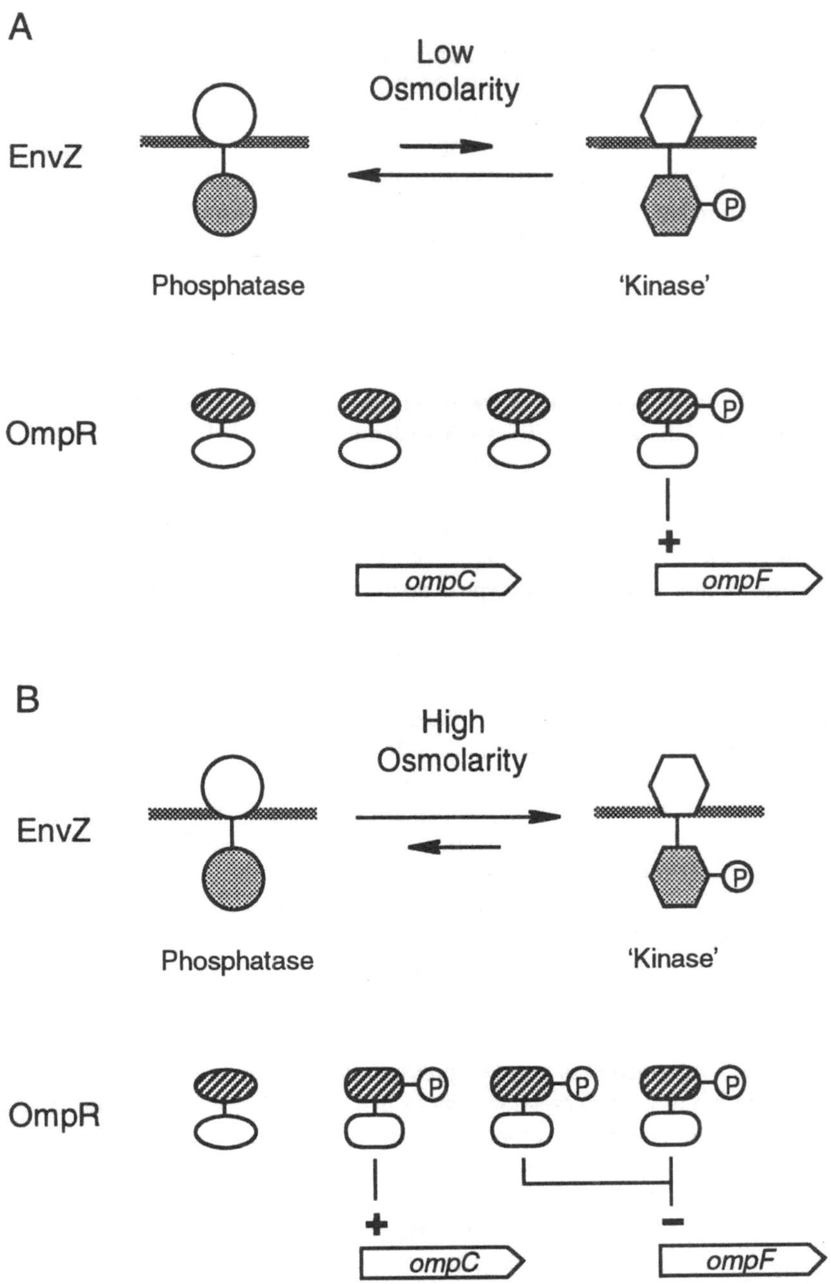

FIG. 3. See adjacent page for legend.

expressed, in high osmolarity, *ompC* is predominately expressed, and in intermediate osmolarity both genes are expressed (Pratt and Silhavy, 1995).

The relative rates of *ompF* and *ompC* transcription are differentially regulated in response to the level of phospho-OmpR. Null alleles of *ompR* result in an OmpF$^-$ OmpC$^-$ phenotype, demonstrating that OmpR is essential for *omp* gene expression. Critically, null alleles of *envZ* demonstrate a similar OmpF$^-$ OmpC$^-$ phenotype, showing that OmpR phosphorylation is essential for expression of both genes. This latter result eliminates one simple model, in which phospho-OmpR would activate one gene, and OmpR (nonphosphorylated) would activate the other (Pratt and Silhavy, 1995).

If phospho-OmpR is required for expression of both *ompF* and *ompC*, then how is differential regulation accomplished? This was determined in a series of elegant genetic experiments, in which missense alleles of *envZ* and *ompR* were isolated and characterized with respect to *ompF* and *ompC* transcription (Slauch and Silhavy, 1989; Russo and Silhavy, 1991; Russo, Slauch and Silhavy, 1993). One class of *ompR* alleles causes constitutive OmpF expression and little OmpC expression (OmpF^c OmpC$^-$ phenotype). A second class exhibits the reciprocal OmpF$^-$ OmpC^c phenotype. Alleles of *envZ* that confer analogous phenotypes have also been identified. Critically, the OmpF^c OmpC$^-$ phenotype is recessive to the wild-type, whereas the OmpF$^-$ OmpC^c phenotype is dominant. This latter observation indicates that OmpR can act to repress *ompF* transcription; the dominant mutations presumably freeze OmpR in its *ompF*-repressing, *ompC*-activating conformation.

These and related observations (Slauch and Silhavy, 1991) led to a model whereby the concentration of phospho-OmpR is the determining factor for *omp* gene expression (Fig. 3). At a low concentration (low osmolarity), phospho-OmpR binds to high-affinity sites upstream of the *ompF* promoter, activating its transcription. At a high concentration (high osmolarity), phospho-OmpR binds also to low-affinity sites upstream of the *ompC* promoter, activating its transcription, and to low-affinity sites near the *ompF* promoter, repressing its transcription. Support for this model comes from in vitro analyses of OmpR-DNA interactions, in which the low- and high-affinity phospho-OmpR binding sites have been directly visualized (Harlocker, Bergstrom and Inouye, 1995; Huang and Igo, 1996).

Thus, control of porin synthesis is regulated by a rheostat rather than a toggle. As the concentration of phospho-OmpR increases, *ompF* expression is decreases simultaneously with increased *ompC* expression (Russo and Silhavy, 1991). This reciprocal pattern of gene expression ensures that the total amount of porin remains relatively constant, but that the properties of the permeability pores are adjusted appropriately in response to growth conditions.

6 NarX-NarL / NarQ-NarP: Control of Anaerobic Respiration

Enterobacteria are facultative aerobes. Although they will use oxygen as a terminal electron acceptor for aerobic respiration, their metabolism is generally geared toward an anaerobic lifestyle, with fermentation and anaerobic respiration providing energy. Distinct respiratory enzyme complexes allow for anaerobic respiration with nitrate, nitrite, *N*- and *S*-oxides, fumarate and other compounds as terminal electron acceptors. Enzyme synthesis is subject to hierarchical control, first by oxygen (the preferred electron acceptor), response to which is mediated by the Fnr protein, and second by nitrate and nitrite (the preferred anaerobic electron acceptors), response to which is mediated by the Nar regulatory system. Aspects of Fnr function are considered elsewhere in this volume.

Nitrate is reduced to nitrite, and nitrite is reduced to ammonium, both in energy-conserving reactions. Thus, a culture growing with nitrate consumes one substrate and signal ligand (nitrate) while simultaneously accumulating another (nitrite); as the nitrate becomes exhausted, the culture begins consuming nitrite. Thus, control of enzyme synthesis is tailored to its physiological role. For example, respiratory nitrate reductase (encoded by the *narG* operon) is synthesized in response to nitrate but not to nitrite, whereas respiratory nitrite reductase synthesis (encoded by the *nrfA* operon) is repressed by nitrate but induced by nitrite.

Anaerobic respiratory gene expression is controlled by dual homologous response regulators, NarL and NarP, the phosphorylated forms of which function as transcriptional activators and repressors (Fig. 4). Both proteins contain a conventional amino-terminal receiver domain and a carboxyl-terminal DNA binding domain. Phospho-NarL and -NarP bind to sites upstream of the relevant promoters to stimulate transcription initiation (Stewart and Rabin, 1995).

The phosporylation states of NarL and NarP are controlled by the dual homologous sensors NarX and NarQ, cytoplasmic membrane proteins (Fig. 4). Both proteins contain a slightly unconventional carboxyl-terminal transmitter domain (lacking the G2 box), located in the cytoplasm, and a unique amino-terminal domain, located in the periplasm. The periplasmic and cytoplasmic domains are separated by a pair of transmembrane helices. In response to nitrate, both the NarX and NarQ sensors autophosphorylate and serve as substrates for NarL and NarP phosphorylation. In the absence of stimulus, NarX acts as a phospho-NarL phosphatase, and NarQ acts as a phospho-NarQ phosphatase. Thus, NarX-NarL and NarQ-NarP form cognate sensor-regulator pairs, although both NarX and NarQ can act to phosphorylate both NarL and NarP (Stewart and Rabin, 1995).

The response to nitrite is more complicated. In this case, as with nitrate, autophosphorylated NarQ serves as a substrate for NarL and NarP phosphorylation. NarX also serves to phosphorylate NarP. However, in response to nitrite, NarX acts primarily as a phospho-NarL phosphatase. Thus, nitrite results in relatively high levels of phospho-NarP but relatively low levels of phospho-NarL (Fig. 4).

Patterns of differential gene expression in response to nitrate and nitrite are therfore set by the number, location and relative affinity of DNA binding sites for phospho-NarL and phospho-NarP. For example, the *narG* operon, which is induced efficiently only in response to nitrate, contains only phospho-NarL binding sites in its upstream control region (Darwin, Li and Stewart, 1996). Thus, in the presence of nitrite, the relatively low concentration of phospho-NarL leads to relatively low level *narG* expression. By contrast, the nitrite-inducible *nrfA* operon is repressed by nitrate. A high-affinity site that binds either phospho-NarP or phospho-NarL suffices for

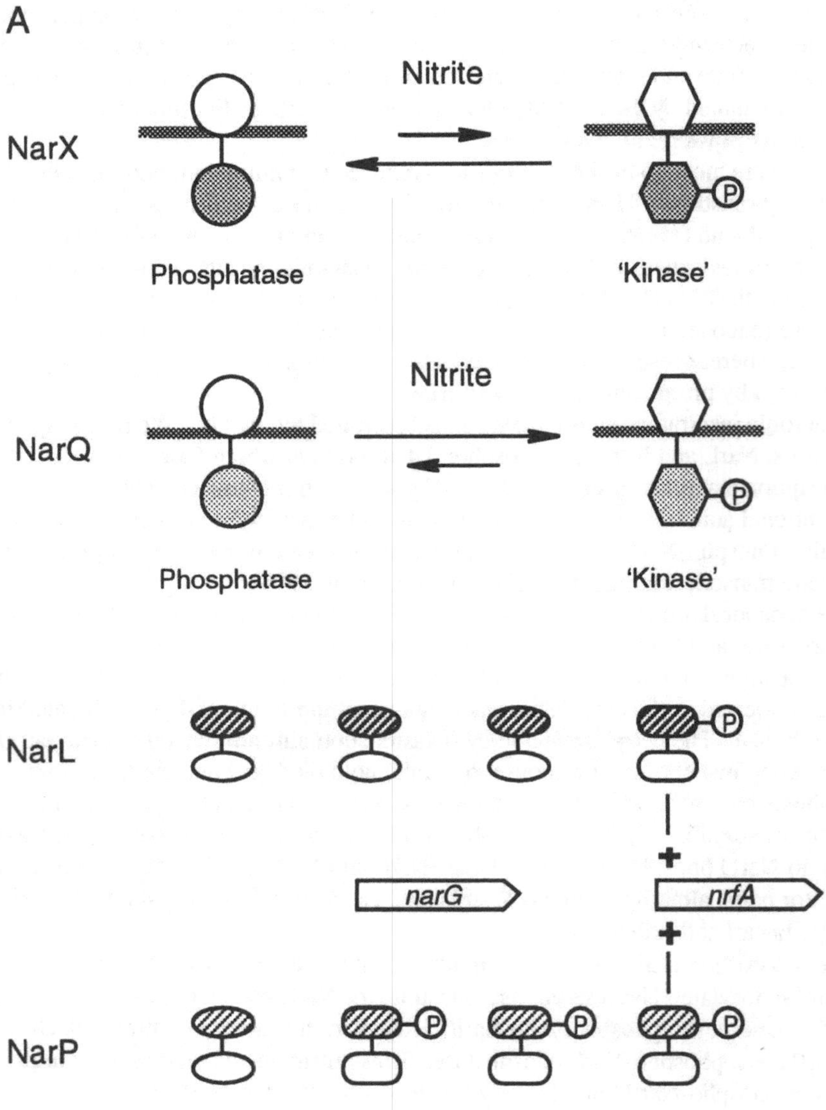

FIG. 4A. NarX-NarL / NarQ-NarP response to nitrite. The NarX equilibrium is shifted toward the phosphatase conformation, whereas the NarQ equilibrium is shifted toward the 'kinase' conformation. The resulting low levels of phospho-NarL are sufficient to fill high-affinity activation sites in the *nrfA* control region, resulting in high-level *nrfA* expression. Phospho-NarP also activates *nrfA* transcription.

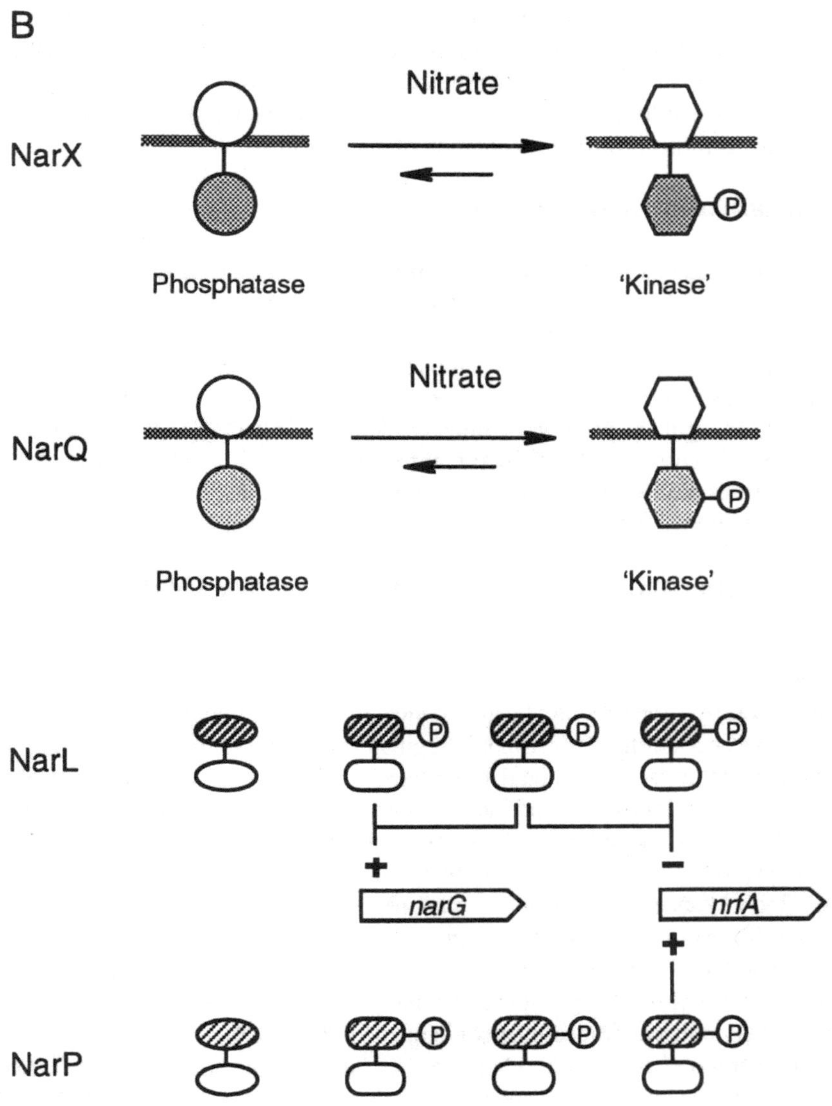

FIG. 4B. NarX-NarL / NarQ-NarP response to nitrate. The NarX and NarQ equilibria are both shifted toward the 'kinase' conformation. The resulting high levels of phospho-NarL are sufficient to fill low-affinity activation sites in the *narG* control region, resulting in high-level *narG* expression., and also to fill low-affinity repression sites in the *nrfA* control region, resulting in low-level *nrfA* expression. Phospho-NarP also activates *nrfA* transcription.

efficient nitrite induction. Lower-affinity sites for phospho-NarL are occupied only during growth with nitrate, when the phospho-NarL concentration is sufficiently high; occupancy of these low-affinity sites results in repression (Tyson, Cole and Busby, 1994). Other operons employ other variations of these themes. Again, as in the case of porin gene expression, it is the relative amount of phosphorylated response regulator that is critical for fine-modulation of gene expression.

7 Crosstalk and Cross-regulation

An important milestone in defining two-component signal transduction was the demonstration that, in vitro, CheA can phosphorylate NtrC, and NtrB can phosphorylate CheY (Ninfa et al., 1988). This experiment provided direct proof that the conserved sequence motifs shared among these proteins reflects a conserved biochemical mechanism, namely transphosphorylation. This interaction between non-cognate sensor-regulator pairs was termed *crosstalk*. These observations invited speculation that sensor-regulator pairs might form a complex network in vivo, whereby various physiological stimuli are coordinated and integrated to ensure a balanced and internally consistent response. This is a very appealing idea that is difficult to test.

Studies of chemotaxis, nitrogen assimilation, osmoregulation and phosphate regulation, among others, have suggested the possibility of non-cognate transphosphorylation in vivo. Invariably, these observations require a null allele of the cognate sensor, in these examples CheA, NtrB, EnvZ and PhoR respectively (Wanner, 1992). A further complication comes from acetyl phosphate, a metabolite that can serve as a substrate for response regulator phosphorylation (Lukat et al., 1992); it is hypothesized that modulation of acetyl phosphate concentrations in vivo might play a physiologically-important role in controlling regulatory output (Kim, Wilmes-Riesenberg and Wanner, 1996; McCleary, Stock and Ninfa, 1993).

As always, well-defined and understood terminology is important for meaningful debate. Barry Wanner has defined *crosstalk* as 'noise' resulting from the common biochemical basis for two-component signalling (Wanner, 1992). In this view, the negative function of sensor proteins (i. e., phosphoprotein phosphatase activity) is critical for dephosphorylating cognate response regulators that may become inappropriately phosphorylated by a non-cognate sensor. Thus, in a wild-type organism, inappropriate phosphorylation of NtrC by autophosphorylated CheA (for example) would be immediately countered by the phospho-NtrC phosphatase activity of NtrB.

The term *cross-regulation* is reserved for wild-type situations in which physiologically-relevant response regulator phosphorylation is mediated by a non-cognate sensor (Wanner, 1992). The search for cross-regulation has been aggressively pursued in the phosphate (Pho) regulatory system, but to date the non-cognate interactions have been observed only in *phoR* null strains and therefore may represent crosstalk instead (Kim, Wilmes-Riesenberg and Wanner, 1996). (The role of acetyl phosphate in this regulation presents an additional complication that I will not discuss here.) Understanding the circumstances under which different two-component systems engage in cross-regulation is therefore an interesting and challenging issue in contemporary bacterial physiology.

Acknowledgements

I thank Andrew Darwin and Stanly Williams from my laboratory for stimulating and thoughtful discussions. My research on two-component signalling is supported by U. S. Public Health Service grant GM36877 from the National Institute of General Medical Sciences.

References

Amsler CD, Matsumura P (1995) Chemotaxis signal transduction in *Escherichia coli* and *Salmonella typhimurium*. 89-103. Two-component signal transduction. Hoch JA, Silhavy TJ, Ed. Washington D. C., ASM Press.

Appleby JL, Parkinson JS, Bourret RB (1996) Signal transduction via the multi-step phosphorelay: not necessarily a road less traveled. Cell 86, 845-848

Baikalov I, Schröder I, Kaczor-Grzeskowiak M, Grzeskowiak K, Gunsalus RP, Dickerson RE (1996) Structure of the *Escherichia coli* response regulator NarL. Biochemistry 35, 11053-11061

Darwin AJ, Li J, Stewart V (1996) Analysis of nitrate regulatory protein NarL-binding sites in the *fdnG* and *narG* operon control regions of *Escherichia coli* K-12. Mol Microbiol 20, 621-632

Harlocker SL, Bergstrom L, Inouye M (1995) Tandem binding of six OmpR proteins to the *ompF* upstream regulatory sequence of *Escherichia coli*. J Biol Chem 270, 26849-26856

Hoch JA, Silhavy TJ (1995) Two-component signal transduction. Washington D. C., ASM Press.

Huang KJ, Igo MM (1996) Identification of the bases in the *ompF* regulatory region, which interact with the transcription factor OmpR. J Mol Biol 262, 615-628

Ikeda TP, Shauger AE, Kustu S (1996) *Salmonella typhimurium* apparently perceives external nitrogen limitation as internal glutamine limitation. J Mol Biol 259, 589-607

Kamberov ES, Atkinson MR, Chandran P, Ninfa AJ (1994) Effect of mutations in *Escherichia coli glnL* (*ntrB*), encoding nitrogen regulator II (NR_{II} or NtrB), on the phosphatase activity involved in bacterial nitrogen regulation. J Biol Chem 269, 28294-28299

Kim SK, Wilmes-Riesenberg MR, Wanner BL (1996) Involvement of the sensor kinase EnvZ in the in vivo activation of the response-regulator PhoB by acetyl phosphate. Mol Microbiol 22, 135-147

Lukat GS, McCleary WR, Stock AM, Stock JB (1992) Phosphorylation of bacterial response regulator proteins by low molecular weight phospho-donors. Proc Natl Acad Sci USA 89, 718-722

Magasanik B (1996) Regulation of nitrogen utilization. 1344-1356. *Escherichia coli* and *Salmonella*. Cellular and Molecular Biology. Neidhardt FC, Curtiss R, III, Ingraham JL et al., Ed. Washington, D. C., ASM Press.

McCleary WR, Stock JB, Ninfa AJ (1993) Is acetyl phosphate a global signal in *Escherichia coli*? J Bacteriol 175, 2793-2798

Merrick MJ, Edwards RA (1995) Nitrogen control in bacteria. Microbiol Rev 59, 604-622

Ninfa AJ (1996) Regulation of gene transcription by extracellular stimuli. 1246-1262. *Escherichia coli* and *Salmonella*. Cellular and molecular biology. Neidhardt FC, Curtiss R, III, Ingraham JL et al., Ed. Washington, D. C., ASM Press.

Ninfa AJ, Atkinson MR, Kamberov ES, Feng J, Ninfa EG (1995) Control of nitrogen assimilation by the NR_I-NR_{II} two-component system of enteric bacteria. 67-88. Two-component signal transduction. Hoch JA, Silhavy TJ, Ed. Washington D. C., ASM Press.

Ninfa AJ, Magasanik B (1986) Covalent modification of the *glnG* product, NR_I, by the *glnL* product, NR_{II}, regulates transcription of the *glnALG* operon in *Escherichia coli*. Proc Natl Acad Sci USA 83, 5909-5913

Ninfa AJ, Ninfa EG, Lupas AN, Stock A, Magasanik B, Stock J (1988) Crosstalk between bacterial chemotaxis signal transduction proteins and regulators of transcription of the Ntr regulon: evidence that nitrogen assimilation and chemotaxis are controlled by a common phosphotransfer mechanism. Proc Natl Acad Sci USA 85, 5492-5496

Nixon BT, Ronson CW, Ausubel FM (1986) Two-component regulatory systems responsive to environmental stimuli share strongly conserved domains with the nitrogen assimilation regulatory genes *ntrB* and *ntrC*. Proc Natl Acad Sci USA 83, 7850-7854

Parkinson JS (1993) Signal transduction schemes of bacteria. Cell 73, 857-871

Parkinson JS, Kofoid EC (1992) Communication modules in bacterial signalling proteins. Annu Rev Genet 26, 71-112

Porter SC, North AK, Kustu S (1995) Mechanism of transcriptional activation by NtrC. 147-158. Two-component signal transduction. Hoch JA, Silhavy TJ, Ed. Washington D. C., ASM Press.

Pratt LA, Hsing WH, Gibson KE, Silhavy TJ (1996) From acids to *osmZ*: multiple factors influence synthesis of the OmpF and OmpC porins in *Escherichia coli*. Mol Microbiol 20, 911-917

Pratt LA, Silhavy TJ (1995) Porin regulon of *Escherichia coli*. 105-127. Two-component signal transduction. Hoch JA, Silhavy TJ, Ed. Washington D. C., ASM Press.

Russo FD, Silhavy TJ (1991) EnvZ controls the concentration of phosphorylated OmpR to mediate osmoregulation of the porin genes. J Mol Biol 222, 567-580

Russo FD, Slauch JM, Silhavy TJ (1993) Mutations that affect separate functions of OmpR the phosphorylated regulator of porin transcription in *Escherichia coli*. J Mol Biol 231, 261-273

Skarpohl K, Waukau J, Forst SA (1997) Role of His243 in the phosphatase activity of EnvZ in *Escherichia coli.* J Bacteriol 179, 1413-1416

Slauch JM, Russo FD, Silhavy TJ (1991) Suppressor mutations in *rpoA* suggest that OmpR controls transcription by direct interaction with the alpha-subunit of RNA polymerase. J Bacteriol 173, 7501-7510

Slauch JM, Silhavy TJ (1989) Genetic analysis of the switch that controls porin gene expression in *Escherichia coli* K-12. J Mol Biol 210, 281-292

Slauch JM, Silhavy TJ (1991) Cis-acting *ompF* mutations that result in OmpR-dependent constitutive expression. J Bacteriol 173, 4039-4048

Stewart V, Rabin RS (1995) Dual sensors and dual response regulators interact to control nitrate- and nitrite-responsive gene expression in *Escherichia coli.* 233-252. Two-component signal transduction. Hoch JA, Silhavy TJ, Ed. Washington D. C., ASM Press.

Stock JB, Ninfa AJ, Stock AM (1989) Protein phosphorylation and regulation of adaptive responses in bacteria. Microbiol Rev 53, 450-490

Stock JB, Surette MG, Levit M, Park P (1995) Two-component signal transduction systems: structure-function relationships and mechanisms of catalysis. 25-51. Two-component signal transduction. Hoch JA, Silhavy TJ, Ed. Washington D. C., ASM Press.

Tyson KL, Cole JA, Busby SJW (1994) Nitrite and nitrate regulation at the promoters of two *Escherichia coli* operons encoding nitrite reductase: identification of common target heptamers for both NarP- and NarL-dependent regulation. Mol Microbiol 13, 1045-1055

Volz K (1993) Structural conservation in the CheY superfamily. Biochemistry 32, 11741-11753

Volz K (1995) Structural and functional conservation in response regulators. 53-64. Two-component signal transduction. Hoch JA, Silhavy TJ, Ed. Washington D. C., ASM Press.

Wanner BL (1992) Is cross-regulation by phosphorylation of two-component response regulator proteins important in bacteria? J Bacteriol 174, 2053-2058

Zhu XY, Amsler CD, Volz K, Matsumura P (1996) Tyrosine 106 of CheY plays an important role in chemotaxis signal transduction in *Escherichia coli.* J Bacteriol 178, 4208-4215

METAL REGULATION OF GENE EXPRESSION IN BACTERIAL SYSTEMS

Nigel L Brown, Kathryn R Brocklehurst, Blair Lawley and Jon L Hobman

School of Biological Sciences, The University of Birmingham, Birmingham B15 2TT, UK

1. Introduction

Metals are important in biochemical processes (da Silva and Williams, 1991). They can be cofactors of enzymatic reactions or they can be the key redox components of electron transport processes. Zinc is an example of a metal whose properties as a Lewis acid are used in the reactions of a wide variety of catalytic processes, and a quick glance through a biochemistry text book will generate a large list of zinc-containing enzymes. The transition metals, iron and copper, can readily lose or gain electrons under physiological conditions and are used in electron transport processes and for some biochemical redox reactions. Metals, such as zinc and magnesium, can also play a structural role in ensuring that enzymes or their substrates maintain the correct atomic and electronic structures. Because they are essential micronutrients to all cells, the intracellular concentrations of metals must be regulated, and in bacteria this appears to be done at the level of transcription of the genes encoding proteins for the uptake and export of the metal. Moreover, because the bacterial cell has no compartments, other than invaginations of the cytoplasmic membrane, and because metals of physiological importance have similar chemical properties, the regulation of expression of the transporter proteins may be controlled specifically by the intracellular concentration of a single metal.

A further complication is that many metals, particularly the 'heavy metals', are toxic to cells. Mercuric ions, for example, react with thiol groups in proteins to form covalent adducts and may inactivate the protein. Arsenate is a physiological mimic of phosphate and can form arseno-derivatives of phosphate-containing compounds; the best-known example is the formation of 1-arseno-3-phosphoglycerate in glycolysis and the subsequent uncoupling of substrate level ATP production. Bacteria need mechanisms for defence against these toxic metals and determinants of bacterial resistance to heavy metals have been found in many organisms (Silver, 1996). Mercury resistance is the single most widespread of all antimicrobial resistances. Expression of these resistance determinants is usually inducible, thus reducing the metabolic load on the cell in the absence of the toxic metal and preventing cross-reaction with other metals except when the toxic threat is present.

The properties of metals that make them essential micronutrients may also cause toxicity when the metal is present in excess. Resistance determinants to some essential heavy metals, such as copper, are known which extend the range of concentrations over which intracellular homeostasis can be maintained (Brown *et al.*, 1993).

NATO ASI Series, Vol. H 103
Molecular Microbiology
Edited by Stephen J. W. Busby,
Christopher M. Thomas and Nigel L. Brown

Unlike antibiotics or nutrients such as sugars and amino acids, metals are widely dispersed in the biosphere and can be mobilised or trapped by geochemical processes (da Silva and Williams, 1991). Bacteria and other microorganisms occupy a wide range of environments in which metals are found at high concentrations, or in which essential metals are sequestered. Most bacteria are thought to have gene regulatory mechanisms for homeostasis of metals, only a few of which have been characterised. This chapter will not be comprehensive, but it will cover examples of gene regulation for homeostasis of an essential metal, for resistance against purely toxic metals, and, finally, the interplay of regulation of homeostasis and resistance for an essential metal which is toxic at high concentrations.

Table 1: EXAMPLES OF GENE REGULATION BY METALS
This not a comprehensive table. Apart from *fur* all examples confer resistance to heavy metals. References are given in the text or in Silver and Phung (1996)

Metal	Bacterial genera	Gene designation
Arsenate	*Staphylococcus, Escherichia*	*ars*
Cadmium	*Staphylococcus, Escherichia,*	*cad*
Copper	*Pseudomonas, Enterococcus, Escherichia*	*cop, pco*
Iron	Various genera	*fur*
Mercury	Various genera	*mer*
Zinc	*Synechococcus*	*smt*

2. Gene regulation by essential heavy metals

2.1. Iron regulation: the Fur protein

Iron metabolism in bacteria is different from that of many other metals in that iron exists in nature as very insoluble ferric salts (da Silva and Williams, 1991). In order for iron to be accumulated it must be scavenged from the environment by small specific iron-chelating compounds called siderophores. In *E. coli* the proteins required for the production of two siderophores, enterobactin and aerobactin, and those required for transport of the chelated iron into the cell (at least 5 systems in *E. coli*) are synthesised under the control of the product of the *fur* (*f*erric *u*ptake *r*egulator) gene.

The product of the *fur* gene, Fur, is an iron-responsive repressor which acts on a variety of promoters. These promoters contain a specific 19 bp DNA operator sequence sometimes called the 'iron box' to which Fur binds (de Lorenzo *et al.*, 1987). This operator is normally found between the -35 and -10 motifs recognised by the σ^{70} factor in RNA polymerase holoenzyme (see Fig 1). At low iron concentration, Fur has a weak affinity for the operator and transcription of the iron-responsive genes occurs. At high intracellular concentrations of Fe(II), Fur binds the ferrous iron and the Fur-Fe(II) complex binds tightly to the operator sequence and blocks the initiation of transcription. Gel mobility shift assays to determine protein binding to a DNA fragment containing the promoter-operator complex from aerobactin synthesis operon (*iuc*) show that either Fur or RNA polymerase holoenzyme can bind to the DNA fragment, but not simultaneously, and that binding by Fur was dependent on a divalent metal ion (Wee *et al.*, 1988). In addition to Fe(II), other ions bind to Fur; but these probably have little physiological significance.

The *fepA-entD* and *fes-entF* operons are required for enterobacterin synthesis in *E. coli* and are divergently transcribed. They are repressed by the Fur protein when iron supply is plentiful. The structure of the control region (Hunt *et al.*, 1994) is shown in Figure 1. There are two main divergent transcripts, as indicated in the diagram. There is a third transcript from a minor promoter for *fepA-entD*, which has its start point within the 'iron box'; this has been omitted from the diagram for clarity. All three transcripts are coordinately regulated by the binding of Fur to a single 'iron box' structure.

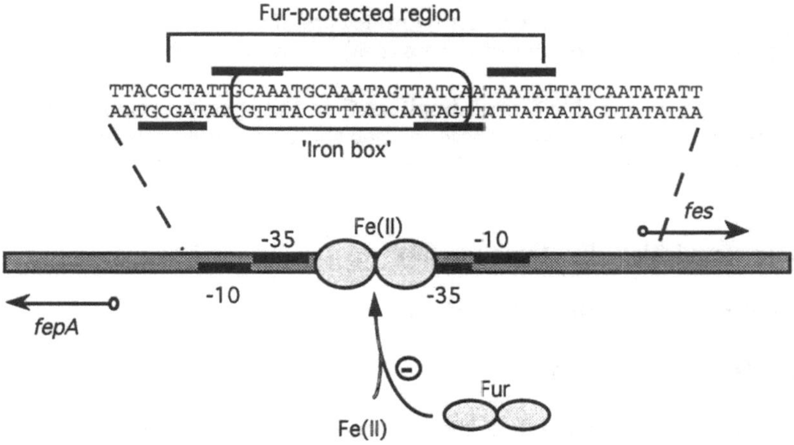

Figure 1. Regulation of the *fepA-fes* control region by Fur. The -35 and -10 regions of the two main divergent promoters are indicated as black bars. The 'iron box' sequence and the extent of protection conferred by Fur against DNaseI digestion is indicated. (Modified from Hunt *et al.*, 1994).

The structure for the Fur protein has been partially derived from NMR studies and mutagenesis, and it is proposed that a series of histidines and cysteines towards the C-terminus of this 148 amino acid protein are involved in Fe(II) binding. DNA protection experiments indicate that Fur wraps around the DNA, but no classical "helix-turn-helix" DNA binding motifs can be identified in the protein sequence. In studies with fusion proteins, the N-terminal region of Fur has been shown to be required for DNA binding (Stojiljkovic and Hantke, 1995).

2.1.1 Multiple systems may regulate uptake genes

Fur can also participate in control alongside other regulatory proteins. This is a common theme in global regulatory systems: the global regulator provides general control related to a general stimulus (in this case iron availability) and a specific regulator controls the specific response of a single operon. The iron-dicitrate transport system of *E. coli* is regulated by two proteins, FecR and FecI. FecR is a periplasmic sensor protein which responds to ferric citrate binding to the outer membrane protein FecA; FecR activates the protein FecI.

FecI is a member of a family of σ^{70}-type factors (ECF sigma factors) that respond to extracytoplasmic stimuli (Lonetto *et al.*, 1994; Angerer *et al.*, 1995). FecI directs RNA polymerase to direct transcription from the *fecAB* promoter, which is itself controlled by *fur*. Synthesis of FecA therefore requires both the activation of FecI (through the presence of external ferric citrate) and the release of repression by Fur (low cytoplasmic concentrations of ferrous iron) (Ochs *et al.*, 1996).

The ECF sigma factors have been implicated in the specific regulation of genes responding to external stimuli. These include the CnrH protein from the cadmium and nickel resistance determinant of *Alcaligenes eutrophus* (Lonetto *et al.*, 1994).

3. Gene regulation by toxic metals

3.1. Mercury regulation: the MerR protein

3.1.1. Mercury resistance

Mercuric ions are highly mobile in the environment and are very toxic to living cells (da Silva and Williams, 1991). About 10^5 tonnes of mercury are released into the biosphere each year by geochemical processes and the global mercury pool is estimated at 10^8 tonnes. Man contributes to the global mercury cycling through industrial processes. It is not surprising that resistance to mercuric ions is the most widespread of all antimicrobial resistances. Mercury resistance (*mer*) genes are found in Gram positive and Gram negative eubacteria (Silver, 1996) and may occur in the Archaea. In eubacteria resistance is conferred up uptake of mercuric ions into the cytoplasm, where they are reduced by the NADPH-dependent flavoenzyme mercuric reductase. It is counterintuitive to bring a toxic metal into the cytoplasm, but this allows reduction by the electron donor which can then be recycled. The mechanism is similar in all eubacteria studied, the major differences being in the number and identity of proteins required to transport mercuric ions into the cell (Hobman and Brown, 1997). The mechanism of resistance and the genes required for this on transposon Tn*501* from *Pseudomonas aeruginosa* are shown in Fig 2. The genes are expressed as an operon under the control of the *merR* gene product in response to the presence of Hg(II).

In addition to resistance to mercuric ions, some *mer* determinants encode an organomercury lyase that cleaves organomercurials. These determinants have similar mechanisms of resistance, the mercuric ion liberated by organomercury cleavage being reduced by mercuric reductase. For these determinants, the genes are expressed in response to organomercurials as well as to mercuric ions. The product of a second gene, *merD*, is also a regulatory protein that is thought to switch off the active *mer* promoter when mercuric ion concentrations are low (Mukhopadhyay *et al.*, 1991).

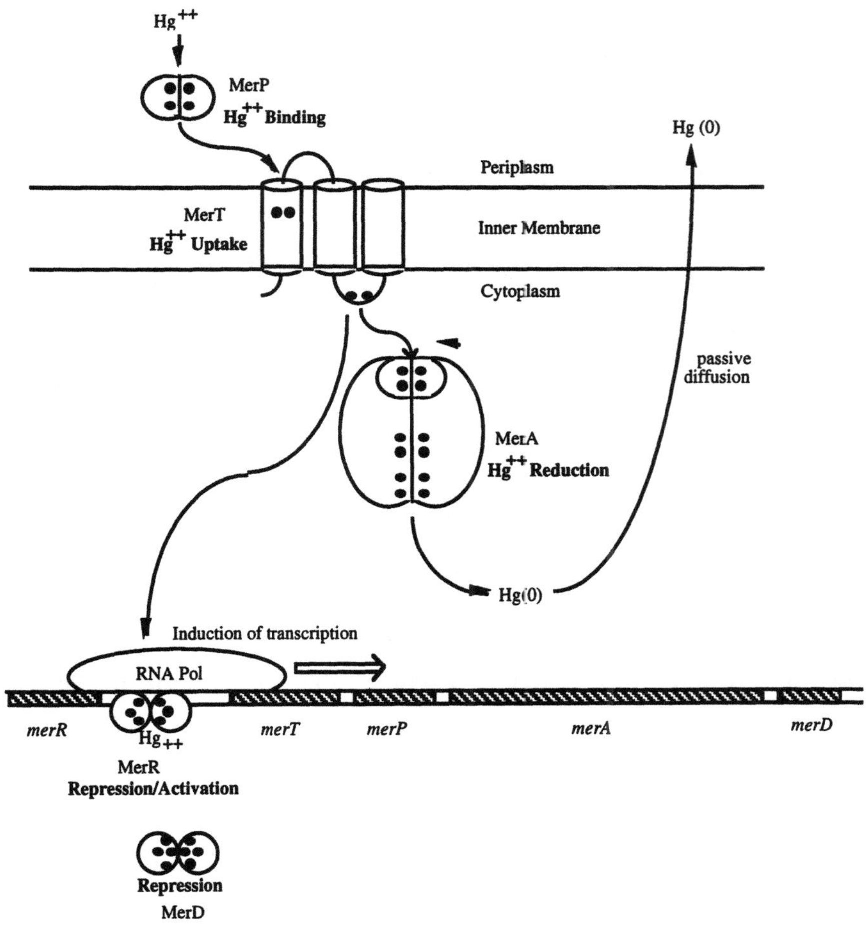

Figure 2: Model of the mechanism of mercuric ion resistance. Mercuric ions are hypothesised to be scavenged in the periplasm of Gram negative bacteria by the periplasmic protein MerP and transferred via the uptake protein MerT to mercuric reductase (MerA). Here the Hg(II) is reduced to Hg(0), which is volatilised from the cell. Cysteine residues are shown as black dots and constitute the Hg(II) binding sites on the proteins. Induction of gene expression occurs by binding of a single mercuric ion to the MerR dimer bound with RNA polymerase at the promoter. MerD is thought to play an unspecified role in down-regulation after induction. (From Hobman and Brown (1997), with permission).

3.1.2. The regulatory region of the *mer* operon

In many Gram negative bacteria the control region of the *mer* operon contains divergent promoters which are both regulated by MerR (Hobman and Brown, 1997). The region from transposon Tn*501* is shown in Figure 3. *merR* gene is transcribed from the *merR* promoter, P_R, and is therefore autoregulated by its own product. This overlaps with the promoter, P_T, responsible for expression of the *merTPAD* genes (and also for expression of transposition genes in Tn*501*). P_R is a normal σ^{70}-dependent promoter with typical -35 and -10 motifs recognised by the σ^{70} factor. P_T is unusual in having good -35 and -10 motifs, but separated by the unusually large distance of 19 base pairs (compared with the normal 16-17 base pairs). This long spacer region results in P_T being a weak promoter. Even though the motifs required for recognition by σ^{70} are good, in B-DNA they would be separated by a further 6.8Å of distance and oriented a further 72° around the helical axis than is normal for strong promoters. This spacer also contains the binding site for MerR, a dyad symmetrical sequence from position -32 to -15 before the P_T transcriptional start. This is also at position +1 to +18 of the P_R promoter.

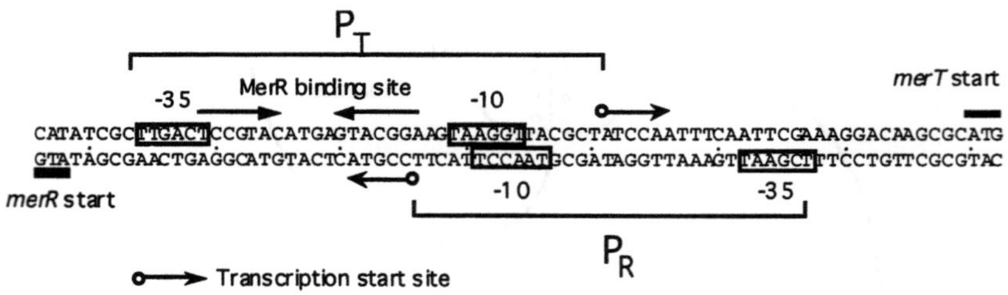

Figure 3: the operator/promoter region of the *mer* operon of Tn*501*.

3.1.3. The MerR protein and the mechanism of transcriptional activation

MerR is a dimeric protein with a subunit size of 144 amino acids. It contains a helix-turn-helix DNA binding motif in the first 30 N-terminal amino acids, and a mercury-binding site in the C-terminal half of the protein. Work on the control region of the Tn*501* and Tn*21* *mer* operons *in vitro* and *in vivo* has shown that MerR is bound to the *mer* control region in the absence of Hg(II). This explains why it acts as a repressor of both P_T and P_R. The mechanism of MerR was last specifically reviewed some time ago (Summers, 1992).

In the presence of mercuric salts a single Hg(II) cation binds to the MerR dimer, being coordinated by three cysteines, Cys82 in one subunit and Cys117 and Cys126 in the other. The Hg-MerR complex binds at the same site on DNA as does the uncomplexed MerR protein. The Hg-MerR complex activates expression from P_T but retains repression and P_R. Activation is due to a distortion of the DNA at the centre of the dyad symmetry of the binding site. This distortion is highly localised and can be detected by chemical nucleases, such as copper phenanthroline. The distortion results in an unwinding of DNA by 55°, presumably orienting the -35 and -10 motifs so that are now recognised by s^{70} in the RNA polymerase holoenzyme, and transcription can be initiated (Ansari *et al.*, 1992; Ansari *et al.*, 1995).

The binding of mercuric ions occurs exclusively in the C-terminal region of the protein. A comparison of predicted MerR sequences from a variety of mercury resistance determinants isolated from Gram negative bacteria shows that the very C-terminal amino acids of MerR differ depending on whether the MerR responds to organomercurials or not. Alteration of these amino acids in MerR from the broad organomercurial resistance determinant of pDU1358 caused loss of response to organomercurials but response to mercuric ions was unaffected (Nucifora *et al.*, 1989).

3.1.4. A family of regulators: the MerR family

The MerR protein shows amino acid similarity to several regulators for other functions unrelated to metal resistance. This similarity is between the DNA-binding region of MerR and the N-terminus of the other proteins. Direct and indirect evidence suggests that these proteins form a family of regulators with similar mechanisms of action, in which the N-terminal region binds DNA and the C-terminal region binds the co-effector. This family includes the regulators shown in Table 2. SoxR from *E. coli* shows some similarity in the C-terminal domain also, which may be due to its binding an iron-sulphur cluster. It is premature to refer to the DNA-binding and effector-binding regions of this class of regulators as domains, as there is little structural information. However, the C-terminal region of $TipA_L$ can exist separately *in vivo* (as $TipA_S$) and the C-terminal region of BmrR has been expressed as a separate rhodamine-binding polypeptide *in vitro*. Recent data from our laboratory (K.R. Brocklehurst and N.L. Brown, unpublished) indicates that MerR may exist in two domains: an N-terminal domain of about 45 amino acids containing the DNA binding site, and a 100 amino acid C-terminal domain (which may be further subdivided).

Table 2 Some members of the MerR family

Protein	Organism	Role	Reference
MerR	Various	Regulation of *mer* genes	(Summers, 1992)
SoxR	*Escherichia coli*	Response to oxidative stress	(Amabile-Cuevas and Demple, 1991)
$TipA_L$	*Streptomyces lividans*	Response to thiostrepton	(Holmes *et al.*, 1993)
NolA	*Bradyrhizobium japonicum*	Nodulation response to plant exudate	(Sadowsky *et al.*, 1991)
BmrR	*Bacillus subtilis*	Response to dyes and drugs	(Ahmed *et al.*, 1995)
BltR	"	"	(Ahmed *et al.*, 1995)

3.2 Arsenate regulation: the ArsR protein

Arsenate toxicity in the cell is prevented by export of the toxic oxyanion. Five arsenic resistance systems have been identified (Silver, 1996; Silver and Phung, 1996). Two from *Staphylococcus* and one from *E. coli* have three genes expressed in an operon and two from *E. coli* have five genes; in all cases the first gene encodes the regulator, ArsR. In addition to the structural genes for the export pump (*arsB* in *Staphylococcus* and *arsAB* in *E. coli*) there is a gene *arsC* which encodes arsenate reductase. This enzyme reduces arsenate (As(V)) to arsenite (As(III)). Arsenite is pumped out of the cell by the export proteins. This is of physiological advantage, as the arsenite pump has no cross specificity for phosphate.

The ArsR protein is a repressor (Wu and Rosen, 1991; Xu *et al.*, 1996) and arsenite, arsenate, antimonite and bismuth act as inducers. The protein contains a putative helix-turn-helix motif just to the C-terminal side of a Cys-Val-Cys sequence that binds the inducer. ArsR binds to an operator sequence which includes the -35 motif in the promoter in both the *E. coli* and *Staphylococcus* operons. Binding of arsenite causes the ArsR protein to be released from the DNA and transcription is induced. In the *E. coli* operon a second regulatory gene, *arsD*, has been identified on the promoter-distal side of *arsR*. This appears to be an inducer-independent *trans*-acting regulator which maintains repression of the downstream structural genes in the *arsRDABC* operon.

3.2.1. The ArsR family of regulators

A number of regulators show amino acid similarity to ArsR. These include CadC, the regulator of the cadmium resistance operon of *Staphylococcus aureus* plasmid pI258; SmtB, the regulator of the metallothionein gene, *smtA*, in the cyanobacterium *Synechococcus*; and MerR from *Streptomyces lividans*. All these regulators are induced by metals (Cd(II), Zn(II) and Hg(II), respectively) and all act negatively.

3.3. Other regulatory systems for toxic metals

A variety of determinants confer resistance to toxic metals and resistance is virtually always inducible. These include resistances to silver, chromate, cadmium, lead, nickel, zinc, and cobalt (Silver, 1996; Silver and Phung, 1996). The Gram negative bacterium *Alcaligenes eutrophus* contains plasmids conferring resistance to a large number of metals (Mergeay *et al.*, 1985). These plasmids confer resistance to mercury (several *mer* determinants), cadmium, zinc and cobalt (the *czc* determinant), cadmium and nickel (*cnr*), copper (*cop*; see below) and lead. The last has not been characterised. The regulation of *cnr* involves both a ECF sigma factor (the *cnrH* gene product) and possibly an anti-sigma factor (the *cnrX* product). The silver resistance determinant found on the *E. coli* plasmid pMG101 shows similarity to the *pco* copper resistance system (section 4.2) and contains two genes (*silRS*) which probably constitute a two-component regulatory system, although there have been no studies as yet of the regulation of the *sil* determinant (S. Silver, personal communication).

4. Gene regulation by essential metals which are toxic in excess

4.1 Regulation of copper homeostasis

Copper is required as an essential component of a number of enzymes, mainly oxidases which use molecular oxygen as an electron acceptor, and of redox-proteins which undertake single electron transfers. In each case the copper is a tightly bound cofactor and its redox properties and substrate accessibility are controlled by the local environment in the protein. However, free copper ions are very toxic. They are thought to be reduced to Cu(I) in the cytoplasm of the bacterial cell, which contains 5-10mM glutathione, and can then cause formation of highly reactive hydroxyl radicals by reaction with hydrogen peroxide generated in the cell:

$$Cu^+ + H_2O_2 = Cu^{++} + OH^- + OH^\bullet$$

The free concentration of copper must be minimised in the cell. This is done in two ways: (i) control of the normal uptake and export of copper from the cell to maintain copper ion homeostasis, and (ii) expression of specific copper resistance genes at high copper concentrations. The copper resistance genes so far identified are plasmid borne and are only found in certain cells. The control of normal copper metabolism has been best studied in the Gram-positive organism *Entercoccus hirae* (formerly *Streptococcus faecalis*), whereas copper resistance has been studied in *E. coli* and *Pseudomonas syringae* (Brown *et al.*, 1993; Silver and Phung, 1996).

4.1 Regulation of copper uptake and export in *Enterococcus hirae*

E. hirae contains two genes, *copA* and *copB* that encode P-type ATPases responsible for copper uptake and copper export, respectively. These are regulated by two genes, *copY* and *copZ*, which respond to both low and high copper concentrations (Odermatt and Solioz, 1995). When copper is low, the copper uptake ATPase is required, when copper is excess, the copper export ATPase is required. Remarkably, the *copA* and *copB* genes are expressed in a single operon of which *copY* and *copZ* are the first two cistrons.

The regulation of the *copAB* genes by CopY and CopZ is actively being investigated. CopY is a 145 amino acid repressor. The working hypothesis is that CopY binds DNA only as a $(CopY\text{-}Cu^+)_2$ complex and inhibits transcription of the promoter. Induction of the operon occur at low copper concentrations because the apoprotein does not bind DNA. The CopZ protein is thought to be a small anti-repressor protein, which binds copper at higher copper concentrations. The CopZ-Cu^+ complex is suggested to bind CopY-Cu^+ to form a complex in which CopY can no longer bind DNA and repress transcription. Therefore, at high copper concentrations expression of the *cop* operon also occurs. Recent data show that CopY does indeed bind to DNA close to the *cop* promoter, but that increasing amounts of copper salts cause loss of DNA binding by CopY; indicating that CopY-Cu^+ does not bind DNA. A full description of the regulation of this system requires further work on CopY and CopZ (M. Solioz, personal communication).

4.2. Copper resistance and management in *Escherichia coli* and *Pseudomonas syringae*: regulation by two-component systems

Plasmid-borne copper resistance determinants have been isolated in *E. coli* from pigs and from the plant pathogenic bacterium *Pseudomonas syringae* pv. *tomato* (Brown *et al.*, 1993; Silver, 1996; Silver and Phung, 1996). The *E. coli* determinant was named *pco* (*p*lasmid *co*pper resistance) and that from P. *syringae* as *cop*. The resistance mechanisms differ, although many of the gene products show similarity of amino acid sequence. In *E. coli* excess copper is exported from the cell; cells containing *pco* accumulate less copper than those which do not. In *P. syringae* excess copper is bound by periplasmic and outer membrane proteins and the cells turn blue. The basis for this difference in mechanism is not understood.

The *pco* determinant (Fig 4) contains seven genes, *pcoABCDRSE*, arranged in at least two copper-inducible transcripts (Brown *et al.*, 1995). The first transcript from a promoter just upstream of *pcoA* encodes *pcoABCDRS*, whereas the second is much stronger and encodes only *pcoE* (Rouch and Brown, 1997). The *pcoE* gene is followed by a ρ-independent terminator. In the *cop* determinant only a single promoter has been identified (Mellano and Cooksey, 1988); this lies upstream of *copA* and expresses *copABCD*, and presumably *copRS*. The regulation of the two systems appears to be very similar.

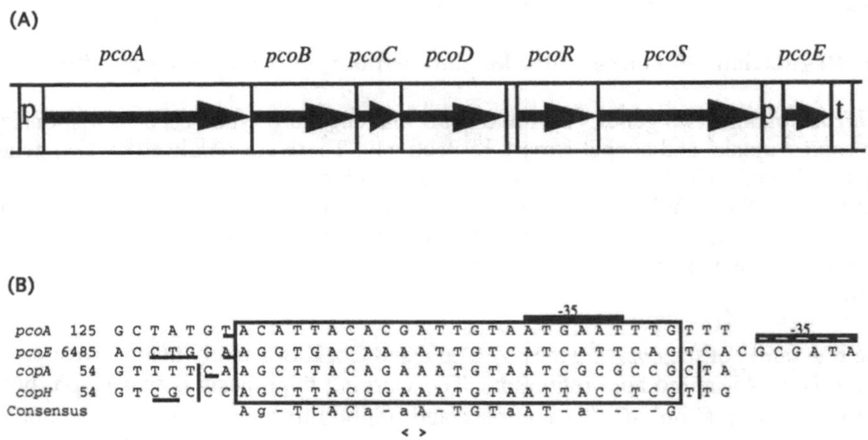

Figure 4. Structural features of *pco* and its promoters.
(A) Genetic structure of the *pco* copper resistance determinant showing the positions of the structural genes, promoters (p) and the ρ-independent terminator (t).. (Brown *et al.*, 1995; Rouch and Brown, 1997).
(B) Alignment of "copper box" sequences from P_{pcoA} and P_{pcoE} from the *E. coli* *pco* determinant, P_{copA} from the *P. syringae cop* determinant and P_{copH} from the *P. syringae* chromosome. The consensus sequence is given underneath and the -35 sequences of the *pco* promoters are marked. The vertical bars show the extent of DNase footprinting in the P_{copA} and P_{copH}. From Rouch and Brown (1997) with permission.

4.2.1. PcoR and PcoS comprise a two-component regulatory system

The two promoters P_{pcoA} and P_{pcoE} show differences in structure. Both have normal -35 and -10 motifs and are induced in the presence of copper (Rouch and Brown, 1997). On high copy number plasmids, at least, zinc is also found to be a weak inducer of the promoters. There is a conserved motif upstream of the transcription start points that is essential for induction of the promoter by copper. This conserved motif has been named the 'copper box' and if deleted, copper regulation of the promoter is lost (Rouch and Brown, 1997). A similar motif occurs in the promoter of the *cop* operon of *P.- syringae*, and CopR protein has been shown to bind to the copper box *in vitro* (Mills *et al.*, 1994).

PcoR (and CopR) show amino acid similarity to the OmpR subgroup of response regulators of the classical two-component gene regulatory systems. PcoS (and CopS) shown similarity to the sensor proteins of these systems. Although not formally proven, it is assumed that the PcoS protein autophosphorylates when copper concentrations are high, and this *trans*-phosphorylates the PcoR protein. The PcoR protein once phosphorylated can bind with high affinity to the copper box operator site and activate transcription.

Transposon insertion and frameshift mutations of *pco* are still regulated by excess copper, and individual promoters are induced in the absence of the *pcoRS* regulatory genes. This is thought to be due to chromosomal homologues of the PcoR and PcoS proteins cross-reacting with the plasmid-borne promoters. A number of candidates for these homologues (CutR and CutS) occur in the *E. coli* genome sequence, but in the absence of further characterisation the homologues have not been identified.

A copper box motif has been identified in the *P. syringae* chromosome and the CopR protein has been shown to bind to this (Mills *et al.*, 1994). Copper box motifs are also found in the *copAB* promoter of *E. hirae*. From these few examples it would be premature to suggest that this is a common motif for regulation of copper resistance and copper management genes in bacteria, but such a possibility is not excluded.

Several other copper resistance determinants have recently been described., such as those from *Xanthomonas campestris* and *Alcaligenes eutrophus*. Little is known about their regulation, but both are inducible. Both determinants have *copABCD* genes showing similarity to the *pcoABCD* genes of *E. coli,* followed by a presumptive P-type copper ATPase, *copF*, in *A. eutrophus*. In the *A. eutrophus* determinant *copRS* regulatory genes lie upstream of *copABCD* (D. van der Lelie, personal communication).

5. The dose-response curve for metal regulated promoters

Transcriptional regulation involving a small effector has a dose-response relationship between the concentration of the effector and the activity of the regulated promoter. Such dose-response relationships differ between different effectors. These have been studied for the *mer* and *pco* promoters.

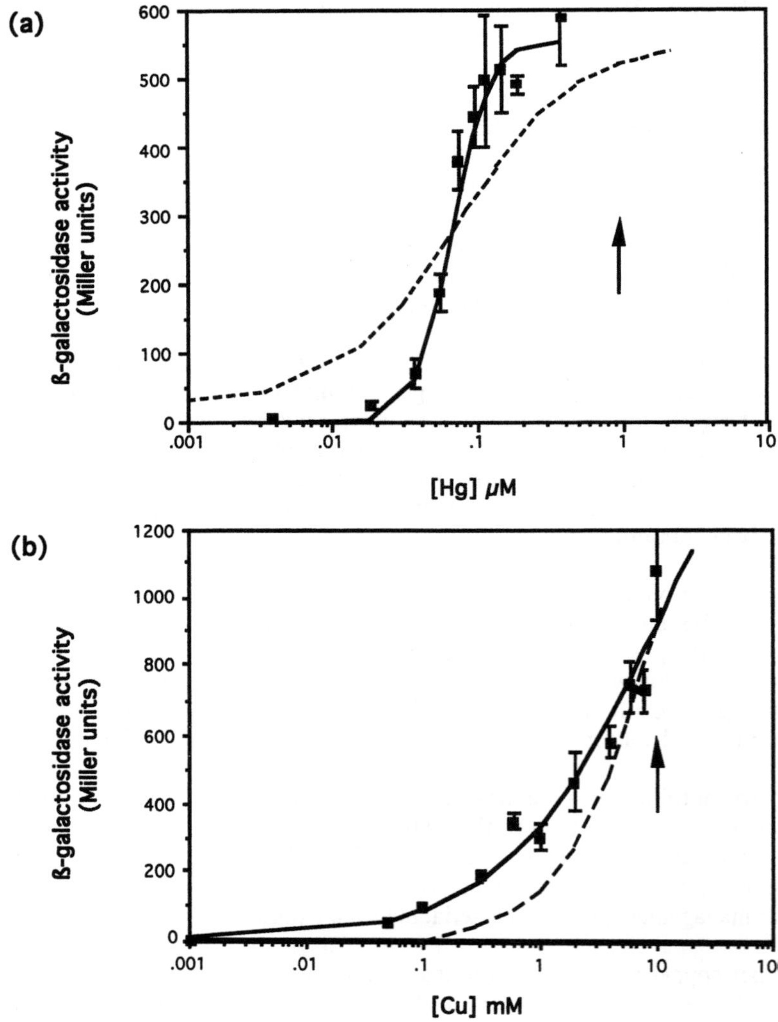

Figure 5 Dose-response curves for the *mer* and *pcoE* promoters. The two graphs show the response of (A) the *mer* promoter and (b) the *pcoE* promoter coupled to the *lacZ* gene; promoter activity (determined from ß-galactosidase activity) is shown against metal ion concentration in each case. The dotted lines show the expected plot for a system obeying Michelis-Menten kinetics. The upwards arrow indicates the metal ion concentration at which the growth rate is half maximal. Modified from Rouch and Brown (1995).

5.1. MerR operates as a hypersensitive biological switch

Studies of the regulation of the *mer* operon *in vitro* and *in vivo* have shown that the rate of increase of expression of P_T with mercuric ion concentration does not follow Michelis-Menten kinetics, but follows a much steeper curve (Rouch and Brown, 1995). This means that the *mer* genes go from fully-repressed to fully-induced across a very narrow range of mercuric ion concentration.

The *mer* promoter is a hypersensitive biological switch. Koshland presented the possibility of a zero-order reaction occurring when all macromolecular components are preassembled and activation of a process requires only the binding of a small ligand. This is the case for the MerR-DNA-RNA polymerase complex preassembled at the repressed *mer* promoter; binding of a single Hg(II) cation activates the promoter. We and others have argued that this has evolved so that the toxic effects of mercury can be avoided by turning on the genes only when necessary to minimise toxicity, without having the *mer* genes expressed at sub-toxic mercury concentrations. Rapid induction of the genes to give maximal expression would remove Hg(II) very quickly. The genes are fully induced well before the toxic effects of Hg(II) affect growth rate.

5.2. PcoR/PcoS operates as a hyposensitive biological switch

In contrast to the *mer* system, the P_{pcoE} promoter has a less sensitive dose-response curve than that predicted by Michelis-Menten kinetics (Rouch and Brown, 1995). Across the range of copper concentrations at which *E coli* will grow, the dose-response curve shows a gradual increase in slope, never reaching the top part of a sigmoidal response. Therefore changes in copper concentration have a small effect on transcription. This may be important in ensuring that increases in copper concentration do not induce expression of the resistance genes, which export copper from the cell, to such an extent that the intracellular copper concentration falls below optimal. By responding in only the bottom part of a sigmoidal response, the slope of the dose-response curve remains shallow.

6. Summary

The regulation of metal metabolism in bacteria is important in providing essential micronutrients and in avoiding the toxic effects of heavy metals. Such regulation occurs at the transcriptional level to control the production of metal transporters and metal binding proteins, and in some cases, enzymes which modify the valence state of toxic metals. This chapter has concentrated on the regulation of genes involved in the management of heavy toxic metals, but other systems for the control of metals which are essentially non-toxic, such as potassium and magnesium have also been studied. Space has not permitted discussion of these.

Acknowledgements

Work from the authors' laboratory has been supported by the Medical Research Council, the Biotechnology and Biological Sciences Research Council (and its predecessors), the Royal Society and the European Commission. The contribution of

Duncan Rouch, Julian Parkhill, Ken Jakeman, and Siobhàn Barrett to work summarised here is gratefully acknowledged. Drs Marc Solioz, Niels van der Lelie, Dietrich Nies and Simon Silver generously informed us of results prior to publication.

References

Ahmed, M., Lyass, L., Markham, P. N., Taylor, S. S., Vazquez-Laslop, N. and Neyfakh, A. A. (1995) Two highly similar multidrug transporters of *Bacillus subtilis* whose expression is differentially regulated. J Bacteriol 177: 3904-10.

Amabile-Cuevas, C. F. and Demple, B. (1991) Molecular characterization of the *soxRS* genes of *Escherichia coli*: two genes control a superoxide stress regulon. Nucl. Acids Res. 19: 4479-84.

Angerer, A., Enz, S., Ochs, M. and Braun, V. (1995) Transcriptional regulation of ferric citrate transport in *Escherichia coli* K-12. Fecl belongs to a new subfamily of sigma 70-type factors that respond to extracytoplasmic stimuli. Mol Microbiol 18: 163-74.

Ansari, A. Z., Bradner, J. E. and O'Halloran, T. V. (1995) DNA-bend modulation in a repressor-to-activator switching mechanism. Nature 374: 371-5.

Ansari, A. Z., Chael, M. L. and O'Halloran, T. V. (1992) Allosteric underwinding of DNA is a critical step in positive control of transcription by Hg-MerR. Nature 355: 87-89.

Brown, N. L., Barrett, S. R., Camakaris, J., Lee, B. T. and Rouch, D. A. (1995) Molecular genetics and transport analysis of the copper-resistance determinant (*pco*) from *Escherichia coli* plasmid pRJ1004. Mol Microbiol 17: 1153-66.

Brown, N. L., Lee, B. T. O. and Silver, S. (1993) Bacterial transport of and resistance to copper. In *Metal ions in biological systems* Sigel, H. and Sigel, A. (eds) New York: Marcel Dekker, 405-430

da Silva, J. J. R. F. and Williams, R. J. P. (1991) *The Biological Chemistry of the Elements*. Oxford: Clarendon Press

de Lorenzo, V., Wee, S., Herrero, M. and Neilands, J. B. (1987) Operator sequences of the aerobactin operon of plasmid ColV-K30 binding the ferric uptake regulation (*fur*) repressor. J Bacteriol 169: 2624-30.

Hobman, J. L. and Brown, N. L. (1997) Mercury Resistance Genes. In *Metal Ions in Biological Systems* Sigel, A. and Sigel, H. (eds) New York: Marcel Dekker Inc., 527-568

Holmes, D. J., Caso, J. L. and Thompson, C. J. (1993) Autogenous transcriptional activation of a thiostrepton-induced gene in *Streptomyces lividans*. Embo J 12: 3183-91.

Hunt, M. D., Pettis, G. S. and McIntosh, M. A. (1994) Promoter and operator determinants for Fur-mediated iron regulation in the bidirectional *fepA-fes* control region of the *Escherichia coli* enterobactin gene system. J Bacteriol 176: 3944-55.

Lonetto, M. A., Brown, K. L., Rudd, K. E. and Buttner, M. J. (1994) Analysis of the *Streptomyces coelicolor sigE* gene reveals the existence of a subfamily of eubacterial RNA polymerase sigma factors involved in the regulation of extracytoplasmic functions. Proc Natl Acad Sci U S A 91: 7573-7.

Mellano, M. A. and Cooksey, D. A. (1988) Induction of the copper resistance operon from *Pseudomonas syringae*. J Bacteriol 170: 4399-401.

Mergeay, M., Nies, D., Schlegel, H. G., Gerits, J. and Charles, P. (1985) *Alcaligenes eutrophus* CH34 is a facultative chemolithotroph with plasmid bound resistance to heavy metals. J. Bacteriol. 162: 328-334.

Mills, S. D., Lim, C. K. and Cooksey, D. A. (1994) Purification and characterization of CopR, a transcriptional activator protein that binds to a conserved domain (cop box) in copper-inducible promoters of *Pseudomonas syringae*. Mol Gen Genet 244: 341-51.

Mukhopadhyay, D., Yu, H. R., Nucifora, G. and Misra, T. K. (1991) Purification and functional characterization of MerD. A coregulator of the mercury resistance operon in gram-negative bacteria. J Biol Chem 266: 18538-42.

Nucifora, G., Chu, L., Silver, S. and Misra, T. K. (1989) Mercury operon regulation by the *merR* gene of the organomercurial resistance system of plasmid pDU1358. J Bacteriol 171: 4241-7.

Ochs, M., Angerer, A., Enz, S. and Braun, V. (1996) Surface signaling in transcriptional regulation of the ferric citrate transport system of *Escherichia coli*: mutational analysis of the alternative sigma factor FecI supports its essential role in *fec* transport gene transcription. Mol Gen Genet 250: 455-65.

Odermatt, A. and Solioz, M. (1995) Two trans-acting metalloregulatory proteins controlling expression of the copper-ATPases of *Enterococcus hirae*. J Biol Chem 270: 4349-54.

Rouch, D. A. and Brown, N. L. (1995) Induction of bacterial mercury- and copper-responsive promoters: functional differences between inducible systems and implications for their use in gene-fusions for in vivo metal biosensors. J Indust Microbiol 14: 249-253.

Rouch, D. A. and Brown, N. L. (1997) Copper-Inducible Transcriptional Regulation at Two Promoters in the *Escherichia coli* Copper Resistance Determinant *pco*. Microbiology 143: 1191-1202..

Sadowsky, M. J., Cregan, P. B., Gottfert, M., Sharma, A., Gerhold, D., Rodriguez, Q. F., Keyser, H. H., Hennecke, H. and Stacey, G. (1991) The *Bradyrhizobium japonicum nolA* gene and its involvement in the genotype-specific nodulation of soybeans. Proc Natl Acad Sci U S A 88: 637-41.

Silver, S. (1996) Bacterial resistances to toxic metal ions--a review. Gene 179: 9-19.

Silver, S. and Phung, L. T. (1996) Bacterial heavy metal resistance: new surprises. Ann Rev Microbiol 50: 753-789.

Stojiljkovic, I. and Hantke, K. (1995) Functional domains of the *Escherichia coli* ferric uptake regulator protein (Fur). Mol Gen Genet 247: 199-205.

Summers, A. O. (1992) Untwist and shout: a heavy metal-responsive transcriptional regulator. J Bacteriol 174: 3097-101.

Wee, S., Neilands, J. B., Bittner, M. L., Hemming, B. C., Haymore, B. L. and Seetharam, R. (1988) Expression, isolation and properties of Fur (ferric uptake regulation) protein of *Escherichia coli* K 12. Biol Methods 1: 62-8.

Wu, J. and Rosen, B. P. (1991) The ArsR protein is a trans-acting regulatory protein. Mol Microbiol 5: 1331-6.

Xu, C., Shi, W. and Rosen, B. P. (1996) The chromosomal *arsR* gene of *Escherichia coli* encodes a trans-acting metalloregulatory protein. J Biol Chem 271: 2427-32.

Regulation of Prespore-Specific Transcription during Sporulation in *Bacillus subtilis*

Jeffery Errington, Richard Daniel, Andrea Feucht, Peter Lewis and Ling Juan Wu

Sir William Dunn School of Pathology, University of Oxford, Oxford, OX1 3RE, UK.

Introduction

During spore formation in the Gram positive bacterium, *Bacillus subtilis*, asymmetric cell division produces a small prespore cell and a much larger mother cell. The two cells then collaborate in an intricate developmental process which culminates with lysis of the mother cell and release of the mature spore (Fig. 1). Many genes involved in sporulation are known and the regulatory pathways controlling their expression are well understood (Errington, 1993; Stragier and Losick, 1996). The main changes in gene expression during sporulation are controlled by four sigma factors, each of which directs RNA polymerase to recognise new promoter sequences and thus turn on new sets of genes. Two of the sporulation-specific sigma factors act successively in the prespore, σ^F and then σ^G: the others act successively in the mother cell; σ^E followed by σ^K (Fig. 1). All four sigma factors are tightly regulated, at both transcriptional and post-translational levels. The complex regulation serves to ensure that the sigma factors only become active at the proper time and place in the developmental process. Although the different regulatory programmes of the prespore and mother cell operate in separate compartments, they are by no means independent. Indeed, in genetic experiments the four sigma factor activities behave as if they operate in a linear dependent sequence, thus:

$$\sigma^F \rightarrow \sigma^E \rightarrow \sigma^G \rightarrow \sigma^K$$

This hierarchy indicates that the mechanisms linking the synthesis or activation of each successive sigma factor involve intercellular signal transduction pathways. The nature of these pathways are not yet understood in detail but will be of considerable interest.

NATO ASI Series, Vol. H 103
Molecular Microbiology
Edited by Stephen J. W. Busby,
Christopher M. Thomas and Nigel L. Brown

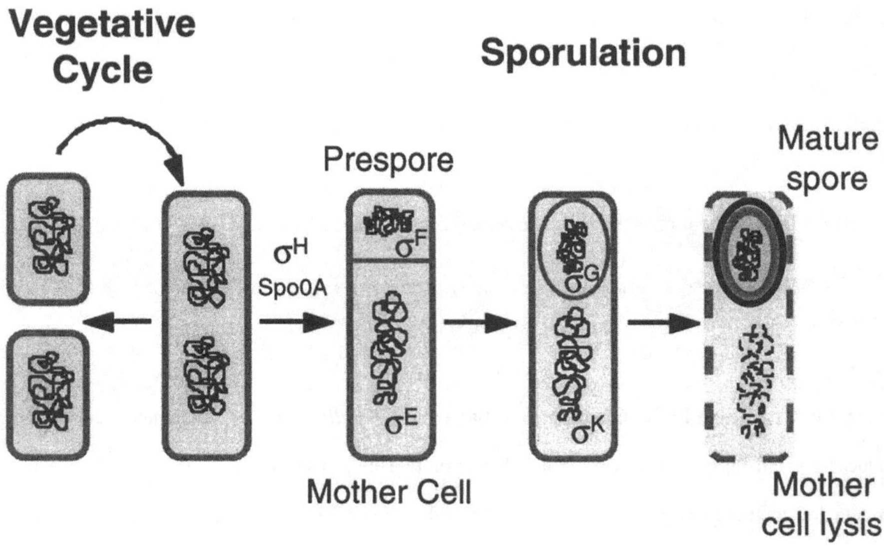

FIG. 1. The life cycle of *Bacillus subtilis* and the major transcriptional regulators of sporulation. In a rich medium the cells grow and divide cyclically in the vegetative state. On starvation, they initiate sporulation, which begins with an asymmetric division. Early sporulation-specific gene expression is controlled by the phosphorylated form of the transcription factor Spo0A in conjunction either with the major vegetative sigma factor, σ^A (not shown) or a stationary phase sigma factor, σ^H. After asymmetric cell division, prespore-specific transcription is directed by σ^F and mother-cell-specific transcription by σ^E. Later, after the prespore has been engulfed by the mother cell, σ^G and σ^K take over as the primary transcriptional regulators in the two cells. Ultimately, the mature spore is released by lysis of the mother cell.

In this essay, we focus on the regulation of the first and probably best understood of the sporulation-specific sigma factors, σ^F.

σ^F is synthesised in the predivisional cell but active only in the prespore

Transcription of the gene encoding σ^F is directed by RNA polymerase containing the stationary phase sigma factor, σ^H, in conjunction with a positive regulator, Spo0A (Trach et al., 1991; Wu et al., 1991). Several lines of evidence show that this transcription occurs early in sporulation, before asymmetric septation (Gholamhoseinian and Piggot, 1989; Partridge and Errington, 1993; Arigoni et al., 1996). Recent immunofluorescence experiments have shown that σ^F protein is indeed present in the pre-divisional cell and that after septation it is distributed into both the large and small compartments (Lewis et al., 1996).

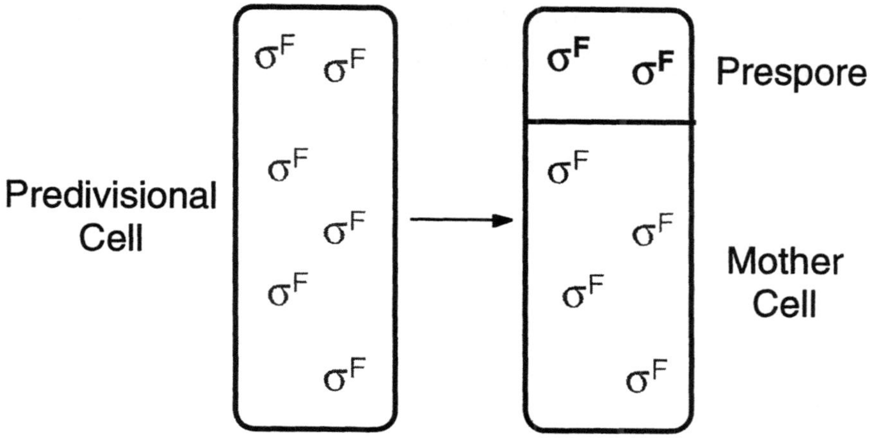

FIG. 2. Cell specificity of σ^F regulation during sporulation. σ^F is synthesised in the predivisional cell but remains latent (grey symbols). After division it is present in both compartments, but it becomes active (black symbols) only in the prespore.

In contrast to the even distribution of σ^F protein, the expression of the genes it controls seems invariably to be restricted to the prespore compartment, on the basis of a variety of experimental approaches. Thus, σ^F is tightly regulated, such that it becomes active only after asymmetric septation and then only in the prespore compartment (Fig. 2). An additional, particularly interesting feature of this regulation is that it imposes a dependence of σ^F activation on formation of the asymmetric septum. Thus, σ^F does not become active if cell division is blocked by use of conditional division mutants, or antibiotics (Levin and Losick, 1994; R.A. Daniel, unpublished results). In the last few years, a great deal of work has been devoted to understanding the nature of the mechanisms controlling the cell-specificity of σ^F activity and its dependence on septation. The remainder of this essay reviews the current state of our understanding.

Gene products involved in σ^F regulation

Genetic analysis was very useful in defining the genes involved in regulation of σ^F. Mutations in a large number of genes can block the expression of σ^F-dependent reporter genes. Most of these mutations turn out to work by preventing the transcription of *spoIIAC* (the gene encoding σ^F). Only two genes specifically required for σ^F activation are known; *spoIIAA* (the first gene in the operon encoding σ^F) and *spoIIE*, an unlinked gene (Schmidt et al., 1990; Margolis et al., 1991; Partridge et al., 1991). The central gene in the *spoIIA* operon, *spoIIAB*, also encodes a crucial regulator of σ^F. Null mutations in this gene cause **increased** expression of σ^F-dependent reporter genes, suggesting that its product is a negative regulator of σ^F (Schmidt et al., 1990; Coppolecchia et al., 1991; Partridge et al., 1991). Experiments with double mutants indicated that the SpoIIE, SpoIIAA and SpoIIAB proteins acted on σ^F in the following hierarchical manner (Schmidt et al., 1990; Margolis et al., 1991; Partridge et al., 1991):

Subsequently, the products of these four proteins were overproduced and purified, and their interactions have been dissected *in vitro* (Fig. 3).

SpoIIAB is an anti-sigma factor

spoIIAB null mutants show deregulated σ^F activity, indicating that this protein is needed to inhibit σ^F before septation and presumably in the mother cell compartment after septation. Purified SpoIIAB was shown to inhibit transcription by RNA polymerase containing σ^F *in vitro* (Duncan and Losick, 1993; Min et al., 1993). Various experimental approaches have shown that the interaction between SpoIIAB and σ^F is a direct one (Duncan and Losick, 1993; Duncan et al., 1996; J.F. Wilkinson and J.E., unpublished results). SpoIIAB thus belongs to an increasingly large family of proteins called anti-sigma factors, which inhibit the action, usually, of a single cognate sigma factor (Brown and Hughes, 1995). Immunofluoresence experiments have revealed that the SpoIIAB protein, like σ^F, is present in the predivisional cell, and that it is apportioned into both the prespore and mother cell after division (Lewis et al., 1996). Given this non-localized distribution, some other factor must be responsible for modulating the action of SpoIIAB on σ^F, to release the sigma factor in the prespore.

The SpoIIAA anti-anti sigma factor and its control by phosphorylation

The genetic experiments mentioned above suggested that inhibition of σ^F by SpoIIAB is antagonised by SpoIIAA. This proved to be the case, but the interactions between SpoIIAB and SpoIIAA turned

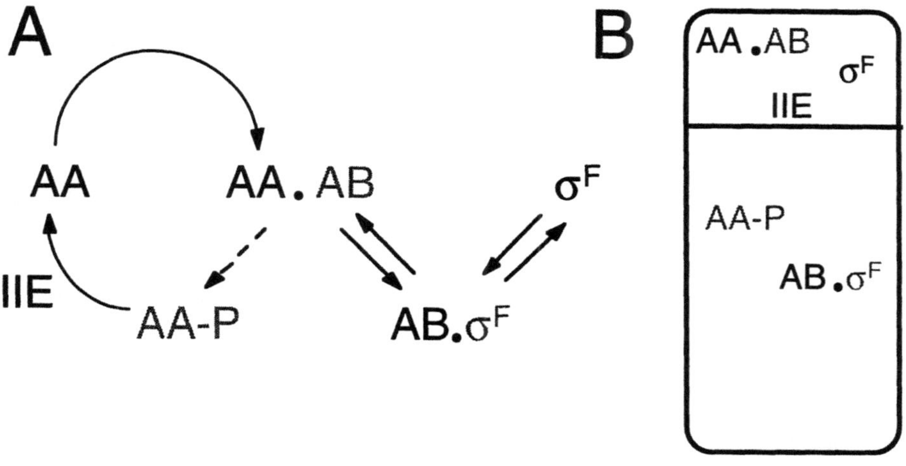

FIG. 3. Regulation of σ^F activity during sporulation. **A**, biochemical interactions between the proteins involved in σ^F regulation. **B**, Predominant interactions in the newly septate sporulating cell. The protein symbols have been abbreviated by leaving off their Spo prefix. Again, grey symbols represent inactive proteins and black symbols, active proteins. The dotted arrow indicates that the dissociation of the AA.AB complex to yield phosphorylated SpoIIAA (or the recycling of SpoIIAB) is slow.

out to be unexpectedly complex. The major complicating factor lies in a second activity possessed by SpoIIAB. In the presence of ATP, SpoIIAB catalyses the phosphorylation of SpoIIAA on a specific serine residue (Min et al., 1993; Diederich et al., 1994; Najafi et al., 1995). This phosphorylation reaction has two important consequences. Firstly, because the turnover rate of this reaction is slow (T. Magnin, S.M.A. Najafi and M.D. Yudkin, unpublished results), and since the binding of SpoIIAA and σ^F to SpoIIAB are mutually exclusive (Alper et al., 1994; Diederich et al., 1994; Duncan et al., 1996), phosphorylation of SpoIIAA can divert SpoIIAB from its role in inhibition of σ^F. However, the product of the reaction, SpoIIAA-P, is inactive, and can no longer interact with SpoIIAB to reverse its anti-σ^F activity *in vitro* or *in vivo* (Alper et al., 1994; Diederich et al., 1994; Magnin et al., 1996). So phosphorylation of SpoIIAA would ultimately favour the inhibition of σ^F. Until recently it was thought that the balance of binding of SpoIIAB to SpoIIAA or σ^F was influenced by the ratio of ATP and ADP, because ADP can stimulate tight association of SpoIIAA and SpoIIAB (Alper et al., 1994; Diederich et al., 1994). However, it now appears that the most important factor controlling this equilibrium is the availablility of non-phosphorylated SpoIIAA (Duncan et al., 1996; T. Magnin, S.M.A. Najafi and M.D. Yudkin, unpublished results).

SpoIIE is a SpoIIAA-P phosphatase

Western blotting (Arigoni et al., 1996; Feucht et al., 1996) and particularly immunofluorescence experiments (Lewis et al., 1996) have shown that most of the SpoIIAA protein is in the phosphorylated, inactive state early in sporulation, presumably as a result of the action of the SpoIIAB kinase. The SpoIIAB protein is thus free to bind to and inactivate σ^F. However, after septation, non-phosphorylated SpoIIAA accumulates in the prespore compartment, in parallel with the release of σ^F activity. Two lines of evidence suggest that the non-phosphorylated SpoIIAA in the prespore is derived by dephosphorylation of SpoIIAA-P rather than synthesised *de novo*. First, genetic mosaic experiments have shown that the intact *spoIIAA* gene does not need to be present in the prespore compartment for sporulation to proceed (Gholamhoseinian and Piggot, 1989). Second, in cells of a *spoIIIE* mutant, in which prespore chromosome partitioning fails and the *spoIIA* locus does not gain access to the prespore compartment (Wu and Errington, 1994), non-phosphorylated SpoIIAA nevertheless accumulates in that compartment (P.J. Lewis, unpublished). It thus appears that the release of σ^F is driven by prespore-specific dephosphorylation of SpoIIAA.

The phosphatase responsible for reactivation of SpoIIAA is encoded by the *spoIIE* gene, as demonstrated by both *in vitro* (Duncan et al., 1995) and *in vivo* (Arigoni et al., 1996; Feucht et al., 1996) experiments. SpoIIE is a large (827 aa) protein with at least two domains (Barak et al., 1996). The N-terminal domain is very hydrophobic and contains up to 12 likely membrane-spanning helices. The remainder of the protein is hydrophilic. Much of the C-terminal domain shows sequence similarity to proteins involved in regulation of another *B. subtilis* sigma factor σ^B (Duncan et al., 1995; Feucht et al., 1996) and it contains sequence motifs found in a family of eukaryotic serine/threonine phosphatases (Adler et al., 1997). Mutations specifically blocking σ^F activation lie in this conserved phosphatase-like domain, and they prevent accumulation of non-phosphorylated SpoIIAA (Feucht et al., 1996). However, SpoIIE also has a second separable function, required for correct formation of the asymmetric septum (Barak and Youngman, 1996; Feucht et al., 1996). This function is probably mediated by a direct interaction with one or more components of the division apparatus, because the SpoIIE protein is targeted to the potential sites of asymmetric division near the cell poles, early in sporulation (Arigoni et al., 1995; Barak et al., 1996).

Spatial and temporal control of σ^F activity

The second, morphogenic function of SpoIIE would seem to make this protein a good candidate for the factor which couples σ^F activation to septation (see above). In accordance with this idea, preliminary results indicate that mutations in *spoIIE* can bypass the dependence on septation (A. Feucht, unpublished results). How might this be achieved? One possibility would be that the protein initially assembles into the division machinery near the cell poles in an inactive state, and that its activity is released by dissociation of the division apparatus when the septum is finished or as it forms. This would explain the dependence of the protein on septation. However, it would not necessarily explain the localization of SpoIIE action in the prespore compartment. We have recently re-examined the localization of SpoIIE protein during septation and preliminary results suggest that it may be sequestered onto the prespore side of the septum as it forms (L.J. Wu and A. Feucht, unpublished results). The effect of this sequestration would be to greatly increase the concentration of the phosphatase in the small prespore compartment (Feucht et al., 1996). Perhaps this pushes the phosphatase activity over a threshold needed to overcome the opposing activity of the SpoIIAB kinase and produce enough non-phosphorylated SpoIIAA to release σ^F.

To conclude, the major factors responsible for proper temporal and spatial regulation of the first sporulation specific sigma factor have now been identified. Their functions and interactions can be reconstructed *in vitro* and their behaviour *in vivo* is well characterized and generally compatible with the *in vitro* data. The remaining challenges mainly lie in understanding how the activity of the SpoIIE protein is controlled so as to render σ^F activation (via dephosphorylation of SpoIIAA-P) dependent on septation and to restrict it to the prespore compartment.

Acknowledgments

Work in the Errington lab is supported by grants from the Biotechnology and Biological Sciences Research Council and the BIOTECH programme of the European Community.

References

Adler, E, Donella-Deana, A, Arigoni, F, Pinna, LA, and Stragier, P (1997) Structural relationship between a bacterial developmental protein and eukaryotic PP2C protein phosphatases. Mol Microbiol *23*, 57-62

Alper, S, Duncan, L, and Losick, R (1994) An adenosine nucleotide switch controlling the activity of a cell type-specific transcription factor in B. subtilis Cell *77*, 195-205

Arigoni, F, Pogliano, K, Webb, CD, Stragier, P, and Losick, R (1995) Localization of protein implicated in establishment of cell type to sites of asymmetric division. Science *270*, 637-640

Arigoni F, Duncan L, Alper S, Losick R, Stragier P (1996) SpoIIE governs the phosphorylation state of a protein regulating transcription factor σ^F during sporulation in *Bacillus subtilis.* Proc Natl Acad Sci USA 93:3238-3242

Barak I, Behari J, Olmedo G, Guzman P, Brown DP, Castro E, Walker D, Westpheling J, Youngman P (1996) Structure and function of the *Bacillus* SpoIIE protein and its localization to sites of sporulation septum assembly. Mol Microbiol 19:1047-1060

Barak I, Youngman P (1996) SpoIIE mutants of *Bacillus subtilis* comprise two distinct phenotypic classes consistent with a dual functional role for the SpoIIE protein. J Bacteriol 178:4984-4989

Brown KL, Hughes KT (1995) The role of anti-sigma factors in gene regulation. Mol Microbiol 16:397-404

Coppolecchia R, DeGrazia H, Moran CP Jr (1991) Deletion of *spoIIAB* blocks endospore formation in *Bacillus subtilis* at an early stage. J Bacteriol 173:6678-6685

Diederich B, Wilkinson JF, Magnin T, Najafi SMA Errington J, Yudkin MD (1994) Role of interactions between SpoIIAA and SpoIIAB in regulating cell-specific transcription factor σ^F of *Bacillus subtilis.* Genes Devel 8:2653-2663

Duncan L, Losick R (1993) SpoIIAB is an anti-σ factor that binds to and inhibits transcription by regulatory protein σ^F from *Bacillus subtilis.* Proc Natl Acad Sci USA 90:2325-2329

Duncan L, Alper S, Arigoni F, Losick R, Stragier P (1995) Activation of cell-specific transciption by a serine phosphatase at the site of asymmetric division. Science 270:641-644

Duncan L, Alper S, Losick R (1996) SpoIIAA governs the release of the cell-type specific transcription factor σ^F from its anti-sigma factor SpoIIAB. J Mol Biol 260:147-164

Errington J (1993) *Bacillus subtilis* sporulation: regulation of gene expression and control of morphogenesis. Microbiol Revs 57:1-33

Feucht A, Magnin T, Yudkin MD, Errington, J (1996) Bifunctional protein required for asymmetric cell division and cell-specific transcription in *Bacillus subtilis.* Genes Devel 10:794-803

Gholamhoseinian A, Piggot PJ (1989) Timing of *spoII* gene expression relative to septum formation during sporulation of *Bacillus subtilis.* J Bacteriol 171:5747-5749

Karow ML, Glaser P, Piggot PJ (1995) Identification of a gene, *spoIIR*, that links the activation of σ^E to the transcriptional activity of σ^F during sporulation in *Bacillus subtilis.* Proc Natl Acad Sci USA 92:2012-2016

Levin PA, Losick, R (1994) Characterization of a cell division gene from *Bacillus subtilis* that is required for vegetative and sporulation septum formation. J Bacteriol 176:1451-1459

Lewis PJ, Magnin T, Errington J (1996) Compartmentalized distribution of the proteins controlling the prespore-specific transcription factor σ^F of *Bacillus subtilis.* Genes to Cells 1:881-894

Magnin T, Lord M, Errington J, Yudkin MD (1996) Establishing differential gene expression in sporulating *Bacillus subtilis*: phosphorylation of SpoIIAA (anti-anti-σ^F) alters its conformation and prevents formation of a SpoIIAA/SpoIIAB/ADP complex. Mol Microbiol 19:901-907

Margolis P, Driks A, Losick R (1991) Establishment of cell type by compartmentalized activation of a transcription factor. Science 254:562-565

Min K-T, Hilditch CM, Diederich B, Errington J, Yudkin MD (1993) σ^F, the first compartment-specific transcription factor of B subtilis, is regulated by an anti-σ factor that is also a protein kinase. Cell 74:735-742

Najafi SMA, Willis AC, Yudkin MD (1995) Site of phosphorylation of SpoIIAA, the anti-anti-sigma factor for sporulation-specific σ^F of *Bacillus subtilis.* J Bacteriol 177:2912-2913

Partridge SR, Foulger D, Errington J (1991) The role of σ^F in prespore-specific transcription in *Bacillus subtilis.* Mol Microbiol 5:757-767

Partridge SR, Errington J (1993) The importance of morphological events and intercellular interactions in the regulation of prespore-specific gene expression during sporulation in *Bacillus subtilis.* Mol Microbiol 8:945-955

Schmidt R, Margolis P, Duncan L, Coppolecchia R, Moran CP Jr, Losick R (1990) Control of developmental transcription factor σ^F by sporulation regulatory proteins SpoIIAA and SpoIIAB in *Bacillus subtilis.* Proc Natl Acad Sci USA 87:9221-9225

Stragier P, Losick R (1996) Molecular genetics of sporulation in *Bacillus subtilis.* Ann Rev Genet 30:297-341

Trach K, Burbulys D, Strauch M, Wu JJ, Dhillon N, Jonas R, Hanstein C, Kallio P, Perego M, Bird T, Spiegelman GB, Foghe C, Hoch JA (1991) Control of the initiation of sporulation in Bacillus subtilis by a phosphorelay. Res Microbiol 142:815-823

Wu JJ, Piggot PJ, Tatti KM, Moran CP Jr (1991) Transcription of the *Bacillus subtilis spoIIA* locus. Gene 101:113-116

Wu LJ, Errington J (1994) *Bacillus subtilis* SpoIIIE protein required for DNA segregation during asymmetric cell division. Science 264:572-575

Quorum Sensing: Bacterial Cell-Cell Signalling from Bioluminescence to Pathogenicity

Simon Swift, John Throup, Barrie Bycroft[1], Paul Williams[1], Gordon Stewart

Department of Applied Biochemistry and Food Science, University of Nottingham, Sutton Bonington Campus, Leicestershire LE12 5RD, UK. [1]Department of Pharmaceutical Sciences, University of Nottingham, University Park, Nottingham NG7 2RD, UK

Abstract

The integration of signals from the bacterial environment, through a network of cellular transduction mechanisms, determines the profile of genes expressed and thereby the bacterial phenotype. Quorum sensing transmits one such signal, i.e. population density, by relying on the accumulation of a small extracellular signal molecule to modulate transcription of target operons.

Keywords. Quorum sensing, environmental sensing, *lux*, *N*-acyl homoserine lactones, population dynamics

1 Introduction

The dynamic nature of the bacterial phenotype depends largely upon selective gene expression from the available genomic pool. To deliver an advantageous phenotype, therefore, controlled gene expression must provide for survival and proliferation within the constraints of the bacterial growth environment. Such evolutionary pressure has ensured that bacteria have developed a network of sensor mechanisms to transduce environmental stimuli into gene expression and hence a phenotype complementary to prevailing conditions. Hence the success of a bacterium in a given environment is dependent upon the range and quality of the information it receives and how effectively it can act upon this information to generate an appropriate phenotype.

In 1992 it was shown that the small molecule *N*-(3-oxohexanoyl)-L-homoserine lactone (OHHL) was responsible for the regulation of synthesis of the β-lactam antibiotic, 1-carbapen-2-em-3-carboxylic acid (carbapenem), in the terrestrial, plant pathogenic bacterium, *Erwinia carotovora* (Bainton et al., 1992a,b). The significance of this discovery lay in the fact that OHHL had already been reported in the scientific literature several years earlier as the autoinducer of bioluminescence in the marine symbiont

NATO ASI Series, Vol. H 103
Molecular Microbiology
Edited by Stephen J. W. Busby,
Christopher M. Thomas and Nigel L. Brown

Photobacterium fischeri (previously classified as *Vibrio fischeri*) (Eberhard et al., 1981; Haygood and Distel, 1993). In the latter organism, OHHL is produced by the action of the LuxI protein which is an enzyme, "autoinducer synthase" (see later). Even in the 1992 literature, OHHL-mediated autoinduction was considered to be uniquely connected with bioluminescence (the "Lux" phenotype) in *P. fischeri* (Shadel and Baldwin, 1992). However the discovery of OHHL elaboration by a terrestrial plant pathogen suggested that production of this, and structurally similar molecules, might be far more widespread than originally supposed (Bainton et al., 1992a). To test this idea a bioluminescence sensor system was developed and used to screen for OHHL production in the spent supernatants of a wide range of bacterial cultures, including strains of *Pseudomonas aeruginosa*, *Serratia marcescens*, *Erwinia herbicola*, *Citrobacter freundii*, *Enterobacter agglomerans* and *Proteus mirabilis* (Bainton et al., 1992a; Swift et al., 1993). The results confirmed, and extended, some earlier indications (Greenberg et al., 1979) that the ability of bacteria to produce such molecules was common.

Other observations made in 1991 and 1992 had reinforced the notion that Lux-type regulation was not unique to bioluminescent *P. fischeri*. In 1991, Gambello and Iglewski identified LasR, a homologue of the *P. fisheri* LuxR transcriptional regulator (see later), involved in the regulation of elastase in *Pseudomonas aeruginosa*. Also, Wang et al. (1991) showed that another LuxR homologue, SdiA (formerly identified as 28K-UvrC, Sharma et al., 1986; Henikoff et al., 1990), was a positive transcriptional activator of a key set of cell division genes (*ftsQAZ*) in *E. coli*. Yet another LuxR homologue, RhiR, responsible for regulation of rhizosphere-expressed genes in *Rhizobium leguminosarum*, was identified in 1992 by Cubo et al. In none of these reports was it specifically suggested that OHHL or structurally related molecules were likely candidates for involvement in the regulation of the various physiological processes investigated. However, in concert with these studies on the LuxR homologues, the discovery of the widespread production of OHHL-like molecules in bacteria (Bainton et al., 1992a) made the general hypothesis of widespread Lux-type regulation in bacteria, intellectually seductive. Since these early observations, it has now become evident that OHHL is only one member of a growing family of *N*-acyl homoserine lactones (*N*-AHLs) which are produced by a wide spectrum of bacteria and are responsible for the regulation of diverse physiological processes in the corresponding producer organisms.

The generic term ‘quorum sensing’ is now used to describes this phenomenon whereby the accumulation of a low molecular weight (c. 170-300 Da) pheromone enables individual cells to sense when the minimal population unit or ‘quorum’ of bacteria has been achieved for a concerted population response to be initiated (Fuqua et al., 1994).

2 The *P. fischeri lux* paradigm

The paradigm for quorum sensing is the regulation of bioluminescence with respect to population density in the marine bacterium *Photobacterium fischeri* (Meighen, 1994; Sitnikov et al., 1995) where the expression of two regulatory genes (*luxI* and *luxR*; signal generator and response regulator) is necessary.

P. fischeri is a free living marine bacterium, strains of which can become symbionts with squid and certain marine fish where they can colonise the corresponding light organs and produce bioluminescence (e.g. see Haygood and Distel, 1993). The bacteria only emit light when they are at high cell density, as found in the contained space of a light organ (or in the artefactual environment of a laboratory culture flask, or in a bacterial colony on an agar plate) (Meighen, 1991; Gray and Greenberg, 1992). The gene cluster responsible for bioluminescence (Fig. 1) contains two transcriptional units. At low cell density there is inefficient production of the rightward transcript (*luxICDABE*) but this is sufficient to allow the low-level expression of LuxI - the protein which is responsible for OHHL synthesis. In a closed system the freely-diffusible (Kaplan and Greenberg, 1985) OHHL accumulates as cell density increases, until a critical OHHL concentration is reached. The molecule is thought to bind to the LuxR protein (Hanzelka and Greenberg, 1995) enabling the latter to then act as a transcriptional activator enhancing expression of the rightward transcript (Nealson et al., 1970; Nealson, 1977; Engebrecht et al., 1983; Engebrecht and Silverman, 1986; Meighen and Dunlap, 1993). As a result of enhanced LuxI expression, more OHHL is produced, more OHHL:LuxR binding occurs and increased transcriptional activation of the *luxI* gene ensues, thereby generating a positive feedback (auto)regulatory loop in which OHHL controls its own synthesis (Stevens and Greenberg, 1997). Clearly, as *luxI* is only the first gene in the operon, the net effect of OHHL autoinduction is enhanced expression of all of the *luxI*-distal genes, the products of which function in bioluminescence. As OHHL autoinduction is cell density-dependent, bioluminescence necessarily becomes cell density dependent. Thus, in a sense, the Lux phenotype of *P. fischeri* is only density dependent because of the genetic location of the *luxCDABE* genes. However, it is important to

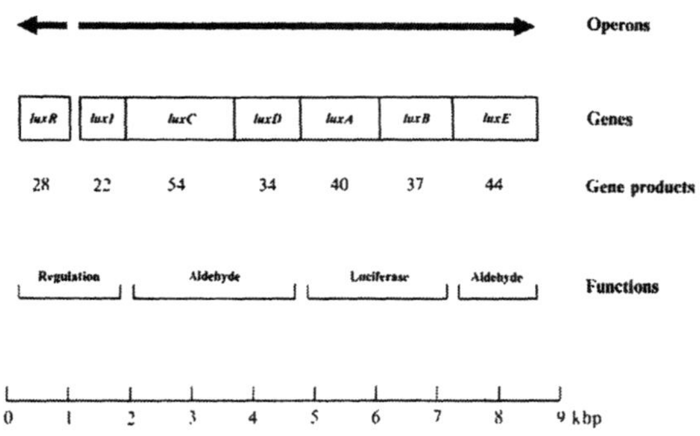

Fig. 1. Organisation and function of *lux* genes cloned from *P. Fischeri*. Arrows denote operons containing *lux* genes. The molecular masses (kDa) of the gene products and their functions are shown below the gene designation.

realise that there is no theoretical requirement for the genes encoding an *N*-AHL-mediated phenotype to be linked to the corresponding regulatory genes. Assuming the *N*-AHL is freely diffusible and can bind to its LuxR homologue, and if the resultant complex then can find its appropriate binding site on the DNA, any form of genetic arrangement is possible.

Before leaving the *P. fischeri* paradigm, it is important to point out that while the LuxR-OHHL complex may be the primary component responsible for *lux* gene induction there is considerable additional regulatory complexity. Expression of the *lux* regulon is modulated by the cAMP-binding protein (CAP)-cAMP system (Dunlap and Greenberg, 1985, 1988). Also, until recently, it was thought that σ^{32} (RpoH), affected *lux* regulon expression (Ulitzur and Kuhn, 1988). However, Dolan and Greenberg (1992) have shown that σ^{32} is not required for transcription of *luxICDABE*. Rather, it appears that the products of *groE* enable folding of LuxR into an active conformation (Dolan and Greenberg, 1992; Adar et al., 1992; Adar and Ulitzur, 1993). Clearly, therefore, *N*-AHL-mediated autoinduction of bioluminescence in *P. fischeri* must ultimately reflect a complex, holistic interaction between several regulatory networks. Consequently, when examining *N*-AHL regulation of gene expression in bacteria other than the *P. fischeri*, this interdependent networking of regulatory circuits should not be overlooked.

A further degree of complexity has emerged recently with the discovery by Kuo et al. (1994) of a second novel autoinduction system, dependent on the *ain* locus, in *P. fischeri* that regulates population desity-dependent signalling through *N*-octanoyl-L-homoserine lactone (see later). Thus there are now multiple, chemically and genetically-distinct but cross-acting, autoinduction systems in *P. fischeri*.

3 Phenotypes associated with *N*-AHL production

To date only a limited group of phenotypes are known to be regulated by *N*-AHLs (for review see Swift et al., 1996). However, based on the results obtained using a *lux* sensor assay (Bainton et al., 1992a; Swift et al., 1993) a far larger spectrum of bacteria make molecules capable of activating expression of the bioluminescence genes. The phenotypes associated with *N*-AHL production are physiologically diverse, apart from bioluminescence these include the synthesis of extracellular enzymes, virulence, antibiotic production and plasmid conjugal transfer. In some bacteria, it is known that multiple phenotypic traits are regulated by the same molecule, e.g. in some strains of the bacterial plant pathogen *Erwinia carotovora*, OHHL regulates synthesis of multiple pectate lyases, polygalacturonase, cellulase, protease and a carbapenem antibiotic (Jones et al., 1993; Pirhonen et al., 1993). In addition, it is now clear that OHHL is only one of a growing family of *N*-acyl homoserine lactones. All of the *N*-AHL regulated phenotypes described so far are related to "secondary metabolism" and to non-essential molecules, including secreted products, although the current list may be simply a reflection of the physiological traits which have been investigated so far.

Outside the laboratory, bacteria live in mixed cultures (e.g. the gut or the rhizosphere) where they can expect to encounter AHLs from a number of rival or symbiotic species. For example, a number of bacteria produce the same AHL signal molecule although in each it is clearly used to regulate the expression of different biological properties. Other bacteria have been shown to produce multiple AHLs, each having different spheres of phenotypic influence (Williams et al., 1996); here possibilities exist for bacteria to regulate genes according to a number of population based parameters by varying the physical and biological properties of the signalling molecule. Further complexity is offered by the potential for cross-talk and interference between the molecules from individual quorum sensing systems.

To add a further complication, dissection of the quorum sensing circuits of *Vibrio harveyi* and *P. fischeri* has identified a second type of AHL synthase activity involving proteins which bear no homology to LuxI [LuxLM (Bassler et al., 1994) and AinS (Gilson et al., 1995) respectively].

4 The Quorum Sensing Signal Generator

Given the importance and range of characteristics positively regulated by the accumulation of different AHL signals, an understanding of the biosynthetic mechanism is important. The recent work of Moré et al. (1996) provides a significant step towards elucidating the underlying biochemistry of AHL synthesis. Purification of the recombinant hexahistidinyl-TraI (H_6-TraI) protein has enabled the synthesis of OOHL from purified substrates *in vitro* (Fig. 2).

4.1 *S*-adenosylmethionine as a substrate

Preliminary experiments with LuxI in crude *P. fischeri* extracts (Eberhard et al., 1991) proposed *S*-adenosylmethionine (AdoMet) as the donor of the homoserine lactone (HSL) moiety, however, other possible substrates have since been considered. For example, HSL accumulation has been proposed as a general signal for starvation (Huisman and Kolter, 1994) allowing the attractive hypothesis of a role for substrate availability in the AHL biosynthetic reaction with HSL providing the amino acid component.

In vitro, H_6-TraI catalyses the biosynthesis of OOHL in the presence of an extract of soluble *Escherichia coli* proteins (S28 extract) and AdoMet (Moré et al., 1996). Substitution of AdoMet by homoserine lactone does not lead to OOHL formation leading to the definitive conclusion that AdoMet provides the homoserine lactone. Consistent with this conclusion, *in vivo* experiments using *E. coli*, methionine and homoserine auxotrophs in the presence of cycloleucine (an inhibitor of AdoMet synthesis) and bearing a plasmid encoding the *luxI* gene, have provided additional support for AdoMet but not HSL as a LuxI substrate (Hanzelka and Greenberg, 1996).

4.2 The fatty acyl moiety is derived from the acyl-ACP

H_6-TraI catalyses the biosynthesis of OOHL in the presence of S28 extract and AdoMet (Moré et al., 1996). In cell-free extracts of *P. fischeri,* both coenzyme A (CoA) and the acyl carrier protein (ACP) adduct of 3-oxohexanoic acid were suggested to be potential acyl side chain donors in OHHL synthesis (Eberhard et al., 1991). To identify the origin of the fatty acyl moiety of OOHL the S28 extract was investigated. Dialysis of the S28 extract abolished any detectable OOHL synthase activity, but this could be restored by the addition of malonyl-CoA (but not acetyl-CoA) and NADPH (and also NADH). A demonstration that ongoing fatty acid metabolism was required for OOHL biosynthesis was given by experiments using cerulenin an irreversible inhibitor of fatty acid biosynthesis. Interestingly, the biosynthesis of a related AHL (*N*-3-(hydroxybutanoyl)-homoserine lactone; HBHL) involved in quorum sensing regulation of bioluminescence in *Vibrio harveyi* requires active fatty acid biosynthesis (Cao and Meighan, 1993). However, it should be remembered that the AHL synthase in this case is not a LuxI homologue and that there is almost certainly at least one alternative mechanism for AHL synthesis.

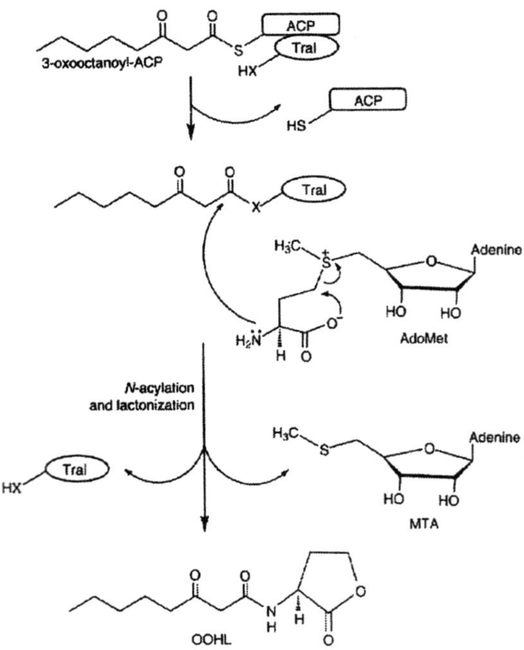

Fig. 2. A model to describe *N*-3-(oxooctanoyl)-homoserine lactone (OOHL) biosynthesis. TraI acquires a 3-oxooctanoyl group from 3-oxooctanoyl-ACP and catalyzes the formation of the amide bond between the amino group of *S*-adenosylmethionine and C-1 of the fatty acid. This reaction is followed by lactonization, creating OOHL and 5'-methylthioadenosine (MTA). Adapted from Moré et al. (1996).

To determine whether the ACP adduct could donate the acyl chain, a quaternary amine derivative of 3-oxo-octanoic acid was used to chemically load the active thiol of the phosphopentathionine group of ACP. The resultant 3-oxo-octanoyl-ACP (OOACP) replaced the requirement for S28 extract or dialysed S28 extract plus malonyl-CoA and NADPH, allowing biosynthesis of OOHL in the presence of AdoMet and H_6-TraI. Hence, it was concluded that the fatty acid moiety derives from the acyl-ACP and not the corresponding acyl-CoA. Significantly, this suggests that the LuxI homologues must selectively interact with the acyl-ACP carrying the specific acyl group for the given AHL. In line with this conclusion, it has been demonstrated that AdoMet and the hexanoyl-ACP, but not hexanoyl-CoA, act as substrates for a maltose binding protein-LuxI fusion *in vitro* (Hanzelka and Greenberg, 1996; Schaefer et al., 1996).

4.3 Perspective

It will be important to confirm that other LuxI homologues utilise the acyl-ACP and AdoMet as substrates and to compare these with the substrate preferences of the LuxLM family of AHL synthases. The challenge remains to elucidate the mechanism of AHL biosynthesis and to identify the LuxI homologue structure/function relationships. An understanding of the mechanism of signal generation may then allow disruption of signal production and the control of expression for a variety of important gene products.

A striking feature of LuxI homologues is that when they are cloned into *E. coli* from the wild type organisms, they retain acyl moiety specificity without the presence of obvious structural motifs that define subfamilies that make specific AHLs. Elucidation of the structural basis for this selectivity is an important question for the future, the answer to which may have relevance to the study of fatty acid biosynthesis.

Finally, the slow rate *in vitro* of the H_6-TraI catalysed reaction using purified substrates suggests that a rate enhancing factor may be missing. It is possible that an NAD moiety may fulfil this role, perhaps as an allosteric modifier of the LuxI homologue. Although the requirement for NADPH in OOHL biosynthesis can be explained by its participation in the elongation reactions of fatty acid biosynthesis, the substitution of NADH to the same effectiveness in this reaction is surprising. It will be interesting, therefore, to see whether the NAD moiety provides any additional functionality.

5 The response regulator

Much is known about the LuxR family, its function and the relationship between structure and function (Slock et al., 1990; Hanzelka and Greenberg, 1995). The proposed two domain receptor/activator modular structure combines a C-terminal DNA binding and transcriptional activator module with an N-terminal AHL-dependent regulator module. Homologies in the N-terminal domain extend to a family of response regulators sensing AHLs, whereas there is homology in the C-terminal with the DNA binding region of the LysR family (where phosphorylation is often the signal for allosteric modification). LuxR is thought to act as a homodimer, with the N-terminal domain membrane associated, where it may bind OHHL partitioned into the membrane. It has been suggested that the N-terminal domain blocks the DNA-binding activity of the C-terminal domain of the intact LuxR protein; on binding to LuxR, OHHL is thought to relieve this inhibition by altering conformation. This appears to facilitate transcriptional activation through binding to a palindromic DNA sequence within the *luxI* operator which has been termed the *lux* box (Stevens et al., 1994).

Recently, the physical interaction of LuxR homologues and their cognate AHLs has been questioned and an intermediary "receptor" protein has been proposed which activates LuxR on sensing the presence of the AHL (Sitnikov et al., 1995). This hypothesis is based upon the differences in transcriptional activation activities seen for LuxR in the presence of a range of exogenously added AHLs when comparing LuxR expressed in *P. fischeri* to a recombinant protein expressed in *E. coli*.

The response regulators LuxN (Meighen, 1994) and AinR (Gilson et al., 1995) appear to autophosphorylate in the presence of their cognate AHLs. Phosphotransfer to a cognate repressor inactivates repressor function thus allowing transcription (Meighen, 1994; Bassler et al., 1994).

5.1 Integration with other mechanisms of cell signalling

Phenotypes controlled through quorum sensing are frequently regulated by additional environmentally responsive circuits. Studies of the expression of many such phenotypes has identified components of what we believe to be a multilayered network of signalling cascades. In some cases many parameters need to be satisfied before expression of a particular phenotype can occur. In these cases population density signals may be overridden or modulated because of other important parameters; for example the effects of iron limitation, oxygen and nutrient starvation, catabolite repression, heat shock, SOS DNA repair inducing agents and molecular chaperones on the expression of *lux* in *P. fischeri* are well documented (see earlier).

Furthermore, population-based signals may modulate the expression of other genes. For example we have already mentioned that in *V. harveyi* and *P. fischeri* an AHL induced phosphoryl transfer cascade may exert regulatory control over bioluminescence. Given the cross talk seen between other sensor kinase regulator systems it is not unreasonable to suggest similar behaviour from these response regulators.

5.2 Co-regulation: The absolute requirement for opines in *Agrobacterium tumefaciens* conjugal transfer

In the plant pathogen *Agrobacterium tumefaciens* conjugal transfer of Ti plasmids is induced by the availability of specific nutrients to populations above a threshold density (as judged by the agrobacterial quorum sensing system comprising TraI, TraR and *N*-3-(oxooctanoyl)-L-homoserine lactone, (OOHL) within the crown gall tumour (Fuqua and Winans, 1994). The important nutrients are opines (octopines or nopalines) which are synthesised by the plant in response to instructions from the infecting bacteria.

In octopine-dependent strains the expression of TraR is activated by the octopine response regulator (OccR) meaning that TraR, and hence conjugation, is only activated in the presence of octopine (Fuqua et al., 1994). In the case of nopaline type strains a similar pattern has emerged, however in this case the expression of *traR* is repressed by the agrocinopine catabolism repressor AccR (Fuqua et al., 1994). In the presence of opines *traR* expression is elevated, the TraR protein is then able to interact with OOHL to activate transcription of the *tra* genes, including *traI*. Since *traI* is regulated via TraR, it seems likely that the formation of an OOHL-TraR complex will result in the formation of a positive feedback loop (i.e. an example of autoinduction) and high levels of OOHL synthesis. This should ensure that once the appropriate conditions are attained, conjugation is strongly stimulated.

Although TraI and TraR are key components in the regulation of Ti plasmid conjugal transfer, other genes are involved in regulatory control. A further level of negative regulation is given by TraM, an 11kDa protein transcribed convergently to *traR*, the disruption of which results in elevated levels of Ti plasmid transfer (Fuqua et al., 1995). Hwang et al. (1995) surmised that TraM is able to sequester TraR from AHLs when opines are absent ensuring a tight repression of conjugation until the appropriate environmental conditions arise.

5.3 Dedicated LuxR homologue regulons: CarR and EcaR(ExpR) in *Erwinia carotovora*

Population density, through quorum sensing, is central to the control of both the expression of plant cell wall degrading exoenzymes and the biosynthesis of carbapenem antibiotics in a second plant pathogen, *Erwinia carotovora* subsp. *carotovora* (Salmond et al., 1995; Jones et al., 1993; Pirhonen et al., 1993; McGowan et al., 1995). Here, the synthesis of the pheromone OHHL is directed by the signal generator CarI (ExpI) and accumulation leads to the activation of a subset of response regulators controlling different regulons.

Two LuxR homologues have been identified in the carbapenem-producing *Erwinia carotovora* ExpR which is convergently transcribed with *carI* (R. Heikinheimo, PhD Thesis, Swedish University of Agricultural Sciences, 1995) and CarR which is part of the *car* operon and unlinked to *carI* (McGowan et al., 1995). CarR is the OHHL response regulator controlling carbapenem biosynthesis, but a role for ExpR has yet to be defined and the proposed OHHL-dependent response regulator for control of exoenzyme biosynthesis has not yet been identified. In a related *E. carotovora* strain, the interruption of *expR* has no significant effect upon virulence, exoenzyme production or OHHL production. Interestingly, however, the over-expression of CarR down-regulates exoenzyme production and this is relieved by additional exogenous OHHL. Two explanations have been offered for these effects, firstly that the LuxR homologue may act to repress exoenzyme transcription, with this repression released upon binding OHHL and second that the LuxR homologue sequesters OHHL from the exoenzyme response regulator(s), preventing transcriptional activation(McGowan et al., 1995).

5.4 Modulation: Positive and negative effectors of exoenzyme synthesis in *Erwinia carotovora*

The phytopathogenicity of *E. carotovora* is highly dependent upon the synthesis and secretion of plant wall degrading enzymes which facilitate maceration of plant tissues and the liberation of nutrients (Barras et al., 1994). Nevertheless, strains constitutively expressing exoenzymes are unable to colonise plant tissue, presumably because at low cell densities the defence reaction mounted by plant hosts will repel these bacteria (Salmond et al., 1995). The tight regulation of exoenzyme production by *E. carotovora* is therefore crucial and requires careful timing of the transition between evasion of the host defences and host attack using the rapid and co-ordinated production of a battery of exoenzymes.

The quorum sensing component of exoenzyme regulation can provide information regarding population density and contribute to the co-ordinate control of the genes in the exoenzyme regulon(s). Other regulators, however, have been identified which both sense other environmental parameters and ensure a tight repression in the absence of the appropriate conditions for exoenzyme production. Accordingly, an important role for a plant extract component has been identified which acts through the *aepABH* regulon (Murata et al., 1991, 1994; Liu et al., 1993). Here three gene products are inducible by plant cell extracts and act co-ordinately to induce exoenzyme biosynthesis and secretion assuming other important parameters have been satisfied. In a basal salts medium, for example, exoenzyme biosynthesis is only induced following the addition of the plant extract (Murata et al., 1991).

A repressor controlling extracellular enzyme synthesis and OHHL production has also been identified in *E. carotovora*. RsmA is able to suppress the synthesis of both exoenzymes and OHHL, with RsmA mutants over-expressing exoenzymes irrespective of the composition of the extracellular milleau (Cui et al., 1995; Chatterjee et al., 1995). Interestingly, RsmA homologues have been reported for most enterobacterial species (Cui et al., 1995) and homologues are also present in *Haemophilus* (GenBank Accession number L45451) and *Bacillus* (SwissProt Accession number P33911). An extensive similarity with the carbon storage regulator CsrA of *E. coli* which mediates gene expression by decreasing RNA stability has allowed the postulation of a similar mode of action for RsmA (Chatterjee et al., 1995).

It has been suggested that AepA detects specific components of the plant cell extract. In the presence of these metabolites, AepABH could act in concert to down regulate *rsmA* transcription. The resultant decrease in RsmA activity could serve to increase *carI* transcription and co-activate exoenzyme synthesis and carbapenem biosynthesis (Barras et al., 1994). The nutrients released by exoenzyme action are usable by both *E. carotovora* and its competitors, but through the co-ordinated production of carbapenem antibiotic and exoenzymes the organism can eliminate these potential competitors.

6 A Role for *N*-acyl homoserine lactones in the Pathogenesis of Infection

Amongst Gram-negative bacteria, bioluminescence and ß-lactam biosynthesis are rather unusual traits. Many different Gram-negative bacteria which do not emit light nor synthesize carbapenems nevertheless produce *N*-AHLs and possess genes analogous to *luxI*. These data suggest that different bacteria employ the same pheromone to control quite different sets of genes. Amongst the *N*-AHL producers identified to date, are several organisms capable of infecting humans, animals and plants. During infection, survival and multiplication in a hostile environment are clearly the priorities of a pathogen which must modulate expression of those genes necessary to establish the organism in a new niche. Parameters such as temperature, pH, osmolarity and nutrient availability are all known to function as environmental signals controlling the expression of co-ordinately regulated virulence determinants in bacteria (Mekalanos, 1992). Such information raises the possibility that*N*-AHLs may also be involved in co-ordinating the control of virulence. A common feature of many bacterial infections is the need for the infecting pathogen to reach a critical cell population density sufficient to overwhelm host immune defences. The ability of a population of bacteria to co-ordinate their attack on the host may therefore be a crucial component in the development of infection.

6.1 Pseudomonas aeruginosa

An indication of a role for *N*-AHLs in the pathogenesis of human infection has come from studies of the opportunistic pathogen *Pseudomonas aeruginosa*. This environmental bacterium can infect almost any body site given the right predisposing conditions. Individuals at greatest risk of *P.aeruginosa* infection include those with burn wounds and cystic fibrosis. *P.aeruginosa* secretes many extracellular toxic factors including an exotoxin which inhibits protein synthesis in mammalian cells and various tissue damaging exoenzymes including an alkaline protease and an elastase. This latter enzyme, which appears necessary for the maximal virulence of *P. aeruginosa*, is a metalloprotease and the product of the *lasB* gene. Elastase synthesis is controlled by a gene termed *lasR*, the product of which (LasR) shares significant amino acid sequence homology with LuxR (Gambello and Iglewski, 1991). When coupled with previous descriptions of the growth phase dependency of elastase production and our discovery that *P.aeruginosa* produces OHHL, there was at this time strong circumstantial evidence for the involvement of OHHL or structurally related compounds in controlling synthesis of this important virulence determinant.

To define a role for *N*-AHLs in controlling elastase expression we employed a similar strategy to that used in our studies of carbapenem biosynthesis in *Erwinia*, i.e. *P. aeruginosa* mutants were selected by chemical mutagenesis which were no longer able to make elastase (Jones et al., 1993). Following the addition of exogenous OHHL, we were able to divide these *P. aeruginosa* mutants into two classes, those which responded to OHHL by synthesizing elastase and those which did not. Confirmation that the OHHL responsive mutants were indeed synthesizing the 33 kDa LasB protein was obtained by sodium dodecyl sulphate-polyacrylamide gel electrophoresis of the protein purified from spent culture supernatants of an OHHL-responsive mutant grown in the presence of OHHL (Jones et al., 1993). These OHHL-responders (one of which was termed PAN067) are therefore likely to have mutations in the *P. aeruginosa* gene analogous to the *P. fischeri luxI* gene. This gene termed *lasI* has been cloned and sequenced (Passador et al., 1993) and shown to be 34.6% identical and 55.9% similar to *luxI*. Passador et al. (1993) also reported that *lasI* was involved in the synthesis of a small diffusible molecule, termed *P. aeruginosa* autoinducer (PAI).

When introduced into *E. coli* on a multi-copy plasmid, *lasI* directs synthesis of predominantly *N*-(3-oxododecanyl-L-homoserine lactone (OdDHL) together with small amounts of molecules exhibiting the same HPLC retention times as OHHL and OOHL (Pearson et al., 1994). Subsequently, Gray et al. (1994) reported that in the presence of LasR, OdDHL but not OHHL could activate a *lasB-lacZ* promoter fusion in an *E. coli* genetic background. However, PAN067 had responded, albeit partially, to the addition of exogenous OHHL but not to OdDHL (Latifi et al., 1995). Since OdDHL failed to stimulate elastase synthesis in PAN067, this suggested that the mutation is unlikely to be in *lasI*. Further examination of PAN067 revealed that it is highly pleiotropic and has an altered colonial morphology, is defective in the production of alkaline protease, haemolysin, pyocyanin, hydrogen cyanide and chitinase (Latifi et al., 1995; Winson et al., 1995) as well as staphylolytic activity (unpublished data). These data suggested that PAN067 must contain a mutation in a global regulatory locus. From a cosmid library of PAO1 chromosomal DNA, we identified a cosmid (pAX27) capable of restoring colonial morphology, autoinducer synthesis and exoproduct production (Latifi et al., 1995).

6.1.1 *P. aeruginosa* PA01 possesses multiple homologues of LuxR and LuxI

To identify the genetic locus on pAX27 capable of complementing PAN067, we employed the strategy first described by Swift et al. (1993) for cloning homologues of *luxI*. Because of the low levels of DNA sequence homology, identifying genes encoding *luxI* homologues by DNA hybridization is not generally possible. An alternative approach is to employ the *E. coli lux* sensor; the introduction of *luxI* or a *luxI* homologue *in trans* on a compatible plasmid should result in the *in vivo* production of an *N*-AHL and consequently, a bioluminescent phenotype. This method does have limitations since LuxR shows some molecular specificity in that the sensor would not detect OdDHL for example. Nevertheless, we identified highly bioluminescent transformants containing sub-clones of pAX27. Spent, cell-free culture supernatants from one of these transformants activated the *lux* sensor suggesting that we had cloned a locus capable of directing autoinducer synthesis. Further analysis of this clone indicated that it contained a 2 kb fragment of PAO1 DNA encoding two genes, the predicted translation products of which were identified as homologues of LuxR and LuxI (Latifi et al., 1995). These homologues are distinct from LasR and LasI and are termed VsmR and VsmI to indicate their role in regulating both *v*irulence determinants and *s*econdary *m*etabolites. Two almost identical genes termed *rhlR* and *rhlI* because of their association with the regulation of ***rh**amno**l**ipid* biosurfactant synthesis have also been identified in another *P. aeruginosa* strain, DSM2659 (Ochsner et al., 1994; Ochsner and Reiser, 1995) as well as in PAO1 (Brint and Ohman, 1995). Disruption of *rhlR* or *rhlI* resulted in a rhamnolipid deficient mutant which also lacked elastolytic activity.

VsmR/RhlR are clearly members of the LuxR family and share around 30% identity with LuxR and LasR as well as the *Pseudomonas aureofaciens* homologue, PhzR (Pierson et al., 1994). This family of transcriptional activators can be considered as modular proteins; the C-terminal region of the protein contains a classical helix-turn-helix DNA-binding motif whilst the N-terminal two thirds of the protein, at least in LuxR, appears to constitute the autoinducer-binding site (Choi and Greenberg, 1992; Stevens et al., 1994). Indeed, Hanzelka and Greenberg (1995) have shown that in whole *E. coli* cells the LuxR N-terminal domain can fold into a polypeptide capable of binding ^{3}H-labelled OHHL in the absence of the C-terminal domain. Therefore, it is perhaps not very surprising to find that the most highly conserved amino-acid sequence stretches amongst the LuxR homologues encompass the DNA- and autoinducer-binding regions. One interesting feature to emerge from sequence comparisons with VsmR is that it is >50% identical to the *E. coli* LuxR homologue, SdiA, which is known to act as a transcriptional *trans*-activator of the cell division genes in the *ftsQAZ* locus (Wang et al., 1991). As yet no *E. coli* equivalent of LuxI or an *N*-AHL has been found.

6.1.2 VsmI Directs the synthesis of two *N*-acylhomoserine lactones

In *P. aeruginosa* PAO1 *vsmI* is located downstream of *vsmR*. VsmI is almost identical to RhlI and is more closely related to the *P. aureofaciens* LuxI homologue PhzI (Wood et al., 1996) than to LasI. Phylogenetic analysis of the LuxI homologues currently deposited in the databases suggest that those from enteric bacteria such *Erwinia*, *Yersinia*, *Serratia* and *Enterobacter* form a group distinct from those of *Pseudomonas, Agrobacterium* and *Vibrio*. Despite the relatively large number of LuxI homologues now known, it is not possible to predict the nature of the *N*-AHL synthesised, even post Moré et al. (1996).

When introduced into *E. coli*, LasI directs the synthesis of OdDHL and also apparently OHHL and OOHL although to a much lesser extent (Pearson et al., 1994). On this basis we assumed that VsmI must direct the synthesis of another *N*-AHL(s) which would probably be responsible for activating VsmR. To explore this possibility, we developed a novel bioassay employing the autoinducer-negative *Chromobacterium violaceum* mutant CV026. This transposon-insertion mutant is white and responds to a range of exogenous *N*-AHLs by producing the violet pigment violacein (Winson et al., 1994). Since PAO1, but neither PAN067 nor OdDHL induced pigmentation, we realised that CV026 could usefully be exploited in the testing of crude extracts and fractions collected from chromatographic separations derived from cell-free culture supernatants. Using this approach, we purified two active compounds from supernatants of *E. coli* transformed with the *vsmRI* locus (Winson et al., 1995). Their structures were deduced from high resolution mass and NMR spectra and confirmed by chemical synthesis. The major compound was identified as *N*-butanoyl-L-homoserine lactone (BHL), the minor as *N*-hexanoyl-L-homoserine lactone (HHL). Neither compound is made by PAN067, and in PAO1, both autoinducers are made in a ratio (BHL:HHL) of approximately 15:1 (Winson et al., 1995; Pearson et al., 1995). These data clearly demonstrate that *P. aeruginosa* produces signal molecules which have either 3-oxo-substituted (OHHL and OdDHL) or unsubstituted (BHL and HHL) *N*-acyl side chains. HHL has also been recently identified in *Y. enterocolitica* (Throup et al., 1995) and in *P. fischeri* (Kuo et al., 1994). Interestingly, neither of these bacteria make BHL and in *Yersinia*, the LuxI homologue, YenI, directs the synthesis of OHHL and HHL in an approximately 1:1 ratio (Throup et al., 1995). This information is all the more remarkable since it is clear that when introduced into *E. coli*, *luxI, lasI, yenI,* and *vsmI* each direct synthesis of the correct *N*-AHL molecules in the same proportions as that found in the parent organism (Kuo et al., 1994; Pearson et al., 1994; Throup et al., 1995; Winson et al., 1995). This must mean that *E. coli* possesses all of the necessary substrates and accessory proteins. Since these signal molecules differ only in the structure of their *N*-acyl side chains, this presumably reflects the capacity of LuxI homologues to "lock into" fatty acid metabolism at different stages to sequester specific *N*-acyl side chains which are enzymatically linked to the homoserine lactone.

6.1.3 Activation of VsmR

A region of dyad symmetry, termed the "*lux* box" centred around residue -40 from the *luxI* transcriptional start site appears to act as the target site for LuxR in *P. fischeri* (Stevens et al., 1994). Related sequences have been found upstream of the *P. aeruginosa lasB* gene as well as upstream of the *A. tumefaciens traA* and *traI* genes (Fuqua et al., 1994). We have also identified a putative *lux* box like element between *vsmR* and *vsmI* which is very similar to that of *lasB* (Latifi et al., 1995). This provided an opportunity to determine whether VsmR was activated by BHL or by OdDHL and to define in more detail the structural requirements for the activation of VsmR by *N*-AHLs. To achieve this we constructed a bioluminescent reporter in which *vsmR* together with the *vsmI* promoter/operator region was linked to the *Photorhabdus luminescens lux* structural operon. Unlike the *lux* reporter described previously which employed the *P. fischeri lux* genes, this new reporter construct, when introduced into *E. coli*, does not require the addition of exogenous long chain aldehyde and functions at temperatures up the 37°C. Using this assay, we observed that BHL and HHL activate this bioluminescent reporter to a similar extent; around 1 μM is required to generate a half maximal response (Winson et al., 1995). Extension of the side chain to the 8 carbon *N*-octanoyl-L-homoserine lactone (OHL) reduced activity by over 50% whilst the 12 C analogue *N*-dodecanoyl-L-homoserine lactone (DHL) was inactive. Introduction of an oxygen at the C3 position of BHL to give *N*-(3-oxobutanoyl)-L-homoserine lactone (OBHL) also reduced activity to around the same level as that observed with OHHL, i.e. about half that obtained with BHL and HHL. OdDHL was slightly active but much less efficient at inducing light emission than any of the other analogues examined (Winson et al., 1995). From this structure/activity analysis, it is clear that optimal activation of VsmR requires a short *N*-acyl side chain without a 3-oxo substituent. In contrast, the 3-oxo substituent and a C12 *N*-acyl side chain have been reported to be essential for the optimal activation of LasR (Passador et al., 1994, 1996). For other LuxR homologues such as LuxR and CarR (from *E. carotovora)* which like VsmR depend on short chain *N*-AHls for activation, the 3-oxo substituent is critical. In *E. carotovora*, carbapenem biosynthesis depends on the activation of CarR by OHHL (Bainton et al., 1992 a, b; McGowan et al., 1995); compounds such as HHL and *N*-(3-hydroxyhexanoyl)-L-homoserine lactone (the *Vibrio harveyi* autoinducer) exhibit less than 1% of the activity of OHHL; BHL, HBHL and OBHL less than 0.15% (Chhabra et al., 1993). In *P. fischeri*, HHL has around 10% the activity of OHHL (Eberhard et al., 1981). Thus although the pattern of substrate specificity emerging for the LuxR homologues is high, they show a certain degree of flexibility.

6.1.4 Restoration of exproduct synthesis in PAN067 by BHL and HHL

Since transformation of PAN067 with a multi-copy plasmid carrying the *vsm* locus restored production of multiple exoproducts and as this mutant had responded to the addition of exogenous OHHL, we added either BHL or HHL to the PAN067 growth medium. Both autoinducers restored elastase production and BHL is more effective than HHL. Further analysis indicated that both chitinase and cyanide synthesis had also been restored, however the concentration of exogenous autoinducer required to restore all three phenotypes to around 50% of that of the parent PAO1 was far greater (100 μM) than that required for activation of the *vsmRvsmI'::lux* reporter (1 μM; Winson et al., 1995). Whilst this may reflect the greater senstivity of the *vsmI* promoter, it also highlights the unknown nature of the mutation in PAN067. Although PAN067 can be complemented by plasmid-borne copies of *vsmRI*, the strain has wild-type *vsmRI* genes.

In *P. aeruginosa* PG201, Ochsner and Reiser (1995) have constructed an *rhlI (vsmI)* mutant in which restoration of rhamnolipid synthesis requires the addition of 0.5 μM BHL to achieve a half-maximal response; a similar response with OHHL requires 10 μM. This information clearly resolves the apparent contradictions in the literature with regard to the regulation of elastase expression by OdDHL and OHHL (Jones et al., 1993; Gray et al., 1994). In particular, the inability of OHHL to activate a reporter system in *E. coli* containing *lasR* and a *lasB::lacZ* fusion (Gray et al., 1994) - whilst being able to induce elastase production in a *P. aeruginosa* genetic background (Jones et al., 1993) - can now be explained given the characterization of the *vsm/rhl* locus and the identification of BHL and HHL. Although OHHL is unable to activate LasR (Gray et al., 1994) it is clearly capable of activating VsmR/RhlR (Winson et al., 1995; Ochsner and Reiser, 1995).

6.1.5 A hierarchical autoinduction cascade regulates the production of virulence determinants and secondary metabolites in *P. aeruginosa*

Thus far, two LuxR homologues (LasR and VsmR[RhlR]) and two LuxI homologues (LasI and VsmI) have been described in *P. aeruginosa* PAO1 and their cognate signal molecules chemically characterized. Both regulatory loci are involved in the expression of alkaline protease and elastase whereas exotoxin A (Latifi et al., 1995) and rhamnolipids (Ochsner and Reiser, 1995) appear to by regulated by only one system, LasR/OdDHL or VsmR/BHL respectively. LasR/OdDHL and VsmR/BHL thus represent two arms of a complex regulatory circuit involved in the control of multiple physiological traits. With regard to *lasB* expression, a picture of the interrelationship between the two quorum sensing circuits is beginning to emerge (Fig. 3). Pearson et al., (1995) have recently reported that although OdDHL is able to activate a *lasB::lacZ* fusion in a *P. aeruginosa lasR* mutant containing a plasmid with *lasR* under the control of the *lac* promoter, it does not activate the reporter when *lasR* is placed under the control of its natural promoter. In fact, BHL was reported to be necessary for the expression of *lasR* in *P. aeruginosa* (Pearson et al., 1995). Whether this BHL-mediated activation of *lasR* requires VsmR was not reported. However, Latifi et al. (1996) obtained evidence to indicate that LasR/OdDHL is required for activation of *vsmR* and that VsmR (in the presence or absence of BHL) has a negative effect on the expression of *lasR*. Thus two interdependent quorum sensing circuits have emerged and although additional work is required to dissect in detail the nature of the inter-relationship between the *lasRI*/OdDHL and *vsmRI*/BHL circuits, it is clear that elastase synthesis depends on both systems. Our current understanding of this hierarchical autoinducer dependent regulatory cascade with regard to *lasB* expression is illustrated by Fig.3. LasR activates *lasI*, such that OdDHL is produced setting up the first autoinduction cascade (Seed et al., 1995). LasR/OdDHL activates expression of *vsmR*, such that VsmR induces *vsmI* expression such that BHL (and HHL) is produced (Winson et al., 1995; Latifi et al., 1996). Since *vsmR* (*rhlR*) mutants are elastase-negative (Ochsner and Reiser, 1995) and as VsmR/BHL activates *vsmI* via an almost identical *lux* box to that located within the *lasB* promoter (Latifi et al., 1995; Winson et al., 1995), we presume that physiologically VsmR/BHL is responsible for *lasB* activation in *P. aeruginosa*. However, Gray et al. (1994) have provided evidence that LasR/OdDHL directly activates *lasB*. Thus although the model presented offers an explanation of why mutations in either *lasR* or *vsmR* lead to the loss of elastase production, it is not yet clear as to why LasR/OdDHL does not activate the *lasB* promoter directly in a *vsmR*(*rhlR*) negative background. Clearly there are many additional layers of regulatory complexity in *P. aeruginosa* which remain to be elucidated. Nevertheless, recognition of the phenomenon of cell-cell communication is beginning to offer new insights into the physiology and molecular biology of *Pseudomonas*. For example, the remarkable link between quorum sensing and the stationary phase sigma factor RpoS in *P.aeruginosa* (Latifi et al., 1996) now opens up a new dimension for the role of cell-cell communication in regulating the survival characteristics associated with entry into and exit from the stationary phase.

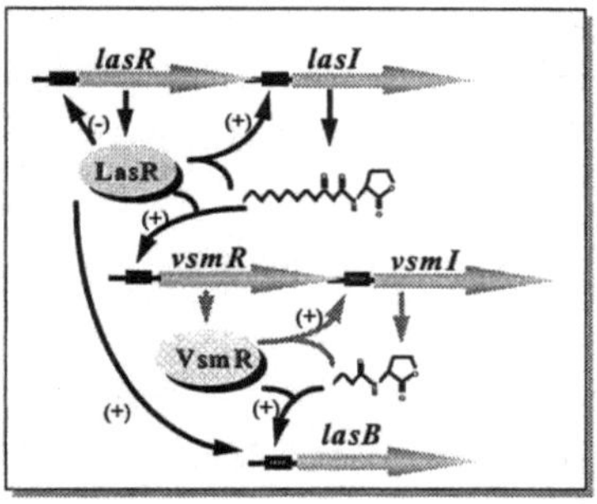

Fig. 3. Quorum sensing in *Pseudomonas aeruginosa*. For details see text.

6.2 Other human pathogens

The above information strongly suggests that bacterial intercellular communication facilitates co-ordination of the expression of multiple virulence determinants thus influencing the outcome of *Pseudomonas* infections. For other human pathogens such as *Yersinia, Aeromonas* ans *Enterobacter*, although we have established that they produce *N*-AHLs and have cloned corresponding *luxI* homologues, the virulence determinants regulated by the pheromone(s) have not yet been identified.

7 Compromising Bacterial Communication Skills - Therapeutic Possibilities

Bacterial pathogens possess distinct genetic properties which confer a significantly greater capacity to compete with other (commensal) bacteria, to gain a foothold within a susceptible host, to multiply in host tissues and to avoid host defences. Pathogens have therefore evolved complex regulatory pathways which control virulence factor expression indicating the crucial importance of being able to maximise survival with the minimum expenditure of metabolic resources when adapting to new and potentially hostile environments. The ability to switch off exogenously virulence gene expression may therefore offer a novel strategy for the treatment or prevention of infection. The recent discovery that a wide spectrum of Gram-negative bacteria employ pheromones to co-ordinate the control of virulence gene expression enables us to envisage such an opportunity. In particular, interference with transmission of the molecular message by an antagonist which competes for the pheromone binding site of the transcriptional activator protein (the LuxR homologue) thereby switching off virulence gene expression so attenuating the pathogen, is an attractive strategy. In this context it is perhaps worth noting that too high a concentration of OHHL in both *P. fischeri* and *E. carotovora* switches off bioluminescence and carbapenem biosynthesis respectively (Eberhard et al, 1986; Bainton et al, 1992a). Several analogues of OHHL have also been shown to inhibit the action of OHHL in *P. fischeri* (Eberhard et al., 1986). The relative ease with which analogues of OHHL can be synthesized (Chhabra et al., 1993) should make it experimentally feasible to evaluate the influence of an OHHL antagonist on the course of infection. Furthermore the observation that *A. tumefaciens* employs an *N*-AHL to promote conjugation and the spread of Ti plasmids throughout the bacterial population is unlikely to be a unique example. It is probable that other bacteria employ pheromone-based self-sensing mechanisms to promote conjugation and the spread of perhaps not only plasmids carrying virulence determinants but also those conferring multiple antibiotic resistant. Pheromone antagonists might therefore also have a role in preventing the spread of antibiotic resistance thus prolonging the useful life of many antimicrobials - including the carbapenems.

The ability of bacteria to modulate phenotype as a consequence of changes in population density through quorum sensing is a phenomenon we are only just beginning to understand. In medicine, quorum sensing offers an attractive target for novel antibacterial agents which jam signal transduction by antagonising the functions of AHLs. Similarly, AHL antagonists may find relevance in agriculture by protecting crops from damage caused by pathogens such as *E. carotovora*. In biotechnology, quorum sensing may be harnessed to control fermentation processes either by triggering early production of a desired metabolite or by making the onset of synthesis of a toxic product dependent on the addition of an exogenous AHL. The ability to detect AHLs sensitively and specifically may also ultimately contribute to the design of novel rapid methods for bacterial detection and enumeration in clinical samples, foods and pharmaceuticals.

Acknowledgements

We would like to thank all our colleagues and in particular Andrée Lazdunski, Amel Latifi, Miguel Camara, Ram Chhabra, Nigel Bainton, Michael Winson, Pan Chan, Philip Hill and Catherine Rees at Nottingham and George Salmond (Dept. Biochemistry, University of Cambridge) who have been involved in the development of our understanding of pheromone-mediated gene expression control. The work described has been funded by grants from the Biotechnology Directorate of the Science and Engineering Research Council, the Wellcome Trust and by Amersham International which are gratefully acknowledged.

References

Adar YY, Simaan M, Ulitzur S (1992) Formation of the LuxR protein in the *Vibrio fischeri-lux* system is controlled by HtpR through the GroESL proteins. J Bacteriol 174:7138-7143

Adar YY, Ulitzur S (1993) GroESL proteins facilitate binding of externally added inducer by LuxR protein-containing *Escherichia coli* cells. J Biolumin Chemilumin 8:261-266

Bainton NJ, Bycroft BW, Chhabra SR, Stead P, Gledhill L, Hill PJ, Rees CED, Winson MK, Salmond GPC, Stewart GSAB, Williams P (1992a) A general role for the *lux* autoinducer in bacterial cell signalling: control of antibiotic synthesis in *Erwinia.* Gene 116:87-91

Bainton NJ, Stead P, Chhabra SR, Bycroft BW, Salmond GPC, Stewart GSAB, Williams P (1992b) *N*-(3-oxohexanoyl)-L-homoserine lactone regulates carbapenem antibiotic production in *Erwinia carotovora*. Biochem J 288:997-1004

Barras F, Van Gijsegem F, Chatterjee AK (1994) Extracellular enzymes and pathogenesis of soft-rot *Erwinia.* Ann Rev Phytopathol 32:201-234

Bassler BL, Wright M, Silverman MR (1994) Sequence and function of LuxO, a negative regulator of luminescence in *Vibrio harveyi.* Mol Microbiol 12:403-412

Brint JM, Ohman DE (1995) Synthesis of multiple exoproducts in *Pseudomonas aeruginosa* is under the control of RhlR-RhlI, another set of regulators in strain PAO1 with homology to the autoinducer responsive LuxR-LuxI family. J Bacteriol 177:7155-7163

Cao JG, Meighen EA (1993) Biosynthesis and stereochemistry of the autoinducer controlling luminescence in *Vibrio harveyi*. J Bacteriol 175:3856-3862

Chatterjee A, Cui YY, Liu Y, Dumenyo CK, Chatterjee AK (1995) Inactivation of *rsmA* leads to overproduction of extracellular pectinases, cellulases, and proteases in *Erwinia carotovora* subsp *carotovora* in the absence of the starvation cell density sensing signal, *N*-(3-oxohexanoyl)-L-homoserine lactone. Appl Environ Microbiol 61:1959-1967

Chhabra SR, Stead P, Bainton NJ, Salmond GPC, Stewart GSAB, Williams P, Bycroft BW (1993) Autoregulation of carbapenem biosynthesis in *Erwinia carotovora* by analogues of *N*-(3-oxohexanoyl)-L-homoserine lactone. J Antibiotics 46:441-449

Choi SH, Greenberg EP (1992) Genetic dissection of DNA-binding and luminescence gene activation by the *Vibrio fischeri* LuxR protein. J Bacteriol 174:4064-4069

Cubo MT, Economou A, Murphy G, Johnston AWB, Downie JA (1992) Molecular characterization and regulation of the rhizosphere-expressed genes *rhiABCR* that can influence nodulation by *Rhizobium leguminosarum* biovar *viciae*. J Bacteriol 174:4026-4035

Cui Y, Chatterjee A, Liu Y, Dumenyo CK, Chatterjee AK (1995) Identification of a global repressor gene, *rsmA*, of *Erwinia cavotavora* subsp *carotovora* that controls extracellular enzymes, *N*-(3-oxohexanoyl)-L-homoserine lactone, and pathogenicity in soft-rotting *Erwinia* spp. J Bacteriol 177:5108-5115

Dolan KM, Greenberg EP (1992) Evidence that GroEL, not sigma-32, is involved in transcriptional regulation of the *Vibrio fischeri* luminescence genes in *Escherichia coli*. J Bacteriol 174:5132-5135

Dunlap PV, Greenberg EP (1985) Control of *Vibrio fischeri* luminescence gene-expression in *Escherichia coli* by cyclic-amp and cyclic-amp receptor protein. J Bacteriol 164:45-50

Dunlap PV, Greenberg EP (1988) Control of *Vibrio fischeri* lux gene-transcription by a cyclic-Amp receptor protein LuxR protein regulatory circuit. J Bacteriol 170:4040-4046

Eberhard A, Burlingame AL, Kenyon GL, Nealson KH, Oppenheimer NJ (1981) Structural identification of the autoinducer of *Photobacterium fischeri* luciferase. Biochemistry 20:2444-2449

Eberhard A, Widrig CA, McBath P, Schineller JB (1986) Analogs of the autoinducer of bioluminescence in *Vibrio fischeri*. Arch Microbiol 146:35-40

Eberhard A, Longin T, Widrig CA, Stranick SJ (1991) Synthesis of the *lux* gene autoinducer in *Vibrio fischeri* is positively autoregulated. Arch Microbiol 155:294-297

Engebrecht J, Nealson K, Silverman M (1983) Bacterial bioluminescence - isolation and genetic-analysis of functions from *Vibrio fischeri*. Cell 32:773-781

Engebrecht J, Silverman M (1986) Regulation of expression of bacterial genes for bioluminescence Genet. Eng. 8:31-44.

Fuqua W C, Winans SC (1994) A LuxR-LuxI type regulatory system activates *Agrobacterium* Ti plasmid conjugal transfer in the presence of a plant tumour metabolite. J Bacteriol 176:2796-2806

Fuqua WC, Winans SC, Greenberg EP (1994) Quorum sensing in bacteria: the LuxR-LuxI family of cell density responsive transcriptional regulators. J Bacteriol 176:269-275

Fuqua WC, Burbea M, Winans SC (1995) Activity of the *Agrobacterium* Ti plasmid conjugal transfer regulator TraR is inhibited by the product of the *traM* gene. J Bacteriol 177:1367-1373

Gambello MJ, Iglewski BH (1991) Cloning and characterization of *Pseudomonas aeruginosa lasR* gene, a transcriptional activator of elastase expression. J Bacteriol 173:3000-3009

Gilson L, Kuo A, Dunlap PV (1995) AinS and a new family of autoinducer synthesis proteins. J Bacteriol 177:6946-6951

Gray KM, Greenberg EP (1992) Physical and functional maps of the luminescence gene-cluster in an autoinducer-deficient *Vibrio fischeri* strain isolated from a squid light organ. J Bacteriol 174:4384-4390

Gray KM, Passador L, Iglewski BH, Greenberg EP (1994) Interchangeability and specificity of components from the quorum-sensing regulatory systems of *Vibrio fischeri* and *Pseudomonas aeruginosa*. J Bacteriol 176:3076-3080

Greenberg, EP, Hastings, JW,Ulitzur S (1979) Induction of luciferase synthesis in *Beneckea harveyi* by other marine bacteria. Arch Microbiol 120:87-91.

Hanzelka BL, Greenberg EP (1995) Evidence that the N-terminal region of the *Vibrio fischeri* LuxR protein constitutes an autoinducer-binding domain. J Bacteriol 177:815-817

Hanzelka BL, Greenberg EP (1996) Quorum sensing in *Vibrio fischeri* - evidence that s-adenosylmethionine is the amino-acid substrate for autoinducer synthesis. J Bacteriol 178:5291-5294

Haygood MG, Distel DL (1993) Bioluminescent symbionts of flashlight fishes and deep-sea anglerfishes form unique lineages related to the genus *Vibrio*. Nature (London) 363:154-156

Henikoff S, Wallace JC, Brown J.P. (1990) Finding protein similarities with nucleotide sequence databases Methods Enzymol 183:111-132.

Huisman GW, Kolter R (1994) Sensing starvation - a homoserine lactone-dependent signaling pathway in *Escherichia coli*. Science 265:537-539

Hwang I, Cook DM, Farrand SK (1995) A new regulatory element modulates homoserine lactone-mediated autoinduction of Ti plasmid conjugal transfer. J Bacteriol 177:449-458.

Jones S, Yu B, Bainton NJ, Birdsall M, Bycroft BW, Chhabra SR, Cox AJR, Golby P, Reeves PJ, Stephens S, Winson MK, Salmond GPC, Stewart GSAB, Williams P (1993) The *lux* autoinducer regulates the production of exoenzyme virulence determinants in *Erwinia carotovora* and *Pseudomonas aeruginosa*. EMBO J 12:2477-2482

Kaplan HB, Greenberg EP (1985) Diffusion of autoinducer is involved in regulation of the *Vibrio fischeri* luminescence system. J Bacteriol 163:1210-1214

Kuo A, Blough NV, Dunlap PV (1994) Multiple *N*-acyl-L-homoserine lactone autoinducers of luminescence in the marine symbiotic bacterium *Vibrio fischeri*. J Bacteriol 176:7558-7565

Latifi A, Winson MK, Bycroft BW, Stewart GSAB, Lazdunski A, Williams P (1995) Multiple homologues of LuxR and LuxI control expression of virulence determinants and secondary metabolites through quorum sensing in *Pseudomonas aeruginosa* PAO1. Mol Microbiol 17:333-343

Latifi A, Foglino M, Tanaka K, Williams P, Lazdunski A (1996) A Hierarchical quorum-sensing cascade in *Pseudomonas aeruginosa* links the transcriptional activators LasR and RhlR (VsmR) to expression of the stationary phase sigma factor RpoS. Mol Microbiol 21:1137-1146

Liu Y, Murata H, Chatterjee A, Chatterjee AK (1993) Characterization of a novel regulatory gene *aepA* that controls extracellular enzyme-production in the phytopathogenic bacterium *Erwinia carotovora* subsp *carotovora*. Mol Plant Microbiol Interactions 6:299-308

McGowan S, Sebaihia M, Jones S, Yu B, Bainton NJ, Chan PF, Bycroft BW, Stewart GSAB, Williams P, Salmond GPC (1995) Carbapenem antibiotic production in *Erwinia carotovora* is regulated by CarR, a homologue of the LuxR transcriptional activator. Microbiology 141:541-550

Mekalanos JJ (1992) Environmental signals controlling expression of virulence determinants in bacteria. J Bact 174:1-7

Meighen EA, Dunlap PV (1993) Physiological, biochemical and genetic-control of bacterial bioluminescence. Advances Microbial Physiol 34:1-67

Meighen 1994 Genetics of bacterial bioluminescence. Annu Rev Gene. 28:117-139

Meighen EA (1991) Molecular biology of bacterial bioluminescence. Microbiol Rev 55:123-142

Moré MI, Finger LD, Stryker IL, Fuqua C, Eberhard A, Winans SC (1996) Enzymatic synthesis of a quorum-sensing autoinducer through use of defined substrates. Science 272:1655-1658

Murata H, McEvoy JL, Chaqtterjee_A, Collmer A, Chatterjee AK (1991) Molecular-cloning of an *aepA*gene that activates production of extracellular pectolytic, cellulolytic, and proteolytic-enzymes in *Erwinia carotovora* subsp *carotovora*. Mol Plant Microbe Ineteractions 4:239-246

Murata H, Chatterjee A, Liu Y, Chatterjee AK (1994) Regulation of the production of extracellular pectinase, cellulase, and protease in the soft-rot bacterium *Erwinia carotovora* subsp *carotovora* - evidence that AepH of *Erwinia carotovora* subsp *carotovora*-71 activates gene-expression in *Erwinia carotovora* subsp *carotovora, Erwinia carotovora* subsp *atroseptica*, and *Escherichia coli*. Appl Environ Microbiol 60:3150-3159

Nealson KH, Platt T, Hastings JW (1970) Cellular control of synthesis and activity of the bacterial bioluminescence system. J Bacteriol 104:313-322

Nealson, K.H. (1977) Autoinduction of bacterial luciferase: occurrence, mechanism and significance *Arch Microbiol* 112:73-79

Ochsner UA, Koch AK, Fiechter A, Reiser J (1994) Isolation and characterization of a regulatory gene affecting rhamnolipid biosurfactant synthesis in *Pseudomonas aeruginosa*. J Bacteriol 176:2044-2054

Ochsner UA, Reiser J (1995) Autoinducer-mediated regulation of rhamnolipid biosurfactant synthesis in *Pseudomonas aeruginosa*. Proc Natl Acad Sci USA 92:6424-6428

Passador L, Cook JM, Gambello MJ, Rust L, Iglewski BH (1993) Expression of *Pseudomonas aeruginosa* virulence genes requires cell-to-cell communication. Science 260:1127-1130

Passador L, Pearson JP, Gray KM, Guertin K, Kende AS, Greenberg EP, Iglewski BH (1994) Use of structural analogs to determine critical features of *Pseudomonas aeruginosa* autoinducer. In Proceedings of the 94th Annual Meeting of the American Society for Microbiology (May 1994, Las Vegas, USA) Abstract D84:111

Passador L, Tucker KD, Guertin KR, Journet MP, Kende AS, Iglewski BH (1996) Functional analysis of the *Pseudomonas aeruginosa* autoinducer PAI. J Bacteriol 178:5995-6000

Pearson JP, Gray KM, Passador L, Tucker KD, Eberhard A, Iglewski BH, Greenberg EP (1994) Structure of the autoinducer required for expression of *Pseudomonas aeruginosa* virulence genes. Proc Natl Acad Sci USA 91:197-20

Pearson JP, Passador L, Iglewski BH, Greenberg EP (1995) A second *N*-acylhomoserine lactone signal produced by *Pseudomonas aeruginosa*. Proc Natl Acad Sci USA 92:1490-1494

Pierson LS, Keppenne VD, Wood DW (1994) Phenazine antibiotic biosynthesis in *Pseudomonas aureofaciens* 30-84 is regulated by PhzR in response to cell density. J Bacteriol 176:3966-3974

Pirhonen M, Flego D, Heikinheimo R, Palva ET (1993) A small diffusible signal molecule is responsible for the global control of virulence and exoenzyme production in the plant pathogen *Erwinia carotovora*. EMBO J 12:2467-2476

Salmond GPC, Bycroft BW, Stewart GSAB, Williams P (1995) The bacterial 'enigma' cracking the code of cell-cell communication. Mol Microbiol 16:615-624

Schaefer AL, Val DL, Hanzelka BL, Cronan JE, Greenberg EP (1996) Generation of cell-to-cell signals in quorum sensing - acyl homoserine lactone synthase activity of a purified *Vibrio fischeri* LuxI protein. Proc Nat Acad Sci USA 93:9505-9509

Seed PC, Passador L, Iglewski BH (1995) Activation of the *Pseudomonas aeruginosa lasI* gene by LasR and the *Pseudomonas* autoinducer PAI: an autoinduction regulatory hierarchy. J Bacteriol 177:654-659

Sharma S, Stark TF, Beattie, WG, Moses, RE (1986) Multiple control elements for the *uvrC* gene unit of *Escherichia coli*. Nucl Acids Res 14:2301-2318

Shadel GS, Baldwin TO (1992) Identification of a distantly located regulatory element in the *luxD* gene required for negative autoregulation of the *Vibrio fischeri luxR* gene. J Biol Chem 267:7690-7695

Sitnikov DM, Schineller JB, Baldwin TO (1995) Transcriptional regulation of bioluminesence genes from *Vibrio fischeri*. Mol Microbiol 17:801-812

Slock J, Van Riet D, Kolibachuk D, Greenberg EP (1990) Critical regions of the vibrio-fischeri luxr protein defined by mutational analysis. J Bacteriol 172:3974-3979

Stevens AM, Dolan KM, Greenberg EP (1994) Synergistic binding of the *Vibrio fischeri* LuxR transcriptional activator domain and RNA polymerase to the *lux* promoter region. Proc Natl Acad Sci USA 91:12619-12623

Stevens AM, Greenberg EP (1997) Quorum sensing in *Vibrio fischeri*: Essential elements for activation of the luminescence genes. J Bacteriol 179:557-562

Swift S, Winson MK, Chan PF, Bainton NJ, Birdsall M, Reeves PJ, Rees CED, Chhabra SR, Hill PJ, Throup JP, Bycroft BW, Salmond GPC, Williams P, Stewart GSAB (1993) A novel strategy for the isolation of *luxI* homologues: evidence for the widespread distribution of a LuxR:LuxI superfamily in enteric bacteria. Mol Microbiol 10:511-520

Swift S, Throup JP, Williams P, Salmond GPC, Stewart GSAB (1996) Quorum sensing - a population-density component in the determination of bacterial phenotype. Trends Biochem Sci 21:214-219

Throup J, Camara M, Briggs GS, Winson MK, Chhabra SR, Bycroft BW, Williams P, Stewart GSAB (1995) Characterization of the *yenI/yenR* locus from *Yersinia enterocolitica* mediating the synthesis of two *N*-acylhomoserine lactone signal molecules. Mol Microbiol 17:345-35

Ulitzur S, Kuhn J (1988) The transcription of bacterial luminescence is regulated by Sigma-32. J Biolum Chemilum 2:81-93

Wang X, de Boer PAJ, Rothfield LI (1991) A factor that positively regulates cell division by activating transcription of the major cluster of essential cell division genes of *Escherichia coli*. EMBO J 10:3363-3372

Williams P, Stewart GSAB, Camara M, Winson MK, Chhabra SR, Salmond GPC, Bycroft BW (1996) Signal Transduction through Quorum Sensing in *Pseudomonas aeruginosa*. In Molecular Biology of Pseudomonads (Nakazawa T, Furukawa K, Haas D. Silver S eds) pp195-206 ASM Press

Winson MK, Bainton NJ, Chhabra SR, Bycroft BW, Salmond GPC, Williams P, Stewart GSAB (1994) Control of *N*-acyl homoserine lactone-regulated expression of multiple phenotypes in *Chromobacterium violaceum*. In Proceedings of the 94th Annual Meeting of the American Society for Microbiology (May 1994, Las Vegas, USA) Abstract H-71:212

Winson MK, Camara M, Latifi A, Foglino M, Chhabra SR, Daykin M, Chapon V, Bycroft BW, Salmond GPC, Lazdunski A, Stewart GSAB, Williams P (1995) Multiple *N*-acyl-L-homoserine lactone signal molecules regulate production of virulence determinants and secondary metabolites in *Pseudomonas aeruginosa*. Proc Natl Acad Sci USA 92:9427-9431

Wood DW, Pierson LS (1996) The *phzI* gene of *Pseudomonas aureofaciens* 30-84 is responsible for the production of a diffusible signal required for phenazine antibiotic production. Gene 168:49-53

STUDYING PROTEIN SYNTHESIS FACTORS IN YEAST: STRUCTURE, FUNCTION AND REGULATION

Mick F Tuite

Research School of Biosciences, University of Kent, Canterbury, Kent CT2 7NJ, UK

Introduction

Once eukaryotic mRNA has been fully processed within the nucleus, it is transported to the cytoplasm where it is translated into its encoded polypeptide chain. The basic components of the protein synthesising machinery have been largely described and functionally defined; nevertheless the list of necessary ancillary translation factors is still expanding. In many cases, these protein factors are essential to facilitate key steps in the process and may themselves be multi-subunit proteins. Translation factors also represent key targets for both global and mRNA-specific control of translation, particularly via phosphorylation of these factors.

A full biochemical description of protein synthesis has come from studies with higher eukaryotic cells through fractionation and purification of the translation factors and their use to reconstitute partial reactions in protein synthesis *in vitro*. While a purely *in vitro* approach has a number of obvious pitfalls, nevertheless we have a more-or-less complete description of the basic mechanisms and factors of the three phases of protein synthesis in eukaryotic cells (Figure 1):

INITIATION: The small (40S) ribosomal subunit is loaded with the initiator $tRNA^{Met}$ to form a 43S pre-initiation complex. This complex then binds to the 5' m^7Gppp 'cap' and 'scans' along the 5' untranslated region (5'UTR) of the mRNA until the initiation codon (usually AUG) is located. This is usually the first AUG codon encountered. Initiation is completed by the joining of the 60S subunit to form a translationally active 80S ribosome. Each step in the initiation phase requires one or more initiation factors (eIFs).

ELONGATION: Aminoacyl-tRNAs are brought to the ribosomal decoding site in a codon-directed manner and the amino acid added to the carboxyl-terminus of the growing polypeptide chain. Following ribosome-directed peptide bond formation, the mRNA and bound peptidyl-tRNA are translocated to bring the next codon in the mRNA to the decoding site . Polypeptide chains grow in eukaryotic cells at the rate of 5-10 amino acids/sec/ribosome and the elongation requires the direct

NATO ASI Series, Vol. H 103
Molecular Microbiology
Edited by Stephen J. W. Busby,
Christopher M. Thomas and Nigel L. Brown

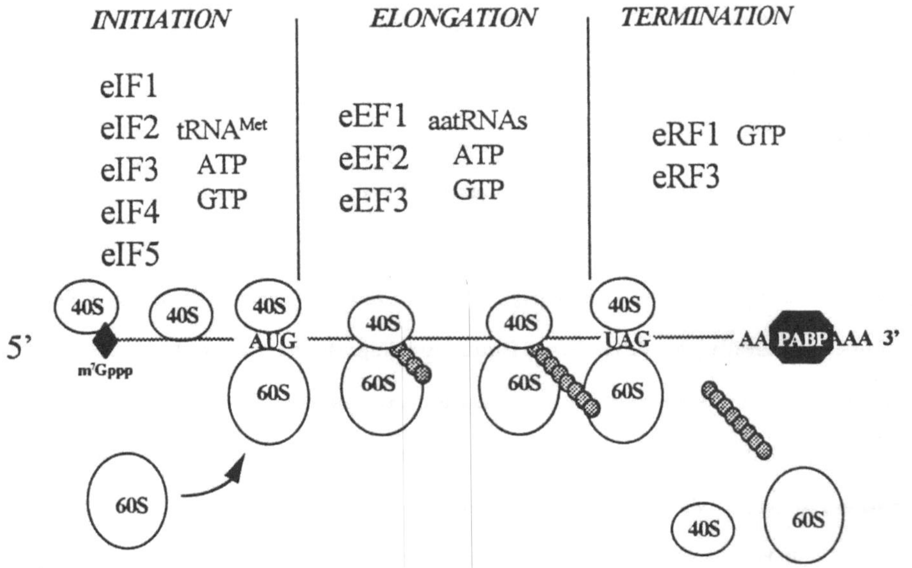

Figure 1. **The translation cycle in eukaryotic cells indicating the three steps in translation and the translation factors involved.**

involvement of two elongation factors (eEFs).

TERMINATION: The arrival of one or other of the termination codons UAG, UAA or UGA at the decoding site of the ribosome signals the termination of polypeptide chain elongation. In the absence of a cognate aminoacyl-tRNA for these three codons, a protein release factor (eRF) binds to the ribosome and promotes the hydrolysis of the polypeptide chain from the last tRNA, and subsequently from the ribosome.

Upon completion of the three phases of protein synthesis, the ribosomal subunits are released from the mRNA and can then either immediately re-enter the translation cycle or remain in storage in the cytoplasm until required.

The translation of an mRNA therefore occurs within the ribosome, a large ribonucleoprotein complex. Within the ribosome, three distinct sites have been biochemically defined:

- A (acceptor) - site, sometimes referred to as the decoding site and to which the incoming aminoacyl-tRNA is brought.
- P (peptidyl) - site contains the previous tRNA covalently attached to the growing polypeptide chain, i.e. peptidyl-tRNA
- E (exit) - site which carries the previous peptidyl-tRNA deacylated during the formation of the last peptide bond.

The existence of these three sites has been firmly established in the bacterial ribosome, (Nierhaus 1993), and evidence has also been provided for the existence of E-site in eukaryotic ribosomes (see below).

A full understanding of the mechanism and control of protein synthesis requires the identification of all the soluble translation factors - eIFs, eEFs and eRFs - which drive the process of protein synthesis, together with a complete description of the specific roles they play. In addition, we must also fully understand how the accuracy of protein synthesis is maintained; what mechanisms are there to ensure that amino acids are covalently bound to their correct, cognate tRNAs? What ensures accurate codon:anticodon interactions at the ribosomal A site? How is the correct reading frame maintained? Some of the answers to these questions have and will come from *in vitro* studies, yet a full understanding can only come from studies with living eukaryotic cells, i.e. *in vivo.*

Yeast as a Model Organism for Studying Protein Synthesis

Following on from the initial success of the *in vitro* biochemical studies with mammalian cells and tissues, recent studies have begun to switch away from complex eukaryotic cells and focus on the more tractable microbial eukaryotic species such as the yeast *Saccharomyces cerevisiae*. This switch has been primarily driven by the opportunity to carry out genetic studies in parallel with biochemical studies with this unicellular eukaryote. A wide range of genetic technologies applicable to yeast are available and include:

- the ability to establish genetic screens for particular classes of mutants with defects in protein synthesis components
- the ease of gene cloning and manipulation including the generation of null mutants for individual genes without altering any other component of the genome

- the potential to create conditional gene expression systems that allow for shutting off the expression of a target gene in normally growing cells
- the complete characterisation of the yeast genome at the nucleotide level and hence identification of all its constitutive genes

In this Chapter, I will demonstrate, through use of specific highly selective examples, how genetic studies on protein synthesis in the yeast *Saccharomyces cerevisiae* have provided important new information on eukaryotic protein synthesis and its control, particularly the identification of key components of both the mechanism and control of protein synthesis. I will also describe aspects of the protein synthetic machinery of yeast which appear to be unique to these fungal species and which therefore may provide potential targets for antifungal drug discovery programmes. For a detailed general overview of eukaryotic protein synthesis and its control, the reader is referred to a recent comprehensive treatise (Hershey *et al.*, 1996).

Studies on Translation Initiation

Overall, the mechanism by which the initiation codon (AUG) is located by the small 40S ribosomal subunit in yeast is very similar, if not identical, to that used by mammalian cells. This includes the repertoire of eIFs. A combination of genetic and subsequent genome sequence studies have identified, almost without exception, yeast homologues of all known mammalian eIFs (Table 1) including the complex multisubunit factors such as eIF2 (3 subunits) and eIF3 (8 subunits).

A number of highly productive genetic screens have been established in yeast as a means of identifying components of the translation initiation machinery. Two will be described here.

Screen 1. ***Selection of sui mutants defective in start codon recognition.*** The scanning 43S preinitiation complex usually initiates at the first AUG codon from the 5' end of an mRNA with the eIF2/GTP/$tRNA^{Met}$ ternary complex playing a central role in codon selection. A genetic screen has been developed to select for yeast mutants that allow the scanning 43S complex to initiate on a non-AUG codon (Figure 2). The screen utilised a strain carrying an *in vitro* engineered *HIS4* gene in which the initiation codon had been mutated to AUU. By selecting for His^+ revertants, three unlinked genes (*SUI1-3*) were identified which, when mutated, allowed translation to initiate from a UUG codon located two codons 3' of the mutated AUG codon (Castilho-Valavicius *et al.*, 1990). Replacing the mutated AUU codon with an UUG codon resulted in translation also being able to initiate at the 5'- most UUG codon.

Table 1. Translation intiation factor genes identifed in yeast by various genetic screens

eIF	Yeast Gene	Mass (kD)
eIFA	*TIF11*	17.4
eIF2α	*SUI2*	34.7
eIF2β	*SUI3*	31.6
eIF2γ	*GCD11*	57.9
eIF3		
p16	*SUI1*	12.3
p62	*GCD10*	54.3
p90	*PRT1*	88.1
(5 others)		
eIF4A	*TIF1, TIF2*	45
eIF4B	*TIF3*	48.5
eIF4E	*CDC33 (TIF45)*	24.3
eIF4G	*TIF4631*	107.1
	TIF4632	103.9
eIF5	*TIF5*	45.2
eIF5A	*TIF51*	17.1

Each of the *SUI* genes identified by this screen encode a component of the translation initiation machinery:

SUI1 : the p16 subunit of eIF3

SUI2 : the α-subunit of eIF2

SUI3 : the β-subunit of eIF2

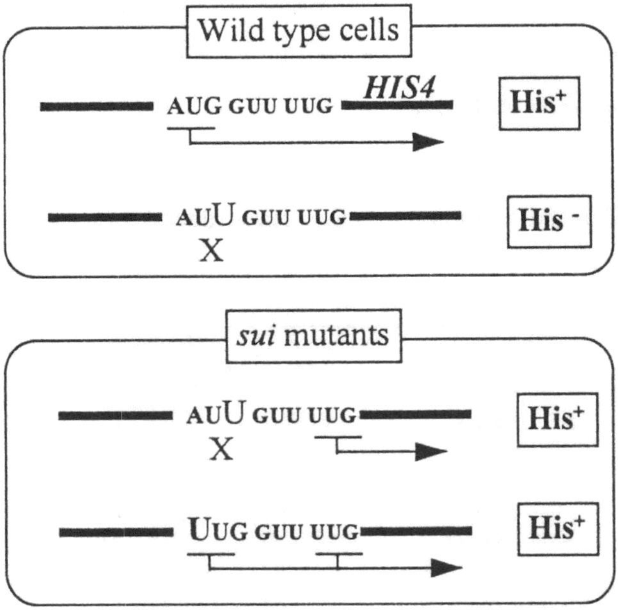

***Figure 2.* A genetic screen for mutants of yeast (*sui*) which allow ribosomes to initiate translation at a non-AUG codon.** It exploits the *HIS4* gene which, when translated either from the authentic AUG codon (in wild type cells) or from the indicated UUG codon in *sui* mutants gives rise to histidine prototrophy (His^+).

Furthermore, the results demonstrated, for the first time, the involvement of these three proteins in AUG codon selection.

Screen 2. Selection of gcd and gcn mutants defective in the post-transcriptional regulation of GCN4. This screen was initially developed to identify how the expression of the gene encoding the Gcn4p transcription factor (*GCN4*) was regulated. Gcn4p is required to activate the transcription of a large number of genes involved in amino acid biosynthesis. During periods of amino acid starvation, the *GCN4* mRNA is translated whereas, in amino acid rich conditions, although the *GCN4* mRNA is synthesised, it is not translated. Unusually, the 5' UTR of the *GCN4* mRNA is long (591 nt) and contains 4 short open reading frames (ORFs) which play a central role in the post-transcriptional regulation of this gene. Essentially, in amino acid-rich conditions, the ribosomes stall at ORF4 and thus fail to initiate on the AUG codon of the *GCN4* coding ORF. In amino acid-starved conditions, the ribosomes bypass ORF4 and are able to initiate on the *GCN4* coding ORF.

Key to *GCN4* regulation is the initiation factor eIF2. Following translation initiation, eIF2 bound GTP is hydrolysed to GDP but, for eIF2 to reenter the translation cycle, the GDP must be exchanged for GTP. This is achieved by a 5 subunit recycling factor eIF2B (Price and Proud, 1994). However, under certain conditions, the α subunit of eIF2 becomes phosphorylated which in turn leads to a much less efficient rate of recycling and hence the pool of eIF2-GTP in the cell declines and translation initiation is impaired. In the *GCN4* regulatory system, there is a kinase that is activated by amino acid starvation and which specifically phosphorylates the α subunit of yeast eIF2 (the *SUI2* gene product - see above). Thus, when eIF2-GTP levels are low (under amino acid starvation conditions) the scanning 40S subunit does not get fully 'loaded' with the eIF2-GTP-$tRNA^{Met}$ ternary complex until it reaches the *GCN4* ORF. If there is a high concentration of eIF2-GTP in the cell (i.e.in amino acid-rich conditions) then the 40S subunit is 'loaded' in time to initiate at ORF4. [For full details of this novel gene-specific post-transcriptional regulatory system, see the recent review by Hinnebusch, 1994.]

By isolating two different types of *GCN4* regulatory mutant (Figure 3) the genes encoding various translation eIF subunits have been identified . The two classes of mutants are:

- *gcd* mutants - which fail the repress translation of the *GCN4* mRNA in amino acid-rich conditions; and
- *gcn* mutants which fail to derepress translation of *GCN4* mRNA under amino acid starvation conditions.

In addition to identifying the γ subunit of eIF2 (*GCD11*) and the p62 subunit of eIF3 (*GCD10*), this screen has also identified the kinase which phosphorylate the eIF2α subunit (*GCN2*) and the 5 subunits of eIF2B, the eIF2 recycling factor (*GCN3*, *GCD7*, *GCD1*, *GCD2* and *GCD6*). Although not originally designed to identify translation initiation factors, it has without question been a very fruitful gene screen for genes encoding these yeast eIFs.

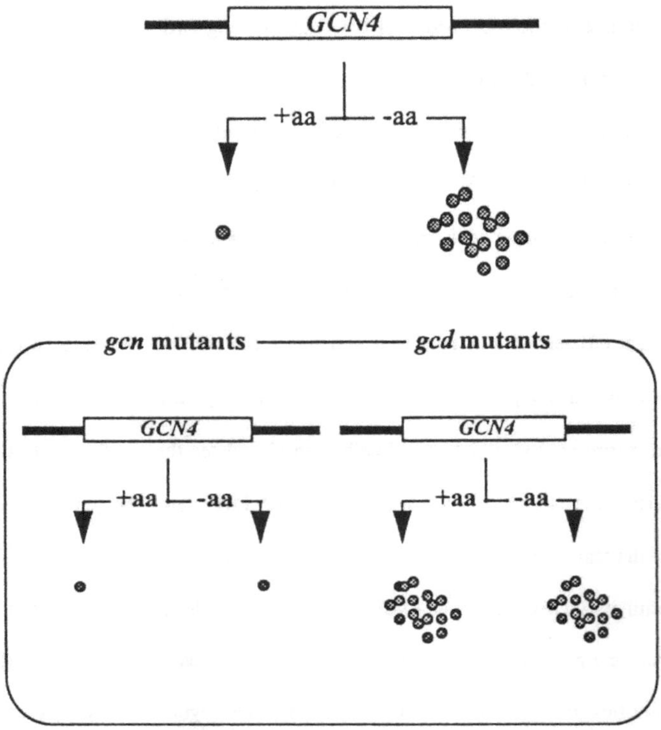

***Figure 3.* A genetic screen for mutants which alter the regulation of translation of the *GCN4* mRNA in response to amino acid starvation (-aa).** The two classes of regulatory mutant are: *gcn*, which fail to derepress translation in response to amino acid starvation; and *gcd*, which fail to repress translation in amino acid rich (+aa) conditions.

It is beyond the scope of this short review to fully discuss how the other initiation factors of yeast have been identified. In some cases, they have been uncovered by design, e.g. the *PRT1* gene (encoding the p90 subunit of eIF3) was identified by the isolation of a temperature-sensitive mutant (*prt1*) that showed a block in translation when shifted to the non-permissive temperature of 37° C. On the other hand, several others have been identified in screens designed to identify entirely different family of genes, e.g. the gene encoding eIF4E (*CDC33*) was identified in a screen for temperature-sensitive cell division cycle mutants.

Translation Elongation in Yeast: Not Exactly Typical Eukaryotic

Translation elongation in most eukaryotic cells requires two eEFs whose functions have been well defined through biochemical studies (Merick, 1992). These are:

- eEF1A (formerly called eEF-1α): promotes the GTP-dependent binding of aminoacyl-tRNAs to the ribosome and is homologous to the bacterial translation elongation factor EF-Tu. Following GTP hydrolysis, the eEF1A-GDP complex leaves the ribosome but, to enter another round of elongation, the bound GDP must be exchanged for GTP. This exchange reaction is catalysed by the three subunit eEF1B recycling factor (formerly called eEF1β/γ/δ).
- eEF2: promotes the GTP-dependent translocation step, placing the peptidyl-tRNA site at the ribosomal P- site, thereby placing the next codon to be translated in the ribosomal A-site.

In vitro translation elongation systems, capable of translating the synthetic template poly(U) into polyphenylalanine, can be readily established with mammalian ribosomes together with eEFlA and eEF2. Supraoptimal Mg^{2+} levels are necessary, however, to 'force' ribosomes to initiate on the AUG-less template. Substitution of the mammalian ribosomes with yeast ribosomes in such an elongation assay results in a complete failure to translate poly(U). This finding provided the first evidence for a third yeast elongation factor which we now call eEF3 and which appears to be a soluble translation factor unique to yeasts and other fungi (Belfield and Tuite, 1993).

The major biochemical properties of eEF3 are:

- it is absolutely required for polyphenylalanine synthesis *in vitro* on yeast ribosomes
- it is a 125kDa soluble protein that cycles on and off ribosomes during the elongation cycle
- it exhibits both a ribosome-dependent ATPase and, to a lesser extent, GTPase activity

The proposed role of eEF3 in the elongation cycle is shown in Figure 4. *In vitro* studies carried out in the laboratory of Neirhaus (Triana-Alanso *et al.*, 1995), using purified ribosomes and other eEFs, suggested that eEF3 promotes the ejection of the 'spent' deacylated tRNA from the E-site. This in turn enhances the binding of the incoming aminoacyl-tRNA: eEF1A:GTP complex to the codon at the ribosomal A-site. This model further implies a degree of cross-talk must take place between the E- and A-sites on the ribosome during translation elongation, possibly via conformational changes within the ribosome mediated through the ribosomal RNA. The eEF3 ejection of the deacylated tRNA at the P-site must be essential for completion of the elongation cycle.

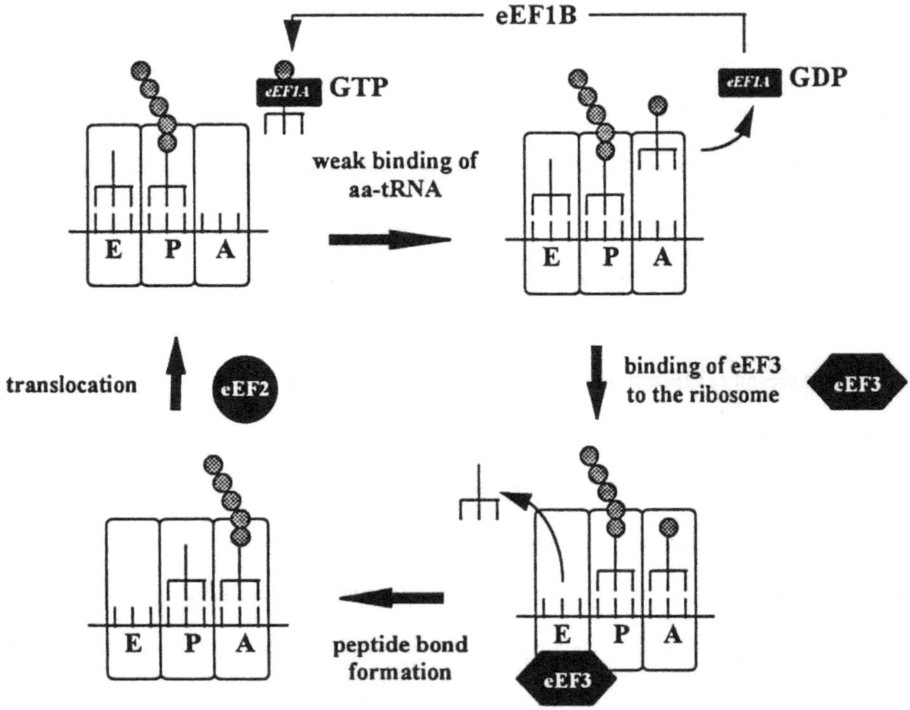

Figure 4. **The translation elongation cycle in yeast depicting the role of the three key elongation factors: eEF1A, eEF2 and eEF3.** The A (acceptor), P (peptidyl) and E (exit) sites of the ribosome are indicated.

In vivo studies have confirmed that eEF3 is indeed an essential protein synthesis factor because:

(a) deletion of the single eEF3-encoding gene (*YEF3*) is haplolethal, and

(b) a temperature-sensitive *yef3* allele results in a cessation of protein synthesis at the non-permissive temperature (37°C) due to a block in translation elongation.

eEF3 has also attracted considerable attention as a potential target for new antifungal drugs since a functionally analogous soluble homologue appears to be lacking from higher eukaryotic cells (Tuite *et al.*, 1995). By either blocking the ribosome - dependent nucleotidase activity of eEF3, or by preventing eEF3 from binding to the ribosome, one could potentially block E-site ejection, thereby leading to a cessation of protein synthesis. The key issue, however, remains the identification of the functional homologue of eEF3 in mammalian cells assuming mammalian ribosomes also have an E-site. We have proposed elsewhere (Belfield *et al.*, 1995) that the nucleotidase activity of eEF3 may be associated with a mammalian ribosomal protein and that eEF3 is simply an 'evolving ribosomal protein'. The recent completion of the cataloguing of all yeast and mammalian ribosomal proteins has confirmed that mammalian ribosomes have one more ribosomal protein (79 vs 78) than *S. cerevisiae* ribosomes (R J Planta, per.comm.). It remains to be demonstrated that this extra mammalian ribosomal protein has the predicted intrinsic nucleolidase activity. Recent studies have however confirmed that washed mammalian ribosomes do have an intrinsic nucleotidase activity that is absent from washed yeast ribosomes (Rodnina *et al.*, 1994).

Table 2. Yeast translation elongation factors and their genes.

Elongation factor	Yeast Gene	Mass (kD)	Bacterial Homologue
eEF1A	*TEF1, TEF2*	50	EF-Tu
eEF1B			EF-Ts
eEF1α	??		
eEF1β	*TEF5*	23	
eEF1γ	*TEF3,TEF4*	47	
eEF2	*EFT1, EFT2*	93	EF-G
eEF3	*YEF3*	125	??

Table 2 summarises the known yeast elongation factors and their genes. The yeast genome sequence has also identified an open reading frame (designated YNL163C) which shows significant homology to eEF2 and may therefore represent a fourth elongation factor (eEF4?).

In mammalian cells, there is evidence that global translation elongation may be regulated via phosphorylation of eEF2. *In vitro*, eEF2 can be phosphorylated on three different threonine resides to varying extents, by a Ca^{2+}-regulated kinase which in turn leads to a block in translation elongation. The physiological significance of this mode of translational regulation in living cells remains to be established. There is, however, no evidence that yeast eEF2 is phosphorylated *in vivo* although, like its mammalian counterpart, it does have a unique post-translational modification, namely a histidine residue that is modified to dipthamide.

Termination: The End of the Cycle

While the structure and function of bacterial RFs have been established for over a decade, the identity of the corresponding eRFs remained elusive until 1995. Genetic studies in yeast had identified a number of genes that, when mutated, gave rise to a defect in translation termination (Stansfield and Tuite, 1994). The genetic screen used exploited a weak nonsense (ochre) suppressor tRNA encoded by the *SUQ5* gene. The *SUQ5*-encoded $tRNA^{Ser}$, although having an anticodon mutation that allows it to translate the UAA codon by a cognate interaction, does so very inefficiently because it is unable to effectively compete with the endogenous eRF-driven termination mechanism. Therefore, when introduced into a strain carrying a defined ochre mutation in a readily scorable gene (e.g. *ade2*-1) no suppression of the mutation can be detected. By screening, post-mutagenesis, for mutations that allowed the *SUQ5* suppressor to suppress the *ade2-1* mutation, five so-called allosuppressor genes (*SAL1-SAL5*) were identified (Cox 1977). A sixth gene, *SAL6*, was identified via a different screen.

Two of the *SAL* genes, so-identified, *SAL3* and *SAL4*, are allelic to the *SUP35* and *SUP45* genes respectively. The latter genes were originally identified by mutations that led to the suppression of a range of nonsense mutations (i.e. as codon non-specific suppressors) in the absence of a defined suppressor tRNA, so-called omnipotent suppressors (Stansfield and Tuite, 1994). The products of the *SUP35* (*SAL3*) and *SUP45* (*SAL4*) genes are the two essential subunits of the yeast eRF; *SUP35* encodes eRF3 and *SUP45* encodes eRF1 (Table 3).

The eRFl and eRF3 proteins physically interact to mediate translation termination at all three stop codons *in vivo*. While the precise roles of the two subunits in translation termination remain to be fully defined biochemically, the following is known:

Table 3. Genes identified by a nonsense suppression-based genetic screen in yeast

Gene	Gene Product	Mass (kD)	Mammalian Homologue?
SAL1 (*UPF2*)	nuclear protein (?)	126	??
SAL2 (*UPF1*)	RNA/DNA helicase	109	Yes
SAL3 (*SUP35*)	eRF3, release factor	79	Yes, eRF3
SAL4 (*SUP45*)	eRF1, release factor	49	Yes, eRF1
SAL5	[not yet cloned]		
SAL6 (*PPQ1*)	ser/thr phosphatase	61	??

- **eRF1**: a 49kDa protein that is tightly associated with the yeast ribosome at a stoichiometry of approximately 1 mol eRF1 to 20 mol ribosomes. *In vivo* depletion of eRF1 leads to a less efficient termination process leading in turn to enhanced nonsense suppression by *SUQ5* (Stansfield *et al.*, 1996). Gene deletion studies confirm that this protein is essential for viability. eRF1 belongs to a family of proteins whose structures mimic that of a tRNA; for example EFG. This would suggest that eRFl recognises the stop codon directly via a novel RNA:protein interaction (Itoh *et al.*, 1996).
- **eRF3** is a 79 kDa that is also tightly associated with the yeast ribosome with a similar stoichiometry to eRF1. It is also essential for viability but, in contrast to eRF1, shows significant amino acid identity to a known translation factor, eEFlA (see above).

Confirmation that the products of the *SUP35* and *SUP45* genes (eRF3 and eRFl respectively) define the essential subunits of the yeast eRF have come from our various studies (Stansfield *et al.*, 1995):

1. A *sal3*, *sal4* (*sup45*, *sup35*) double mutant is inviable (Cox, 1977).

2. Only overexpression of both the *SUP35* and the *SUP45* genes give rise to antisuppression, i.e. inhibition of the activity of an otherwise efficient nonsense suppressor rRNA to translate a known stop codon (Stansfield *et al.*, 1995). Expression of either *SUP35* or *SUP45* alone does not lead to antisuppression.

3. The Sup35p and Sup45p proteins directly interact as demonstrated *in vivo* - using the two hybrid system - and *in vitro* (Stansfield *et al.*, 1995).

Translation termination can therefore be viewed as a defective translation elongation step. Instead of an aminoacyl-tRNA being brought to the ribosomal A-site by eEFlA, eRFl is brought to the ribosomal A-site by eRF3 in a protein complex that 'mimics' an aminoacyl-tRNA - eEFlA complex. With no amino acid bound to eRFl, this leads to an aborted elongation step. This must be followed by peptidyl-tRNA hydrolysis, polypeptide chain release and finally release of the ribosome from the mRNA. Little is known about these latter steps although, by analogy with *E. coli*, one would expect there to be a RRF (ribosome recycling factor) that facilitates ribosome release. Given that eRF3 has four potential GTP binding sites, it is likely that it is eRF3 that provides the GTP hydrolysing activity known from biochemical studies to be required for translation termination.

One of the most remarkable features of translation termination in yeast is that eRF3 is a prion protein. In certain strains of yeast, designated [*PSI*$^+$], the efficiency of nonsense suppression by the *SUQ5*-encoded suppressor tRNA^{Ser} is enhanced but the responsible 'mutation' does not segregate according to Mendelian rules. In other words, a [*PSI*$^+$] (mutant) by [*psi*$^-$] genetic cross gives a [*PSI*$^+$] diploid, and when sporulated, this diploid gives rise to tetrads consisting only of 4 [*PSI*$^+$] spores. If a nuclear gene mutation were responsible, then a 2 mutant : 2 wild-type segregation pattern would be expected. Since the [*PSI*$^+$] determinant does not map to any known cytoplasmically-located nucleic acid genome, another explanation must be found to describe the unusual genetic behaviour (Cox *et al.*, 1988).

Based on both genetic and biochemical data (reviewed in Tuite and Lindquist, 1996) it has been proposed that the eRF3 (Sup35p) protein can behave in the same way the PrP prion protein does during the development of the group of mammalian neurodegenerative diseases known as the transmissible spongiform encephalopathies (Wickner, 1994). This model fully explains all of the currently available data. One important issue that remains, however, is whether the [PSI^+] state is simply a mutant 'disease' state, or actually represents a novel post-translational mechanism for generating additional genetic diversity. The efficiency of translation termination in [PSI^+] cells is significantly lower than in [psi^-] cells and one might therefore get C-terminal extensions to certain yeast proteins if translation bypassed the natural termination codon. While this possible role for the [*PSI*] prion remains to be proven, the genetic screens available to probe such a phenomenon in yeast will allow us to shed new light on one of the most unusual of biological phenomenon, namely the inheritance and replication of a protein without an underlying nucleic acid genome.

Conclusion

The object of this article is to illustrate the potential that genetic analysis in *S. cerevisiae* provides us with to fully elucidate the role of the large number of translation factors in this fundamental step in gene expression. Only following the full cataloguing of these factors will we begin to understand the interplay and cross-talk that occurs between the physiological state of a cell and its ability to translate specific and/or the total mRNA population in a cell and how this interplay is regulated. With the completion of the *S. cerevisiae* genome sequence, we now know the amino acid sequences of all of the translation factors required, but it will still take a significant amount of effort to both identify these genes and to elucidate their roles in translation. Genetic screens such as those described in this article will be invaluable in such analyses in the post-genome era.

Acknowledgements

Work on yeast protein synthesis factors in the author's laboratory has been funded by the BBSRC, the Wellcome Trust and Glaxo-Wellcome and carried out by a large number of excellent postdoctoral and postgraduate researchers.

References

Belfield GP, Tuite MF (1993) Translation elongation factor 3 : a fungus-specific translation factor? *Mol. Microbiol* **9**, 411-418.

Belfield GP, Ross-Smith NJ, Tuite MF (1995) Translation elongation factor 3 (EF-3) : an evolving eukaryotic ribosomal protein? *J.Mo.l.Evol.* **41**, 376-387.

Castilho-Valavicius B, Yoon H, Donahue TF (1990) Genetic characterisation of the *Saccharomyces cerevisiae* translational initiation suppressors *sui1*, *sui2* and *SUI3* and their effects on *HIS4* expression. *Genetics* **124**, 483-495.

Cox BS (1977) Allosuppressors in yeast. *Genetical Res.* 30, 187-205.

Cox BS, Tuite, MF, McLaughlin CS (1988) The [*PSI*] factor of yeast: a problem of inheritance. *Yeast* 4, 159-178.

Hershey JWB, Matthews MB and Sonenberg N (1996) eds *'Translational Control'*, Cold Spring Harbor Laboratory Press, NY.

Hinnebusch A (1994) Translational control of *GCN4*: an *in vivo* barometer of initiation factor activity. *Trends Biochem. Sci.* **19**, 409-414.

Itoh K, Ebihara K, Uno M, Nakamura Y (1996) Conserved motifs in prokaryotic and eukaryotic polypeptide release factors: tRNA-protein mimicry hypothesis. *Proc. Natl. Acad. Sci. USA* 93, 5443-5448.

Merick W (1992) Mechanism and regulation of eukaryotic protein synthesis. *Microbiol. Rev.* **56**, 291-315.

Nierhaus KH (1993) Solution of the ribosome riddle : how the ribosome selects the correct aminoacyl-tRNA out of 41 similar contestants. *Mol. Microbiol.* **9**, 661-669.

Price N, Proud CG (1994) The guanine nucleotide exchange factor elF-2B. *Biochimie* 76, 748-760.

Rodnina MV, Serebryanik AI, Ovcharenko GV, El'Skaya AV (1994) ATPase strongly bound to higher eukaryotic ribosomes. *Eur. J. Biochem.* **225**, 305-310.

Stansfield I, Tuite MF (1994) Polypeptide chain termination in *Saccharomyces cerevisiae*. *Curr Genetics* **25**, 385-395.

Stansfield I, Jones KM, Tuite MF (1995) The end in sight: terminating translation in eukaryotes. *Trends Biochem. Sci.* 20, 489–491.

S tansfield I, Eurwilaichitr L, Akhmaloka, Tuite MF (1996) Depletion in the levels of the release factor eRF1 causes a reduction in the efficiency of translation termination in yeast. *Mol. Microbiol.* 20, 1135-1144.

Triana-Alonso FJ, Chakraburtty K, Nierhaus KH (1995) The elongation factor 3, unique in higher fungi and essential for protein synthesis, is an E-site factor. *J. Biol. Chem.* 270, 20473-20478.

Tuite MF and Lindquist SL (1996) The maintenance and inheritance of yeast prions. *Trends Genet.* 12, 467–471.

Tuite MF, Belfield GP, Colthurst DR and Ross-Smith NJ (1995) Defining a new molecular target in fungal protein synthesis: eukaryotic elongation factor 3 (eEF-3). In '*Antifungal Agents: Discovery and Mode of Action*' (LG Copping et aL, eds.), Bios Scientific Publishers, Oxford, pp119-129.

Wickner RB (1994) [*URE3*], as an altered Ure2p protein: evidence for a prion analogue in *Saccharomyces cerevisiae*. *Science* 264, 566-569.

Part 4

Microbial Cell Biology

The Roles of Molecular Chaperones in the Bacterial Cell

Peter A. Lund[1]

[1] School of Biological Sciences, University of Birmingham, Birmingham B15 2TT, UK

1 Introduction

In this chapter, I shall look at the major bacterial molecular chaperones, and summarise the evidence for what we know about their cellular role. Molecular chaperones have risen from obscurity to near superstar status in the last few years, as we have learned that many fundamental aspects of protein function in all cells require the help of other proteins in order to take place. These include the correct folding of proteins as they are translated, their delivery to and passage across membranes, and the repair of proteins which have become inactive because of exposure of the cells to stresses such as an increase in temperature. "Molecular chaperone" is the generic name given to a protein that assists in these processes. The review will be from the point of what is known about what these proteins do in the cell, although I will also describe what is currently understood about the mechanisms by which some of the chaperones work, and how their diverse functions can sometimes be understood in terms of a single mechanism. Finally, I will try to summarise some of the areas where I think key questions about the cellular role of molecular chaperones remain unanswered.

Molecular chaperones could be defined as proteins that help other proteins to fold into their correct native conformation, although this is rather too strict a definition that excludes proteins such as SecB (see below) involved in secretion rather than protein folding per se, but generally agreed to be chaperones. Perhaps a better definition is by mechanism rather than role: molecular chaperones share the common property of being able to bind unfolded or partially folded protein in such a way as to prevent incorrect interactions taking place that prevent that protein from eventually carrying out its cellular function. Different research groups use this term in rather different ways, so for the purposes of this chapter I will restrict myself as much as possible to describing only those molecular chaperones that interact non-covalently with their substrates. and that act with a range of substrates rather than one specific protein. This excludes enzymes which have a direct effect on bonds in proteins (such as catalysts of disulphide bond formation and rearrangement); however, as will be seen this is not a complete exclusion as some evidence exists that some chaperones may act both covalently and non-covalently.

NATO ASI Series, Vol. H 103
Molecular Microbiology
Edited by Stephen J. W. Busby,
Christopher M. Thomas and Nigel L. Brown

Molecular chaperones were discovered relatively recently. This was because most experiments on protein folding had been done on purified proteins under conditions where molecular chaperones were not necessary. It is thus ironic that many of the detailed studies on molecular chaperones which are now being carried out are done in biochemistry laboratories, using purified proteins. Although these studies can tell us a great deal about the way in which molecular chaperones work, and although this in turn can suggest how they may act inside the cell, such experiments must be interpreted carefully if we are to understand the precise cellular role of molecular chaperones. For this it is necessary to study the genetics of the molecular chaperones, which mainly means looking at the effects of mutation or deletion of the genes altogether. As will become clear, this is not always easy, partly because many molecular chaperones are central to cell growth, and partly because much overlap of function exists between different families of molecular chaperones. It is thus also important to study non-mutated molecular chaperones in a cellular context, for example by looking at proteins which are thought to interact directly with them. In addition, we can make many inferences by looking at the conditions under which molecular chaperones are expressed.

We can best understand why molecular chaperones are needed by the cell by imagining ourselves being able to observe the whole process of protein folding as it actually takes place. Within a rapidly dividing *E. coli* cell, there are approximately 60,000 proteins synthesised per minute. A protein as it begins to be made is a simple polymer of different amino-acid residues with no defined structure, but in order to take on its final role in the cell it must fold into a unique conformation and (in some cases) traverse one or more lipid membranes. The forces that drive proteins to fold into their final shapes are many, but initially the "hydrophobic effect" is the most important, which refers simply to the fact that amino-acid residues with hydrophobic side chains prefer to be in environments where they are not exposed to water. The optimum environment for these is on the inside of a protein molecule, and the first step in protein folding is sometimes referred to as a hydrophobic collapse. In this, the protein forms a structure where the amino-acid side chains on the outside are charged or hydrophilic, with most of the hydrophobic side chains buried on the inside. However, there are several problems for a protein which is folding up inside the cell. One is that at the high concentration of folding proteins present in a rapidly growing cell, hydrophobic side chains from different protein molecules may start to interact with each other, leading to aggregates of protein with no activity, rather than to individual folded protein molecules. Another is that folding of protein, if too rapid, will inhibit the ability of that protein to cross membranes, yet many proteins made in bacteria have destinations other than the cytoplasm. Finally, most proteins in their active form are not particularly rigid molecules, their flexibility being needed for their function. However, the price paid for this flexibility is often low thermal stability, and thus bacteria need ways of keeping proteins active and folded as temperatures start to rise. As we start to look at the different families of molecular chaperones, it will become clear how they address some of these problems, and how some of the same chaperones can also help the cell remove proteins which have become damaged and cannot be restored to function.

The precise definition of a protein as a molecular chaperone is not easy. Ideally there will be data from both *in vitro* and *in vivo* experiments to show that a given protein which is a candidate for this designation interacts non-covalently with a range of different substrates in unfolded or partially folded form, and through this interaction helps them to reach their final destination in an active state. The picture is complicated, however, by the fact that many proteins that appear to act as chaperones can also have a role in cellular proteolytic pathways, and that some chaperones can also influence protein folding through covalent interaction with their substrates. Moreover, some proteins are "candidate chaperones" but their case is not yet proven: the designation may rest mainly on sequence homologies with known chaperones, or on *in vitro* data only. Indeed, despite the explosion of high quality research in this field in recent years, most attention has focused on a small number of chaperone families. The two most studied are the GroE proteins (GroES and GroEL) and the DnaK/DnaJ/GrpE family. I will thus consider these two groups first, and then look at some of the other chaperones whose functions are beginning to be studied in bacteria. It seems likely that more chaperones remain to be discovered, and it is beyond question that much research remains to be done before these fascinating and important groups of proteins can be said to be well understood.

2 The GroE proteins

There are two GroE proteins, GroES and GroEL, and both are essential for the growth of *E. coli* at all temperatures. They were originally discovered by the isolation of mutants in *E. coli* that prevented the growth of bacteriophage lambda, and it was noticed that many of these mutants also made cells temperature sensitive (Zeilstra-Ryalls et al., 1991). Other observations further linked the GroE proteins with the effects of temperature on growth. First, although they are present in the cell under all conditions of growth, their expression is strongly induced by heat shock. Second, mutants of *E. coli* that cannot undergo the heat-shock response because of deletion of the *rpoH* gene are extremely temperature-sensitive, growing only below 20°C. However, revertants of these strains can be selected by simply looking for growth at higher temperatures, and many of these revertants prove to be mutants that over-express the GroE proteins, with the degree of over-expression correlating neatly with the maximum growth temperature (Kusukawa and Yura, 1988). Third, over-expression of the GroE proteins can suppress many different temperature sensitive (ts) mutations in a whole variety of different genes (Van Dyk et al., 1989). This last observation also points to a major role of the GroE proteins as being in some aspect of protein folding, as ts mutations are often caused by steps in the pathway to the folded protein being temperature sensitive, rather than the final folded protein itself.

Definitive proof of this has come from a variety of sources. One is to show that over-expression of GroES and GroEL aids in the folding of proteins which normally do not fold well in *E. coli*. This was first demonstrated for bacterial Rubisco (Goloubinoff et al., 1989), and has been shown for several other examples since. Both GroES and GroEL are required for this, and the same effect has been demonstrated *in vitro* with many different protein substrates which are first denatured and then allowed to refold either spontaneously or in the presence of GroEL, GroES, and MgATP. In many cases, much higher yields of folded protein can be obtained when the GroE proteins are present. Inspection of strains carrying mutations in the *rpoH* gene shows that they are generally poor at folding protein, and that many *E. coli* proteins become insoluble when RpoH is not fully functional. All the components of the heat shock regulon are of course expressed at low levels in such strains, but the protein folding defect can be largely suppressed by over-expression of the GroE proteins alone (Gragerov et al., 1993). The converse experiment is to inactivate the GroE proteins at a normal growth temperature by selection of highly temperature sensitive GroE mutants. This has been done for the GroEL protein, and these experiments show that when GroEL is inactivated, a number of *E. coli* proteins and co-expressed heterologous proteins now fail to reach their active state (Horwich et al., 1993). Depletion of wild-type GroEL from cells by placing the *groEL* gene under the control of a tightly regulated promoter also leads to the formation of inclusion bodies of insoluble protein (Ivic et al., 1997). Studies have also been done on the original ts mutants which were isolated in the lambda screen, and while these tend to support the notion that the *groE* genes have central roles in cellular growth, these studies can be difficult to interpret, as the *groE* mutants are very pleiotropic, and it is difficult to distinguish direct from indirect effects. One result that does seem clear is that both the GroE proteins have a role in the export of β-lactamase from the cell, as pulse-chase experiments showed that its export was much reduced at 42°C in strains carrying ts mutations in *groES* or *groEL*; this was not true for a number of other secretory proteins tested in the same experiment (Kusukawa et al., 1989).

GroEL and GroES levels are much higher in the cell after heat shock, for example to 43 °C, amounting to up to 12% of total cellular protein, and it is generally assumed that part of their role in cells growing at higher temperatures is the protection of proteins which are less stable at these higher temperatures, or the refolding of proteins which have been partially denatured by heat shock once the temperature of the cells returns to normal. (In this context, it is significant that the whole heat shock regulon can be induced by the presence of unfolded proteins within the cell, even at normal temperatures of growth). Direct *in vivo* evidence for this with *E. coli* is surprisingly scant, although it has been shown that strains which carry a mutated *groEL* or *groES* gene cannot reactivate firefly luciferase which has lost its activity after heat shock (Schroder et al., 1993). However, Hsp60 (a close GroEL homologue found in mitochondria) has been shown to be capable of protecting dihydrofolate reductase imported into mitochondria at high temperatures (Martin et al., 1993). The GroE proteins may act as part of a quality control system in heat shocked cells, assisting proteins which have become partially denatured to refold, but also possibly directing some proteins to the pathways of protein degradation. There is good evidence from looking at model proteins expressed *in vivo* that the GroE proteins have a role in proteolysis as well as in protein refolding, but it is unknown if, or how, they are able to distinguish refoldable proteins from those which are irreparably damaged. Surprisingly, it is still not known precisely which proteins in *E. coli* are the actual substrates on which GroEL acts, or how this spectrum may be changed at different temperatures or under different growth conditions. Certainly not all the proteins present in the cell require GroEL for their folding, and one calculation suggests that only about 5% of all cellular proteins do require GroEL in order to fold (Lorimer, 1996) although other estimates put the figure at nearer 30%. A number of proteins were identified as being misfolded in the study using a highly ts mutant of GroEL, (Horwich et al., 1993); an alternative study using a depletion approach identified a different set (Kanemori et al., 1993. This difference may reflect the different experimental approaches used. It is also surprising that at normal growth temperatures, the levels of GroEL protein can be reduced quite substantially without any severe effects on strain growth (Kanemori et al., 1993, Ivic et al., 1997).

The ability of GroES and GroEL to apparently fold proteins from the unfolded to the native state has been the subject of a tremendous amount of interest, and the reaction has been intently studied from a biochemical and biophysical viewpoint. The fascination with the GroE proteins stems in part from their remarkable structure. GroEL is a homotetradecamer, with two rings of seven subunits each stacked back to back in what is sometimes referred to as a double doughnut structure (Braig et al., 1994). An obvious feature of the structure is that each ring has a large central cavity, which is thought to be the site where proteins may bind and fold. GroES is a single ring of seven subunits, and can be shown to interact with GroEL during the folding reaction. The development of models to explain how the GroE proteins work has been an exciting and sometimes stormy journey, and is not yet over. A review of all the biochemical and structural information about these proteins is beyond the scope of this chapter, but one currently favoured model (Hartl, 1996) for how they may function is as follows (Figure 1).

First, the protein substrate (shown as a filled circle) is thought to bind to the end of one of the two rings (shown as the back-to back horseshoe-shaped structures). The substrate binds by virtue of hydrophobic interactions between amino-acid side chains on the substrate and in the GroEL protein, and it follows that the substrate itself must be partially or fully unfolded. Both GroES (shown as a triangle) and ATP subsequently bind to the same ring of the complex (step 1), and the ATP is hydrolysed to form a GroEL/ADP/substrate/GroES complex (step 2). Mutagenic studies have shown that some of the sites occupied by GroES are the same as those occupied by the protein substrate, so the net effect of GroES binding is to displace the bound substrate into the cavity, and to simultaneously cap the cavity which prevents the substrate protein from diffusing away from the GroEL. This means that GroES is not strictly a chaperone as it has no protein folding ability on its own, and it is more correct to refer to it as a co-chaperone. While in the cavity, the substrate protein is free to fold, and as there is insufficient space for other proteins to be present, this folding can proceed without fear of any unwanted aggregation reactions occurring. ATP is bound and hydrolysed by the opposite ring (step 3), rather slowly, and this hydrolysis in turn displaces the bound ADP and GroES from the top ring. GroES will come away from the GroEL slowly - a consequence of the weak ATPase activity of GroEL under these conditions - which gives plenty of time for folding to occur and the partially or fully folded protein to be released (the open circle in step 4). However, if the substrate has not buried its hydrophobic residues in the time available, it can rebind to the GroEL ring and the whole process can start again. This model is often referred to as the Anfinsen cage model (Ellis, 1994), since the GroEL-GroES complex effectively forms a cage wherein proteins may fold in the thermodynamically spontaneous fashion first described by Christian Anfinsen, as there are no other proteins nearby with which the folding protein might interact to form aggregates. The protein may eventually be released in a folded and active form, or at least in a form which now has the bulk of its hydrophobic residues buried and thus has no propensity to aggregate. It should be noted that this model, while attractive, is by no means universally accepted.

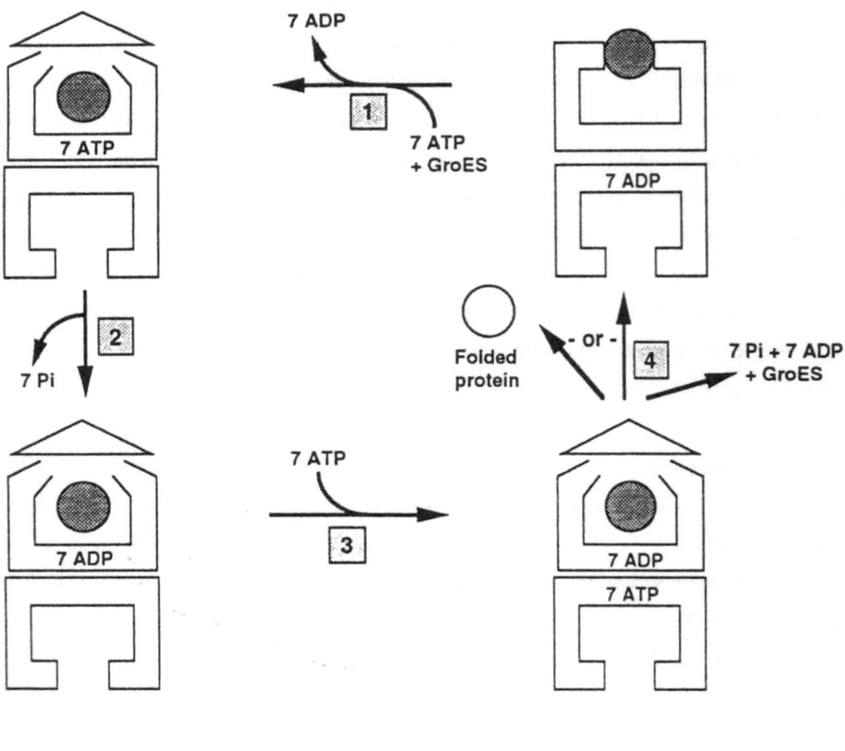

Figure 1. Proposed reaction cycle for GroEL and GroES

To summarise: GroEL and GroES have multiple roles in the bacterial cell and are essential to its growth. They appear to act by helping an as yet incompletely defined subset of proteins to fold to their correct final conformation, probably by blocking off-pathway steps that would normally lead to aggregation. They are also of crucial importance in helping bacteria to survive heat-shock, perhaps by preventing or reversing heat-shock induced aggregation. Although the GroE proteins have been to date the most intensively studied of the molecular chaperones, there is still a great deal to be learned about them. Not only do the details of the remarkable mechanism by which they help proteins to fold remain to fleshed out, but many aspects of their biology are still obscure. For example, there is some evidence that the specificity of GroEL may vary depending on other factors such as the degree of phosphorylation of the protein, or the presence of other protein cofactors (see below under trigger factor). Moreover, although most work has focused on their role as chaperones of protein folding, and to a lesser extent on their role in protein degradation, other evidence suggests that they also have an important role in mRNA stability (perhaps as an "RNA chaperone") (Georgellis et al., 1995). In bacterial species other than *E. coli* there are further enigmas to be tackled. Why do several bacterial species possess more than one copy of the *groE* genes, for example (the current record holder is *Bradyrhizobium japonicum*, which has five (Fischer et al., 1993), and what significance attaches to the persistent reports that *groE*L can be found outside the cell in some species (e.g., Ensgraber and Loos, 1993)?

3: The DnaK/DnaJ/GrpE family

As with the GroE proteins, the above proteins were originally identified by looking for mutants of *E. coli* that failed to plate bacteriophage lambda, and, like the GroE proteins, the DnaK/DnaJ/GrpE family are all stress-induced. In *E. coli*, *dnaK* and *dnaJ* are cotranscribed, while *grpE* is unlinked, but in many organisms all three genes are found in the same operon, sometimes with other genes largely of unknown function. Their association in the cell can be demonstrated to some extent by the isolation of suppressor mutants (thus for example suppressors of a *grpE* ts mutation have been shown to map in the *dnaK* gene), and also by co-imunoprecipitation. As with the *groE* genes, the over-expression of DnaK and DnaJ can suppress the general protein aggregation seen in some *rpoH* mutants (Gragerov et al., 1992). More effective suppression of this protein aggregation phenotype is seen when both DnaK/DnaJ and GroES/GroEL are over-expressed simultaneously, suggesting that the two families may differ in substrates or modes of action along the protein folding pathway. At least one specific cases have been described of a protein which is wholly or partially aggregated in *dnaK*, *dnaJ* or *grpE* mutant strains but not in *groE* mutants, also supporting the idea that these two classes of molecular chaperone are distinct in some ways (Thomas and Baneyx, 1996). Over-expression of DnaK alone or in concert with DnaJ and GrpE has also in some cases been shown to enhance the folding of some heterologous proteins. In the experiments described in the GroE section where suppressers of *rpoH* deletions that allowed cells to grow at normal temperatures were isolated. up-promoter mutants in the *dnaK/dnaJ* operon were found in a second round of mutagenesis following the isolation of mutants in the *groE* promoter: these allowed the cells to grow at up to 42 °C. All these lines of evidence point to a role of this trio of proteins in aspects of protein folding in the cell. It is rather surprising then to find that although GrpE is essential to the cell, strains of *E. coli* can be constructed expressing no DnaK or DnaJ protein. DnaK null mutants grow poorly and only at a restricted range of temperatures; they also show defects in cell division, DNA replication and chromosome segregation (Bukau and Walker, 1989). DnaJ null mutants also grow relatively poorly although the defects in growth are not as marked as in *dnaK* mutants. Both null mutants rapidly accumulate secondary suppressing mutations which give normal rates of growth - this makes their detailed study rather tricky. However, a range of ts mutants is also available in all three genes, and numerous studies have looked at the effects on various cellular processes of these mutations.

The results of such studies show that the DnaK/DnaJ/GrpE trio have fingers in many pies, and it is at first not easy to understand their broad roles in terms of a single, unifying mechanism. The evidence suggesting that they have an important role in protein folding has already been referred to. Moreover, they have been shown with one heterologous *in vivo* substrate (firefly luciferase) to be capable of reactivation of heat inactivated protein, a reaction that can be reproduced *in vitro* but which requires the presence of the chaperones before inactivation takes place: it is a protection, rather than a dissolution of aggregated material, that occurs (Schroder, 1993). Not only are the genes members of the heat shock regulon, but they are (at least in *E. coli*) also closely implicated in its regulation. Briefly, the heat shock sigma factor $\sigma 32$ is an unstable protein and the instability is much reduced in strains mutated in the *dnaK*, *dnaJ* or *grpE* genes (Straus et al., 1990). The trio may thus have a role in presenting $\sigma 32$ to proteases in the cell, perhaps by stabilising it in a protease-sensitive conformation. This is only one aspect of a broader role of the trio in proteolysis: as with the GroE proteins, proteolysis of several abnormal proteins has been linked with either DnaK or DnaJ, although the effects seen vary depending on the nature of the substrate protein and the particular allele of the chaperone gene studied.

DNA replication is also clearly affected in some ways by the action of this trio of proteins. Indeed, the block on lambda growth in strains mutated in *dnaK*, *dnaJ* or *grpE* is due to a block on lambda DNA replication (unlike GroEL, where assembly of the phage head is altered). The effects of these proteins on replication have been studied in a variety of model systems: chiefly bacteriophage lambda and P1, and the F plasmid. The mechanisms of action are different in each case but generally consist of activation of one or more components of the replication complex: in the case of P1, by the conversion of a replication protein from inactive dimers to active monomers (Skowyra et al., 1995). Some effect on chromosomal replication also seems likely given the abnormal segregation patterns of chromosomes in *dnaK* null strains.

One area of chaperone involvement in the cell which is particularly hard to untangle is that of protein secretion across the cytoplasmic membrane. As proteins are known to be in an unfolded form when they cross membranes, the involvement of chaperones in holding proteins in this form is unsurprising, but different proteins appear to require different chaperones to act upon them, and there is also a good deal of degeneracy in the system. This is well illustrated by the DnaK/DnaJ/GrpE trio. DnaK over-expression alone can improve the export of lacZ hybrids from the cytoplasm (as can GroEL alone, although less efficiently; Phillips and Silhavy, 1990). More significantly, strains which are deficient in the secretion chaperone SecB (see below) are non-viable on rich medium, but this can be suppressed by over-expressing DnaK and DnaJ (Wild et al., 1992). More recent studies have confirmed and extended this result to GrpE too, and shown that all three proteins can play a role in the secretion of both SecB-dependent and independent proteins (Wild et al., 1996).

How can these various effects of the DnaK/DnaJ/GrpE trio, plus others not described here, be understood in terms of a common mechanism? Several studies strongly suggest that the action of this set of chaperones occurs at an earlier stage in protein folding that that of the GroE proteins. *In vivo* studies have shown evidence for association of DnaK and DnaJ with nascent polypeptide chains bound to ribosomes (Gaitanaris et al., 1994), and nascent chains on ribosomes *in vitro* have been shown to be releasable to give active protein when incubated with DnaK/DnaJ/GrpE plus GroES/GroEL (Kudlicki et al., 1994). *In vitro* studies with purified protein initially also supported the idea that the GroE proteins acted subsequently to the DnaK/DnaJ/GrpE proteins, although subsequent experiments have shown that there is not absolute obligation for the interaction to be in this order (Buchberger et al., 1996). Much work has been carried out on the nature of the bound substrate of DnaK, and this suggests a model where DnaK can bind certain sequences of amino-acid residues if they are in an extended conformation, unlike the binding of GroEL to proteins which have already partly collapsed into a globular form. Much less work has been carried out on the nature of protein binding to DnaJ. Recent work has started to focus on the action of the three proteins together, and as with the GroE system a model has been proposed for a cycle of binding and release which accommodates most of the data to date.

In this model (Figure 2, with the unfolded or nascent protein shown as a wavy line), the cycle is initiated by DnaJ binding to exposed amino-acid side chains, a step which increases the affinity of DnaK-ATP for the polypeptide (step 1). This interaction between DnaJ and DnaK also stimulates the hydrolysis of the DnaK-bound ATP (step 2), and the resulting complex is stable unless the bound ADP is removed, a reaction which is favoured by GrpE (step 3). (Thus as with GroES, GrpE is a co-chaperone rather than a chaperone in its own right). The step where DnaJ leaves the complex is unclear, but DnaK apparently releases the polypeptide when it rebinds ATP. Thus the whole cycle leads to a transient protection of a section of protein, by the binding of DnaK/DnaJ to exposed hydrophobic regions; GrpE acts as a nucleotide exchange factor to promote the release of this complex. How can this model explain the many different *in vivo* functions of DnaK/DnaJ/GrpE? Protection of exposed hydrophobic regions while a protein is elongated on the ribosome may prevent those regions either from aggregating with other nearby nascent chains or forming incorrect interactions with other parts of the growing protein until all the protein has been synthesised. Whether the protein will then require the action of GroES and GroEL to fully fold or not presumably varies with different proteins. Equally, binding of the DnaK/DnaJ complex to a protein could prevent it from folding until (in the case of a secretory protein) it has been delivered to the membrane translocation complex. It is not so easy to see how this model fits in with the role of these chaperones in DNA replication or proteolysis, but it may be as a result of stabilisation by DnaK/DnaJ of transiently exposed hydrophobic regions in the proteins involved in this case: this will tend to make proteins unfold or possibly undergo other changes in conformation, and thus raises the possibility that the chaperones' role in this case may be an active one (altering the state of a protein within the cell) rather than the more passive one of merely preventing aggregation.

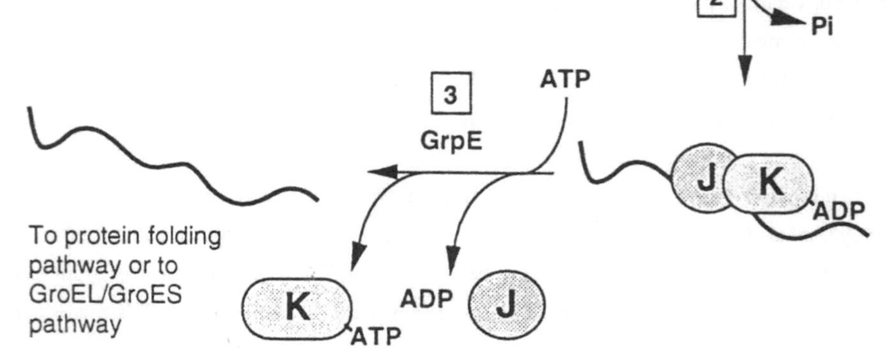

Figure 2: Proposed reaction cycle for DnaK, DnaJ and GrpE

A new twist has been added to the story of this particular set of chaperones by the discovery, both through genetic approaches and genomic sequencing, of the existence of multiple homologues of *dnaK* and *dnaJ* in *E. coli*. This raises the possibility of degeneracy within chaperone families as well as between them, and this has already been experimentally demonstrated: for example, one of the homologues of DnaJ (CbpA) can complement in part for defective DnaJ function (Ueguchi et al., 1994). Exploring the role of these homologues, as well as describing the reaction cycle in more detail, remain important challenges in this area of the molecular chaperone story.

4 The SecB protein and trigger factor

Because of the need for proteins to be kept in an unfolded state in order to cross membranes, molecular chaperones are often implicated in secretion, and some of the *sec* genes which have been identified by the use of various mutant screens do appear to have chaperone activity. The best studied of these is the SecB protein, mutants of which accumulate the precursors of proteins that otherwise would have been secreted from the cytoplasm (Kumamoto and Beckwith, 1985). SecB appears to lack the broad specificity of DnaK/DnaJ or of the GroE proteins, but is not restricted in its role to a single substrate: genetic and immunoprecipitation experiments suggest that it recognises a relatively small set of substrates including maltose-binding protein (MBP), LamB, OmpF, and OmpA. Unlike the proteins mentioned above, it is not a component of the heat shock regulon, nor is it essential, although deletion of the gene is lethal on rich media. However, there is good evidence that like the above families of molecular chaperones, SecB acts by binding to a protein and maintaining it an unfolded but non-aggregated state. Moreover, it has been shown that SecB overlaps to some extent with both GroEL and DnaK/DnaJ in function. The evidence for the latter has already been mentioned: over-expression of DnaK and DnaJ can complement a *secB* mutant and permit its growth on rich medium. The evidence for the former is based on *in vitro* data, and show that SecB and GoEL can both maintain a secretory protein in a secretion competent state when it is diluted from the unfolded form into a system containing vesicles that can import protein - a process analogous to secretion in the cell (Lecker et al., 1989). GroEL does not however have any reported genetic overlap with SecB, and experiments with different substrate have suggested that maintaining substrates in export-competent states is not the only role of SecB (Francetic and Kumamoto, 1996). SecB has been proposed to work by a "kinetic partitioning" mechanism, similar in principal to that proposed for the DnaK/DnaJ molecular chaperones (Randall and Hardy, 1995). By this model, it acts by sequestering regions of the protein substrate that would tend either to aggregate or to fold prematurely, and only release them when the protein is correctly positioned - in the case of SecB, for unloading into the membrane translocase complex. A question which is still unanswered is how SecB recognises those proteins with which it interacts, given that they have no sequence homology. It is easy to see, however, how over-expression of DnaK/DnaJ could suppress SecB defects to some extent, since both proteins are proposed to interact with their substrates at an early stage in protein folding - including even while the substrate is ribosome bound. However, the interactions between substrate and chaperone are not necessarily the same in the two cases.

A less clearly defined component of the secretory system is the protein known as trigger factor. Trigger factor was initially identified biochemically, and shown to interact with unfolded precursors of secretory proteins to keep them in a translocation competent state (Lecker et al., 1989). Although this implied a chaperone-like role in protein secretion similar to that seen for SecB and DnaK/DnaJ, the depletion of trigger factor from cells does not lead to any accumulation of secretory precursors or to protein aggregation (Guthrie and Wickner, 1990). Another intriguing observation is that trigger factor can interact with GroEL, an interaction which has been suggested to be involved in some way in GroEL's role in proteolysis (Kandror et al., 1997). The final twist comes from the demonstration that trigger factor has a highly efficient prolyl cis-trans isomerase activity: this makes it appear more like an enzyme that covalently interacts with proteins to speed them in folding (Hesterkamp et al., 1996). The precise role of trigger factor in vivo still is unclear, and it is on the edge of proteins that may be unequivocally defined as being molecular chaperones.

5 Other molecular chaperones

The overlap between degradative and repair functions has already been discussed both with respect to the GroES/EL and DnaK/DnaJ/GrpE chaperones. Another family of proteins which appear in this dual guise are the Clp ATPases, a family of proteins which bind to a common subunit with serine protease activity, ClpP. Various lines of evidence suggest that these proteins (ClpA, ClpB and ClpX) can also act as molecular chaperones, although these are so far almost entirely based on in vitro data. For example, ClpA can both activate the P1 RepA replication protein (as can DnaK/DnaJ/GrpE) and protect firefly luciferase from heat inactivation (Wickner et al., 1994). ClpX can both protect and rescue lambda O protein from heat aggregation (Wawrzynow et al., 1995), and a homologue of ClpB in yeast (Hsp104) can disaggregate preformed complexes in vivo. More detailed study on mutants in these genes is needed to confirm that they do show molecular chaperone activity in bacteria in vivo, and this is likely to be complicated by their dual role.

There is an unfortunate tendency in the literature to label proteins as molecular chaperones before evidence has come to light which can justify this. Nevertheless, although it is probably true to say that the major cellular molecular chaperones of bacteria have been discovered, there are undoubtedly many more promising candidates which may yet turn out to have significant roles in some aspects of bacterial growth. Initially such proteins may be identified only in terms of their expression (i.e. they are induced under conditions where chaperones are thought to be likely), or by homology with proteins with known chaperone function in eukaryotes. Two particular cases may be mentioned in *E. coli*. The first is the HtpG protein, which has an essential homologue in yeast and homologues which have shown by in vitro studies to have molecular chaperone activity in other eukaryotes. Yet its deletion produces only a very mild temperature sensitivity in *E. coli* (Bardwell and Craig, 1988), and its role in bacteria is still not known. The second are two low molecular weight proteins IbpA and IbpB, which were additionally identified by their co-purification with inclusion bodies (Allen et al., 1992), and which are also known to be very strongly induced on heat shock. No mutants have yet been reported in the genes for these proteins, but they seem likely to have some chaperone function by virtue of the fact that they share sequence homology with a range of low molecular weight proteins in eukaryotes (including α-crystallin, the major component of the mammalian eye lens) which have been shown to posses molecular chaperone activity *in vitro* (Jakob et al., 1994).

6 Where next in chaperone biology?

The whole area of molecular chaperone biology and biochemistry has developed so rapidly that predicting areas where new developments may soon be expected is not an easy task: events are likely to overtake the prophet before the ink is dry on the page. However, some new trends are certainly appearing in the field. Perhaps the most fascinating area from the bacterial point of view which is now receiving a lot of attention from several groups is, what happens to proteins in the bacterial periplasm? Given that proteins must be unfolded to pass through the membrane, and equally must fold up to become functional, it seems reasonable to suggest that there are chaperones in the periplasm. Certainly a lot of work has recently been published on the role of the Dsb proteins, which are important in the catalysis of the formation and isomerisation of disulphide bonds. Whether there are "true" molecular chaperones in the periplasm - in the sense of having no covalent interaction with their substrate - remains to be seen. Genomic sequencing is also likely to play an increasing role, through the identification of homologues and conserved domains in different species, and this may help us to establish the importance of candidate chaperones such as the Ibp proteins mentioned above. With the publication of structures for GroEL and for the protein-binding domain of DnaK, the structure-function studies are likely to continue, as is controversy over models for the action of the different chaperones. But perhaps some more attention will be paid to the definition of the chaperones' cellular role, particularly in species other than *E. coli* where the possibility that chaperones have functions associated with events in bacterial development have hardly been explored. Bacteria and archaea that grow at extremes of temperature or salinity present a particularly fascinating problem as far as protein folding is concerned, and the study of these may also be relevant in one aspect of the field which has not been discussed above but which is frequently used to justify a lot of grant proposals: will our understanding of the chaperones lead to many applications in biotechnology and medicine? If the pace of recent work can be maintained, there is no reason to doubt that this will remain an exciting field for some time yet.

References

Allen SP, Polazzi JO, Gierse JK, Easton AM (1992) Two novel heat shock proteins produced in response to heterologous protein expression in *Escherichia coli*. J Bacteriol *174*, 6938-6947

Bardwell JCA, Craig EA (1988) Ancient heat-shock gene is dispensable. J Bacteriol 170, 2977-2983.

Braig K, Otkinowski Z, Hegde R, Boisvert DC, Joachimiak A, Horwich AL, Sigler PB (1994) The crystal structure of the bacterial chaperonin GroEL at 2.8-angstrom. Nature 371, 578-586.

Buchberger A, Schroder H, Hesterkamp T, Schonfeld HJ, Bukau B (1996) Substrate shuttling between the DnaK and GroEL systems indicates a chaperone network promoting protein folding. J Mol Biol 261, 328-333

Bukau B, Walker GC (1989) Δ-dnaK52 mutants of Escherichia coli have defects in chromosome segregation and plasmid maintenance at normal growth temperatures. J Bacteriol 171, 6030-6038

Ellis RJ (1994) Molecular chaperones - opening and closing the Anfinsen cage. Current Biol 4, 633-635

Ensgraber M, Loos M (1992) A 66kDa heat shock protein of Salmonella typhimurium is responsible for binding of the bacterium to intestinal mucus. Infect Immun 60, 3072-3078

Fischer HM, Babst M, Kaspar T, Acuna G, Arigoni F, Hennecke H. (1993) One member of a *groESL*-like chaperonin multigene family in *Bradyrhizobium japonicum* is coregulated with symbiotic nitrogen fixation genes. EMBO Jou 12, 2901-2912

Franetic O, Kumamoto C (1996) Escherichia coli SecB stimulates export without maintaining export competence of ribose-binding protein signal sequence mutants. J Bacteriol 178, 5954-5959

Gaitanaris GA, Vysokanov A, Hung SC, Gottesman ME, Gragerov A (1994) Successive action of *Escherichia coli* chaperones in vivo Mol Micro 14, 861-869

Georgellis D, Sohlberg B, Hartl FU, Von Gabain A (1995) Identification of GroEL as a constituent of an mRNA protection complex in Escherichia coli. Mol Micro 16, 1259-1268

Goloubinoff P, Gatenby AA, Lorimer G (1989) GroE heat shock proteins promote assembly of foreign ribulose bisphosphate oligomers in *Escherichia coli*. Nature 337, 44-47

Gragerov A, Nudler E, Komissarova N, Gaitanaris GA, Nikiforov V (1992) Co-operation of GroEL/GroES and DnaK/DnaJ heat shock proteins in preventing protein misfolding in *Escherichia coli*. Proc Natl Acad Sci USA 89, 10341-10344

Guthrie B, Wickner W (1990) Trigger factor depletion or over-production causes defective cell division but
does not block protein export. J Bacteriol 172, 5555-5562

Hartl FU (1996) Molecular chaperones in cellular protein folding. Nature 381, 571-580

Hesterkamp T, Hauser S, Lutcke H, Bukau B (1996) *Escherichia coli* trigger factor is a prolyl isomerase that associates with nascent polypeptide chains. Proc Natl Acad Sci USA 93, 4437-4441

Horwich AL, Low KB, Fenton WA, Hirshfield IN, Furtak K (1993) Folding in vivo of bacterial cytoplasmic proteins: role of GroEL. Cell 74, 909-917

Ivic A, Olden D, Wallington EJ, Lund PA (1997) The groEL deletion of Escherichia coli is complemented by a Rhizobium leguminosarum groEL homologue at 37oC but not at 43oC. Gene, in press.

Jakob U, Gaestel M, Engel K, Buchner J. (1994) Small heat shock proteins are molecular chaperones. J Biol Chem 268, 1517-1520

Kandror O, Sherman M, Moerschell R, Goldberg (1997) Trigger factor associates with GroEL in vivo and promotes its binding to certain polypeptides. J Biol Chem 272, 1730-1734
Kanemori M, Mori H, Yura T. (1994) Effects of reduced levels of GroE chaperones on protein metabolism: enhanced synthesis of heat-shock proteins during steady state growth. J bacteriol 176, 4235-4242.
Kudlicki W, Odom OW, Kramer G, Hardesty B (1994) Activation and release of enzymatically inactive, full length rhodanese that is bound to ribosomes as peptidyl-tRNA. J Biol Chem 269, 16549-16553.
Kumamoto CA and Beckwith J (1985) Evidence for specificity at an early step in protein export in *Escherichia coli*. J Bacteriol 163, 267-274
Kusukawa N, Yura T (1988) Heat shock protein GroE of *Escherichia coli*: key protective roles against thermal stress. Genes Dev 2, 874-882
Kusukawa N, Yura T, Ueguchi C, Akiyama Y, Ito K (1989) Effects of mutations in the heat shock genes *groES* and *groEL* on protein export of *Escherichia coli*. EMBO Jou 8, 3517-3521
Lecker S, Lill R, Ziegelhoffer T, Georgopoulos C, Bassford PJ, Kumamoto CA, Wickner W (1989) Three pure chaperone proteins of *Escherichia coli*- secB, trigger factor, and GroEL - form soluble complexes with precursor proteins *in vitro*. EMBO Jou 8, 2703-2709
Lorimer GH. (1996) A quantitative assessment of the role of chaperonin proteins in protein folding in vivo. FASEB Jou 10, 5-9
Martin J, Horwich AL, Hartl F-U. (1992) Prevention of protein denaturation under heat stress by the chaperonin Hsp60. Science 258, 995-998
Phillips GJ, Silhavy TJ (1990) heat-shock proteins DnaK and GroEL facilitate protein export of LacZ-hybrid proteins in *Escherichia coli*. Nature 344, 882-884
Randall LL, Hardy SJS (1995) High selectivity with low specificity: how SecB has solved the paradox of chaperone binding. Trends Biochem Sci 20, 65-69
Schroder H, Langer T, Hartl F-U, Bukau B (1993) DnaK, DnaJ and GrpE form a cellular chaperone machinery capable of repairing heat-induced protein damage. EMBO Jou 12, 4137-4144
Skowyra D, McKenny K, Wickner SH (1995) Function of molecular chaperones in bacteriophage and plasmid DNA replication. Seminars Virol 6, 43-51
Straus D, Walter W, Gross CA (1990) DnaK, DnaJ and GrpE heat-shock proteins negatively regulate heat-shock gene expression by controlling the synthesis and stability of sigma-32. Genes Devel 4, 2202-2209
Thomas JG, Baneyx F (1996) protein folding in the cytoplasm of *Escherichia coli*: requirements for the DnaK-DnaJ-GrpE and GroEL-GroES molecular chaperone machines. Mol Micro 21, 1185-1196
Ueguchi C, Kakeda M, Yamada H, Mizuno T (1994) An analogue of the DnaJ molecular chaperone in *Escherichia coli*. Proc Natl Acad Sci USA 6, 1165-1172.
Van Dyk TK, Gatenby AA, laRossa RA (1989) Demonstration by genetic suppression of interaction of GroE products with many proteins. Nature 342, 451-453.
Wawrzynow A, Wojtkowiak D, Marszalek J, Banecki B, Jonsen M, Graves B, Goergopoulos C, Zylicz M (1995) The ClpX heat shock protein of *Escherichia coli*, the ATP-dependent substrate specificity component of the ClpP-ClpX complex, is a novel molecular chaperone. EMBO Jou 14, 1867-1877
Wickner S, Gottesman S, Skowyra D, Hoskins J, McKenny K, Maurizi MR (1994) A molecular chaperone.
ClpA, functions like DnaK and DnaJ. Proc Natl Acad Sci USA 91, 12218-12222
Wild J, Altman E, Yura T and Gross CA (1992) DnaK and DnaJ heat-shock proteins participate in protein export in *Escherichia coli*. Genes Devel 6, 1165-1172
Wild J, Rossmeissl P, Walter WA and Gross CA (1996) Involvement of the DnaK-DnaJ-GrpE chaperone team in protein secretion in *Escherichia coli*. J Bacteriol 178, 3608-3613
Zeilstra-Ryalls J, Fayet O, Georgopoulos C (1991) The universally conserved GroE (Hsp60) chaperonins. Annu Rev Microbiol 45, 301-325

Protein Traffic in Bacteria

Anthony P. Pugsley

Unité de Génétique Moléculaire (CNRS UMR321), Institut Pasteur, 25, rue du Dr Roux, 75724 Paris Cedex 15, France.

Keywords. Protein export, protein secretion, general secretory pathway, Sec proteins, pili, ABC transporters, contact secretion pathway

1 Introduction

The bacterial plasma (cytoplasmic) membrane, essentially equivalent to the plasma membrane of any living cell, is a lipid bilayer with integral and peripheral membrane proteins that perform a variety of functions ranging from solute transport to gene regulation. Gram-negative bacteria have a second bilayer membrane, the outer membrane, which, unlike the cytoplasmic membrane, contains glycolipids (lipopolysaccharide) located exclusively in the outer leaflet. Both Gram-positive and Gram-negative bacteria possess a peptidoglycan layer that lines the outer surface of the cytoplasmic membrane, and many bacteria have a protein layer (the S-layer), composed of one or a limited number of protein species, that covers the cell surface.

In Gram-positive bacteria, proteins are found in the cytoplasm, the plasma membrane and outside of the cell. In Gram-negative bacteria, proteins are found in the cytoplasm, the plasma and outer membranes, the periplasm between the two membranes and outside of the cell. By convention, proteins that are located beyond the outermost lipid bilayer of the cell, irrespective of whether they are assembled into surface appendages such as pili, flagella or S-layers, anchored to the cell surface by fatty acids or other linkages or released into the growth medium, are called secreted- or exoproteins and the process by which they reach their final destination is referred to as secretion. Proteins located within the cell envelope are referred to as exported. The purpose of this article is to provide a brief overview of the ways in which these proteins reach their final destinations. Further details and a more extensive list of citations can be found in several recent reviews (Cornelis and Wolf-Watz 1997; Pugsley 1993; Simonen and Palva 1993; Wandersman 1996).

2 Basic Questions

Protein traffic in bacteria involves the relocalisation of proteins synthesised in the cytoplasm to another cellular location or to the extracellular milieu. At most, 20 % of the total mass of protein synthesised by the cell is usually non-cytoplasmic. One major question concerns the nature of information that distinguishes these proteins from those that remain in the cytoplasm, and the way this information is recognised and decoded. The following step in the process concerns the

NATO ASI Series, Vol. H 103
Molecular Microbiology
Edited by Stephen J. W. Busby,
Christopher M. Thomas and Nigel L. Brown

way in which the polypeptide carrying the targeting signal(s) is transported through one or both membranes. What, for example, are the energetic requirements for polypeptide translocation, what is the conformational state of the protein before and during translocation and which membrane components are directly involved in polypeptide movement through the cell envelope?

3 Basic Tools

Two basic approaches have been used to study protein traffic in bacteria. The genetic approach involved the development of procedures for selecting mutations affecting protein traffic. Some of the selection procedures, designed to permit the identification of genes involved in major pathways for protein traffic, are particularly elegant (Pugsley 1989; Pugsley 1993), while others, affecting a narrower spectrum of non-cytoplasmic proteins whose traffic relies on more-or-less specialised pathways, could be identified simply by their ability to impede the correct localisation of a particular reporter protein (Pugsley et al. 1990). The biochemical approach involves the development of systems to analyse and reconstitute protein traffic *in vitro*. Clearly, the two systems go hand-in-hand: the identification, cloning and high level expression of genes coding for proteins involved in protein traffic have provided important clues to the identity of interacting partners and have permitted the ready purification and reconstitution of these components *in vitro*. Conversely, genetic approaches, while extremely powerful, failed to identify important components of major protein traffic machineries that had been characterized biochemically.

4 Pathways for Protein Traffic in Bacteria

The vast majority of non-cytoplasmic proteins produced by bacteria are transported to and through the cell envelope by a common pathway called the General Secretory Pathway (GSP). Because of the importance of this pathway, it will be given particular emphasis in this review. The GSP is composed of two steps, one for each membrane in the Gram-negative cell envelope (obviously, only the first step exists in Gram-positive bacteria). The first step involves recognition of a protein export signal, the signal sequence located at or close to the N-terminus of the secretory protein. Polypeptides bearing this signal are translocated across the cytoplasmic membrane by the preprotein translocase composed of integral and peripheral membrane Sec proteins. Secretory proteins are released into the periplasm/external milieu upon completion of translocation and specific proteolytic processing of the signal sequence (unprocessed secretory proteins remain anchored in the plasma membrane). In Gram-positive bacteria, the GSP ends here, whereas in Gram-negative bacteria, secretory proteins may remain in the periplasm or they may insert into the outer membrane or be shunted into one of several terminal branch pathways that lead them across the outer medium to the external milieu. Collectively, these branch pathways, which are specific for particular proteins or groups of proteins, form the second step in the GSP (Fig. 1)

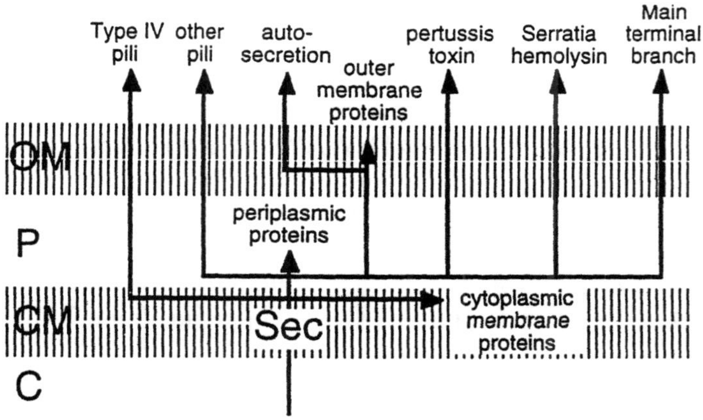

Fig. 1. Branches within the general secretory pathway (GSP) in Gram-negative bacteria.

In addition, a number of extracellular proteins bypass the GSP completely (Fig. 2). Two major bypass pathways have been characterized in some detail: one, called the ABC pathway because of its dependence on an integral plasma membrane protein with a characteristic ATP binding cassette motif, is present in both Gram-positive and Gram-negative bacteria, while the other, the contact secretion pathway, is found only in Gram-negative bacteria. Details of these two bypass pathways will be discussed at the end of this chapter.

5 The First Step in the GSP

This section of the review describes components of the first part of the GSP, often called the Sec pathway, which has been most extensively characterized in *Escherichia coli*.

5.1 The Signal Sequence and Signal Peptidases

Proteins entering this pathway bear a signal sequence composed of an N-terminal hydrophobic segment usually located just downstream from one or more positively-charged amino acids. The hydrophobic segment, which usually comprises 8 or more amino acids, is able to form an alpha-helix when mixed with organic solvents or detergents that mimic the environment within the lipid phase of the plasma membrane. This behaviour of the isolated signal sequence does not appear to be coincidental: amino acid changes that are predicted to disrupt helix formation or which reduce signal sequence hydrophobicity prevent insertion into lipid bilayers, prevent helix formation in apolar environments and diminish or abolish signal sequence function (Hoyt and

Gierasch 1991). These results suggest very strongly that the signal sequence comes into direct contact with the fatty acyl chains of the lipid bilayer following its association with the cytoplasmic membrane.

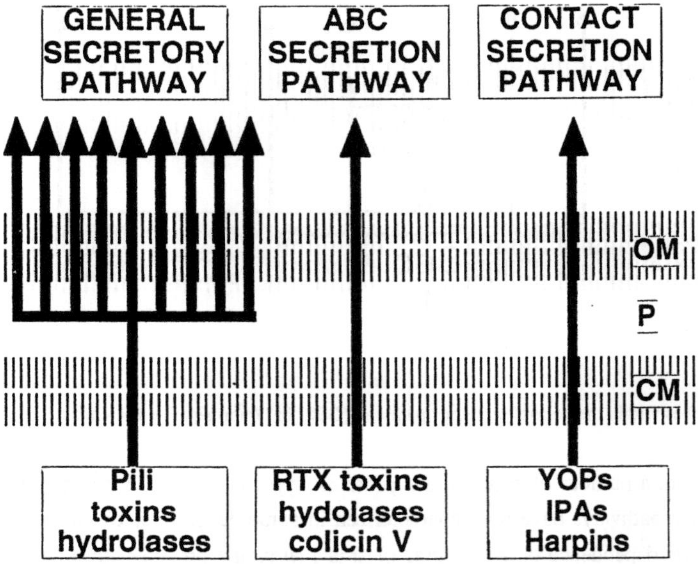

Fig. 2. Three major protein secretion pathways in Gram-negative bacteria. OM, outer membrane; P, periplasm; CM, cytoplasmic membrane. The names of the three pathways are given at the top of the figure and examples of proteins secreted are given at the bottom.

The hydrophobic segment of most signal sequences is followed by a turn-inducing amino acid or motif and then, 3-5 amino acids further downstream, by a cleavage site recognised by a specific protease called signal peptidase. In fact, bacteria have three signal peptidases. One of them, called leader peptidase, recognises the sequence alanine-X-alanine (where X is almost any amino acid) and cleaves immediately after the second alanine residue. Other amino acids such as glycine are frequently found in place of alanine, but cleavage sites can be predicted with high precision in most cases (von Heijne 1986). The second signal peptidase, lipoprotein signal peptidase, recognises the sequence leucine-alanine-glycine-cysteine in which the cysteine residue is modified by a diacylglyceride. Cleavage occurs between the glycine and the cysteine. Proteins that are processed by lipoprotein signal peptidase are less numerous than those processed by leader peptidase (approximately 30 in *E. coli)* and can be readily identified by labelling with radioactive palmitate. Most lipoproteins are anchored by their fatty acids in periplasmic leaflet of the cytoplasmic or outer membranes. Both of these signal peptidases are integral plasma membrane proteins with their catalytic sites facing the periplasm. This implies that the precursor polypeptides are inserted in the membrane in such a way that the signal sequence is perpendicular to the plane of the membrane. The third class of signal peptidase, prepilin peptidase, is described in section 8.

Surprisingly, signal sequences do not appear to be recognised by a specific membrane receptor. Although bacteria have a cytosolic complex composed of a 48 kDa protein linked to a 4.5S RNA that is structurally equivalent to a primative eukaryotic signal recognition particle (SRP) and, like eukaryotic SRP, is able to bind to signal sequences as they emerge from ribosomes (Valent et al. 1995), it does not appear to exert a dramatic effect on protein export (Phillips and Silhavy 1992). Instead, the major role of the signal sequence might be to retard the folding of precursor polypeptides (see below) and to insert into the plasma membrane at the interface between the preprotein translocase and the fatty acyl chains of the lipid bilayer.

5.2 Sec Proteins

In contrast to the bacterial SRP, many of the identified components of the translocase are absolutely essential for translocation of almost all presecretory proteins. One of these essential components is a ca 200 kDa homodimeric protein called SecA, an ATPase that is found associated with the plasma membrane and free in the cytoplasm. Depletion of SecA or conditional mutations that abolish its activity are lethal and cause the accumulation of precursor proteins in the cell. Mutations that diminish the levels of the integral membrane proteins SecE, SecY, SecD and SecF have similar effects, indicating that these proteins are probably essential translocase components. SecE and SecY proteins are particularly interesting because they are the structural and functional equivalents of proteins required for preprotein translocation across the membrane of the endoplasmic reticulum in eukaryotes (Finke et al. 1996). Furthermore, preprotein translocation can be reconstituted *in vitro* with purified, denatured or chaperone-complexed presecretory protein (see below) and SecA, SecY and SecE proteins (Brundage et al. 1990). SecY protein has 10 transmembrane segments (Ito 1992) and, thus, resembles many other integral membrane proteins that are involved in solute transport. It is not surprising, therefore, that the bulk of the evidence published to date, and in particular cross-linking data showing that presecretory proteins are in intimate contact with SecY, indicate that SecY probably forms the walls of a channel through which proteins are translocated (Joly and Wickner 1993). This does not exclude the possibility that the single essential transmembrane segment of SecE protein or even the signal sequence could contribute to channel formation.

SecE and SecY form a complex that can be readily purified or co-immunoprecipitated after solubilization of the plasma membrane in non-ionic detergents. In most cases, another integral membrane protein, initially called band 1 or P12 and now called SecG, remains associated with SecE and SecY throughout their purification (Hanada et al. 1994). Although this suggested that SecG might be an integral component of the translocase, *secG* mutations did not turn up in any of the screening or selection procedures designed to identify genes affecting protein export. In fact, it turns out that SecG is not essential except under some rather unusual growth conditions, but it does improve the efficiency of translocase *in vitro*. Recent studies of SecG reveal two interesting properties: it flips across the membrane during preprotein translocation and its function is dispensable when the acidic phospholipid composition of the membrane is high. These two observations led to the conclusion that the function of SecG is to facilitate the movement of another molecule across the lipid bilayer (Nishiyama et al. 1996). There are two candidate molecules. One is obviously the presecretory protein itself; the other is SecA protein, a large C-terminal segment of which actually penetrates through the plasma membrane as part of an ATP-driven insertion/desinsertion cycle that accompanies- and likely provides part of driving force for preprotein translocation (Economou et al. 1995). Thus, the translocase seems to function in a way that is at least partially analogous to a sewing machine: the thread is the presecretory protein and the needle is SecA, which picks up the presecretory protein and pushes it into the membrane (the fabric) at specific sites formed by the integral components of the translocase. Once the translocated segment of presecretory protein is on the trans (periplasmic) side of the membrane, SecA is withdrawn from the translocase (and may even be released from the membrane) and the whole cycle is repeated until the entire polypeptide chain is translocated.

The analogy between the sewing machine and the translocase is not complete, because in the former, the needle is never released from the thread and in the latter, the presecretory protein is not inserted back and forth across the membrane. Translocase may have an equivalent of the sewing machine shuttle, however, in the form of SecD and SecF proteins. The exact role of these proteins is still an open question, but early indications that they might help release translocated polypeptides from the trans side of the membrane could indicate an involvement in a late step in translocation (Matsuyama et al. 1993), possibly the release of SecA from the translocated polypeptide.

5.3 Protein Conformation and Energy Dependence

Although *in vitro* translocation can be driven solely by repeated cycles of step-wise translocation by SecA ATPase, the protein motive force (pmf) is absolutely required for translocation *in vivo* and *in vitro* when SecA or ATP is limiting. The way in which the pmf drives translocation remains unclear (Schiebel et al. 1991).

One important aspect of preprotein translocation is the conformation of presecretory proteins prior to and during translocation. Sewing machines become jammed if the thread has a knot in it. Likewise, presecretory proteins cannot be translocated if they have extensive tertiary structure. This means that precursors have to be maintained in a loosely-folded configuration before they are translocated. The maintenance of this so-called translocation-competent state depends on molecular protein chaperones. *In vitro*, any one of a number of chaperones seems able to maintain translocation competence, but one particular chaperone, SecB, seems to play a predominant role *in vivo* (Lecker et al. 1990). Indeed, the role of SecB seems to be restricted to the recognition of slowly-folding nascent polypeptides (remember, signal sequences are reported to retard folding) and targeting them to the translocase via interactions between SecB and SecA, and then SecA and the translocase and acidic phospholipids in the plasma membrane (Hartl et al. 1990).

SecB protein is not essential for viability except under optimal growth conditions, and many proteins can be exported efficiently in its absence. There are several explanations for these two phenomena. One is that some proteins are exported co-translationally so that very little of the precursor polypeptide emerges from the ribosome before it is translocated. The ribosome might also exert an anti-folding (chaperone-like) activity. Surprisingly, however, the absence of SecB reduces co-translational export, which is not what one would expect if this scenario were correct in all cases. In some cases, the bacterial SRP might bind to signal sequences and reduce tertiary structure formation; indeed there does seem to be at least a degree of complementarity between SRP and SecB since these proteins whose export is markedly affected by the absence of SRP tend to be those whose export is SecB-independent (Phillips and Silhavy 1992). A third possibility is that other chaperones are able to replace SecB (Altman et al. 1991). Finally, some presecretory proteins may have a strong intrinsic incapacity to fold tightly in the cytoplasm due, at least in part, to the presence of the signal sequence. It is worth stressing that the idea that the signal sequence is the major SecB binding site (Watanabe and Blobel 1989) has now been discarded, even though SecB might bind to some signal sequences (Randall et al. 1990). Instead, SecB binding sites are located at several positions along the length of the polypeptide (de Cock et al. 1992) and the signal sequence retards folding sufficiently to allow SecB to bind to these sites.

6 Post-translocation Folding

It is now widely accepted that although polypeptides contain all the information necessary for them to adopt their final, uniform conformation, they often cannot achieve this state without assistance from molecular chaperones and other proteins that are often referred to as foldases. Proteins extruded across the cytoplasmic membrane are no exception: the periplasm (or the outer surface of the plasma membrane) contains a set of enzymes whose function varies from disulphide bond oxido-reduction (isomerase) activity through peptidyl-proline isomerase and possibly chaperone activities to proteases that collectively act to stimulate proper, stable folding or the destruction of proteins that are incorrectly folded (Lazar and Kolter 1996; Missiakis et al. 1996). Interestingly, proteins secreted by Gram-positive bacteria rarely contain disulphide bonds and these bacteria produce very few, if any soluble multimeric proteins. The first of these two features might be explained by the absence from these bacteria of enzymes that can catalyse the formation of- or rearrange disulphide bonds, while the latter reflects the absence of a compartment (the periplasm) in which components of multimeric proteins can accumulate to the high concentrations that would favour their oligomerisation.

7 Integral Plasma and Outer Membrane Proteins

Outer- and plasma membrane proteins are readily distinguished by the absence, from the former, of potential membrane-spanning segments composed of 20-or-so mainly hydrophobic amino acids. In plasma membrane proteins, these transmembrane segments display a polarised organisation of basic amino acids around their two ends, with positive charges being more abundant on the cytoplasmic side of the membrane. This charge distribution is the major determinant of plasma membrane protein topology.

There appear to be two different modes of plasma membrane protein insertion. In some cases, the presence of a large periplasmic/cell surface loop or domain necessitates translocation of the polypeptide via the Sec machinery. The simplest case is-that of bitopic proteins, which differ from a periplasmic/secreted proteins only by the fact that the signal sequence is not processed. In this case, the signal sequence also acts as a membrane anchor that retains the protein embedded in the lipids of the plasma membrane. Some polytopic membrane proteins also contain large extracytoplasmic loops that are probably translocated across the cytoplasmic membrane in a Sec-dependent manner (Anderson and von Heijne 1993). In other cases, the extracytoplasmic loops can cross the plasma membrane in a Sec-independent fashion, with pmf-dependent insertion (Bassilana and Gwizdek 1996; Cao et al. 1995) occurring by progressive, pair-wise integration of adjacent hydrophobic transmembrane segments into the lipid phase of the bilayer (Pugsley 1993).

Topological- and, more recently, high resolution crystallographic analysis of outer membrane proteins reveal that their transmembrane segments are short amphipathic ß-strands of polypeptide, the extremities of which are often marked by aromatic amino acids that form a pair of belts around the protein in its final, inserted state (Cowan et al. 1992). Folding and oligomerisation of outer membrane proteins occur in the periplasm and are assisted by periplasmic foldases (Missiakis et al. 1996) but their insertion into the membrane is probably a gradual, unassisted process that might require prior interaction with newly-synthesised LPS. The apparent absence of a receptor protein that would ensure correct insertion into the outer (rather than the plasma) membrane is particularly noteworthy.

8. Terminal branches of the GSP

The GSP has several terminal branches whose specific function is to direct one or a limited number of secretory proteins across the outer membrane (Fig. 1). Two of these pathways are relatively simple. The autosecretory pathway, of which the *Neisseria gonorrhoeae* IgA protease is the best example, involves the insertion of the C-terminal segment of the polypeptide into the outer membrane as described in section 7, followed by translocation of the N-terminal segment, probably in a loosely-folded configuration, through a pore formed by the C-terminal segment.

Autoproteolysis or cleavage by non-specific outer membrane proteases lead to release of the catalytic N-terminal domain into the medium (Klauser et al. 1992).

Little is known about the secretion of the *Serratia* haemolysin except that the sole protein required for transport of the toxin across the outer membrane protein is an integral outer membrane protein that is also involved in the posttranslational activation of the toxin (Schiebel et al. 1989). Other proteins, such as the *Bordetella* filamentous haemagglutinin and surface-exposed proteins in *Haemophilus influenzae*, are probably secreted in a very similar manner. *Bordetella* possess yet another terminal branch of the GSP for the secretion of pertussis toxin, a hexameric structure composed of several different subunits that probably fold and oligomerize in the periplasm before the toxin is transported across the outer membrane by a complex pathway comprising several components. These components bear a striking resemblance to proteins required for the conjugal transfer of T DNA from *Agrobacterium* to plant cells and for the interbacterial transfer other conjugal plasmids, raising the possibility that protein secretion and conjugation might be mechanistically similar processes (Weiss et al. 1993).

8.1 The Secreton

The major pathway for extracellular protein secretion by the GSP is probably present in the vast majority of Gram-negative bacteria. Indeed, recent studies show that it even exists in *E. coli* K-12 and other *E. coli* strains, hitherto thought unable to secrete proteins by this pathway. This secretion pathway exhibits several interesting characteristics:

a. Exoproteins adopt their final conformation in the periplasm, prior to translocation across the outer membrane (Hardie et al. 1995; Hirst and Holmgren 1987; Pugsley 1992).

b. At least 14 integral- and peripheral membrane proteins are specifically required for secretion. The genes coding for these proteins are clustered (most of them in an operon), often together with genes coding for the exoproteins. These proteins are thought to be directly or indirectly involved in the formation of a transenvelope complex, the secreton, that recognises secretion signals in the exoproteins and directs them to a specific gated pore (secretin) in the outer membrane (Pugsley 1993).

c. Many secreton components are interchangeable between different species. They also bear a striking resemblance to proteins involved in the formation of type IV pili in Gram-negative bacteria (Pugsley 1993). These common components include proteins with signature sequences found in the type IV pilus subunits (pilins; see section 8.2).

d. Secretin is the only integral outer membrane component of the secreton (Hardie et al. 1996a). It forms characteristic multimeric structures (probably composed of 12 subunits) that resist dissociation by heating in SDS. Secretins resist the action of periplasmic proteases and are able to insert into the outer membrane through the action of a specific periplasmic chaperone-like protein that probably remains bound to the secretin after it has inserted into the membrane. Thus, secretins are different from most other outer membrane proteins (Hardie et al. 1996b).

e. Both ATP and the pmf are required for secretion (Letellier et al. 1997; Possot et al. 1997). ATP is probably required for secreton assembly, since both secretion and piliation pathways include a member of a unique family of peripheral plasma membrane ATPase/kinase proteins. The pmf is probably required to direct the opening and closing of secretin pore in the outer membrane, presumably via a novel signal transduction pathway mediated by secreton components that are embedded in the plasma membrane and establish pmf-dependent contacts with the secretin. Recent studies in *Aeromonas* indicate that ATP might also be required for the actual process of protein translocation, at least in some bacteria (Letellier et al. 1997).

8.2 Type IV Pili

The similarity between the main branch of the GSP and type IV piliation has already been mentioned. Type IV pilins, also called mePhe pilins, exhibit several unique characteristics compared with other pilins (Pugsley 1993):

a. They have unusually long, very hydrophobic signal sequences

b. Signal sequence processing occurs on the N-terminal side of the hydrophobic segment, rather than on its C-terminal side (see section 5), after the glycine residue in the consensus cleavage site Gly-Phe/Met/Leu-X-X-X-Glu, where X is a hydrophobic amino acid. The polypeptide sequence upstream from the glycine residue is usually 5-7 amino acids long and contains one or more arginine residues and often a glutamine, but some upstream sequences can be over 20 amino acids long.

c. Processing is carried out by a special integral membrane signal peptidase called prepilin peptidase. Unlike other signal peptides, the catalytic site is located in the cytoplasmic side of the plasma membrane.

d. Prepilin peptidases also have N-methyl transferase activity that catalyses the S-adenosyl methionine-dependent methylation of the newly exposed N-terminal amino acid. Proteolysis and methylation are independent reactions (Strom et al. 1993).

e. Type IV pilins characteristically have an intramolecular disulphide bond

All but the last of these features are shared with the pilin-like components of secreton (see above). Interestingly, the site in the envelope to which type IV pili are anchored has not been determined but the retention of the strongly hydrophobic signal sequence core, which is essential for filament formation, suggests that they might remain anchored in the plasma membrane.

Type IV pili are commonly produced by major pathogens such as *Neisseria*, *Pseudomonas aeruginosa*, and *Vibrio cholerae*. Type IV piliation appears to be an extremely complex process; in *P. aeruginosa*, the only bacterium in which a systematic search for piliation mutants has been carried out, at least 25 and probably as many as 40 genes seem to be involved in pilus assembly and its regulation. Many of these genes exist in *E. coli* K-12 and in other *E. coli* strains, which are not thought to produce type IV pili (Francetic and Pugsley, in prepartation). This raises that possibility that type IV piliation might be more common than hitherto thought likely and that the pili might be involved in processes other than adsorption to epithelial cells, the role usually attributed to these appendages in the pathogens mentioned above. Interestingly, type IV pili of both *Pseudomonas* and *Neisseria* are capable of movement, possibly through successive cycles of polymerisation and depolymerisation, causing the bacteria to undergo twitching or other forms of mobility over short distances. This unique form of motility is dependent on a signal transduction pathway not unlike that involved in that the more conventional, flagella-mediated motility of enteric and other bacteria. Indeed, in *Myxococcus xanthus*, type IV-dependent chemotaxis is a major element in the unique social mobility exhibited by this bacterium (Wu et al. 1997).

Type IV pilin-like proteins are not exclusive to Gram-negative bacteria. The Gram-positive bacterium *Bacillus subtilis* has a set of proteins, including 4 type IV pilin homologues, a prepilin peptidase and a putative assembly ATPase, that are all closely-related to the components of the secreton. However, they are not involved in protein secretion but perform an as-yet undetermined role in the import of DNA by naturally transformation-competent bacteria. The transformation system in *Bacillus* lacks one key element that is common to both the type IV piliation system and the secreton, namely the outer membrane secretin. *B. subtilis*, of course, does not have an outer membrane.

8.3 Other Pili

The assembly of other pili appears to be far less complex and is much better understood than that of type IV pili. Although several different pilins are assembled into the pilus filament, where they perform different functions (see below), assembly depends on only two specific components, a periplasmic chaperone that forms a complex with pilins in the periplasm, protects them from proteolysis by periplasmic proteases and guides them to the outer membrane, and an integral outer membrane assembly protein (often called the usher), that causes the pilins to be released from their chaperone and assembled into the pilus filament (Hultgren et al. 1991; Hung et al. 1996). The chaperone binding site is located near the C-terminus of the pilus subunits. Pilus assembly is energy-independent. Some pili, such as the well-characterized P or Pap pilus, are composite structures comprising structurally-different types of filaments. Specific minor pilins are required to form the cores of these filaments and the junctions between different segments of the filament or between the filament and the outer membrane (Jacob-Dubuisson et al. 1993), or have receptor binding activity that allows the pilus to adhere to target cells.

9 The ABC secretion pathway

The ABC secretion pathway (also called the type I secretion pathway) is used by a wide variety of Gram-negative and Gram-positive bacteria to secrete extracellular proteins and peptides. The best studied examples are the alpha hemolysin secretion pathway of *E. coli*, the metalloprotease secretion pathway of *Erwinia chrysanthemi* and the secretion of a heme binding protein by *Serratia marcescens*. In this review, I will describe mainly those features found in these systems, all of which come from Gram-negative bacteria, namely a polytopic plasma membrane protein with a characteristic ATP binding cassette motif (from which this secretion pathway derives its name), an integral outer membrane protein that probably forms a specific pore and a protein that is embedded in the plasma membrane by its N-terminal unprocessed signal sequence and which probably creates specific contacts with its cognate outer membrane protein (the so-called membrane fusion protein). These three proteins form a complex when presented with a translocation substrate (Létoffé et al. 1996), leading to the idea that exoproteins are translocated across both membranes in a single step driven by ATP hydrolysis (Koronakis et al. 1991). The repeated failure to demonstrate the presence of a periplasmic intermediate in this pathway, even when the outer membrane component is removed, supports this idea.

Although the assembly of the secretion machinery is Sec-dependent, the translocation process itself is not known to require any of the Sec proteins. In most cases, the signal that drives secretion is an unprocessed peptide of approximately 40 amino acids located at the extreme C-terminal end of the polypeptide (see below)(Wandersman 1996). The way in which this peptide is recognised by the secretion machinery, initially by the ABC protein itself, is unknown, although it appears that structural motifs are more important than precise sequence recognition. The secretion signal can be transplanted into other polypeptides to promote their secretion, although the presence additional sequences upstream of the secretion signal in the exoprotein usually favours the secretion of heterologous proteins by ensuring correct signal presentation. These unusual features raise interesting questions regarding the conformation of the exoprotein prior to translocation: since exoproteins must be released from the ribosome before they can be recognised. Do they fold and are they then unfolded before translocation, or are they complexed with a specific cytoplasmic chaperone? Interestingly, exoproteins that are not secreted (because they carry a defective secretion signal or because components of the secretion machinery are absent) are unstable, suggesting that they do not fold into a protease-resistant conformation before translocation and that any interaction with a protective chaperone can only be transitory.

The secretion signal of many of the peptide bacteriocins that are secreted by the ABC pathway by Gram-positive bacteria is a processed N-terminal leader peptide with a characteristic double glycine motif. Such a motif is also found in colicin V, a peptide bacteriocin produced by certain strains of *E. coli.* Colicin V appears to be the only the protein secreted by the ABC pathway in Gram-negative bacteria that does not have a C-terminal secretion signal (van Belkum et al. 1997).

10 The Contact-dependent Secretion Pathway (CDSP) and Flagellum Assembly

The contact-dependent secretion pathway (originally called the type III secretion pathway) was discovered in *Yersinia* but is now known to be present in a wide variety of animal and plant pathogens. Secretion was first observed in *Yersinia* grown in medium without calcium. The proteins (YOPs), and other proteins secreted by the same pathway in other bacteria, tend to form structured aggregates that are readily pelleted when cultures are centrifuged, which complicates analysis of their secretion (Parsot et al. 1995). However, it is now recognised that their secretion occurs by a unique pathway that, like the ABC pathway described in section 9, is not directly dependent on the Sec machinery.

Although in vitro-cultured bacteria secrete YOPs and other proteins directly into the growth medium, this is probably an artefact caused by the use of conditions that mimic those prevailing in the natural environment that cause bacteria to inject their proteins directly into the cytosol of the epithelial cells to which they adhere. Thus, the secreted proteins have to cross three lipid bilayer membranes, two belonging to the bacterium that is producing and secreting them and one belonging to the cell into which they are to be injected (Boland et al. 1996; Cornelis and Wolf-Watz 1997; Hakansson et al. 1996; Rosqvist et al. 1994).

At least 20 different gene products are involved in the CDSP. Many of these genes code for the secreted proteins, including some that form the channel that spans the plasma membrane of the target cell, while others form the secretion machinery in the envelope of the producing bacterium and still others are chaperones that prevent illicit interactions between secreted proteins prior to their exit from the cell (Wattiau et al. 1996). Many of the components of the CDSP are similar to those involved in the assembly of the flagellum in both Gram-positive and Gram-negative bacteria, leading to the suggestion that the two processes are fundamentally similar.

Different bacteria seem to have adopted different strategies to controlling protein efflux by the CDSP. Some bacteria, including *Yersinia*, control secretion at the level of gene expression; the presence of an appropriate signal induces expression of the entire system, leading to co-ordinated synthesis and secretion of the YOPs. In other bacteria, e.g., *Shigella*, proteins (called IPAs) that are to be secreted are stockpiled, apparently in the cytoplasm whence they are rapidly released when the bacteria detect appropriate signals. The way in which these signals are recognised and transduced to the secretion/regulation machinery is not fully understood but in *Shigella*, it is proposed that two of the secreted IPAs block the secretion pore in the cell envelope until the signal is detected (Ménard et al. 1994). The CDSP has only a single outer membrane protein that could form this pore. Interestingly, this protein, which belongs to the secretin family of proteins described in sections 8.1 and 8.2, is not a component of the flagellum assembly pathway.

The kinetics and energy requirements for secretion by the CDSP or flagellum assembly have not been determined. However, one of the components shared by both systems is an integral membrane ATP binding protein, which raises the possibility that secretion or assembly of the secretion machinery are driven by ATP hydrolysis. The ease with which the CDSP can be manipulated in *Shigella* should facilitate further studies of this important aspect of this fascinating secretion pathway.

The nature of the secretion signal is better understood. Studies with hybrid proteins indicate that the extreme N-terminal 20-or-so amino acids of the YOPs are absolutely required for secretion *into the growth medium*. Additional sequences are required for injection into the target cell. These sequences, located just downstream from the secretion signal, are also the sites recognised by the specific chaperone (Sory et al. 1995). Thus, it is proposed that the function of the chaperone is to prevent premature interactions between the YOPs that have to be injected and those that form the channel by which they will be injected into the target cell.

11 Concluding remarks

Curli, a novel type of cell surface appendage, were discovered about 5 years ago in *E. coli*. This pilus-like structure differs fundamentally from other types of pili in its unique mode of assembly (Hammar et al. 1995; Hammar et al. 1996). Instead of being assembled at a specific assembly platform in the outer membrane or within the periplasm, curli subunits are translocated across the cell envelope, probably via a periplasmic intermediate, to the outside of the cells were contact with a specific protein called the nucleator causes the formation of the appropriate filament. Indeed, the nucleator need not be furnished by the cells that produce the curli subunits, since it can occur in trans. This novel discovery discloses just how little we know about the diversity of mechanisms used by bacteria to secrete and assemble proteins.

All Gram-negative bacteria that have been studied so-far have at least two systems for extracellular protein secretion, and most have far more. For example, *P. aeruginosa* has functional GSP secreton, ABC and CDSP systems for the secretion of toxins and enzymes, a type IV pilus assembly system and a flagella assembly system. Other pathways for protein secretion remain to be discovered and characterized. For example, *S. marcescens* secretes a wide variety of extracellular enzymes and toxins. Among these, a lipase, a protease and a heme-binding protein are secreted by apparently independent ABC pathways, a serine protease is secreted by the autosecretion pathway and a haemolysin is secreted by a unique branch of the GSP (see section 8), but the mechanism by which other proteins, including a chitinase and a nuclease, are secreted, remains to be elucidated. Both of these proteins are translocated to the periplasm by the Sec pathway when their genes are expressed in *E. coli* K-12, implying that this bacterium lacks a specific GSP branch pathway that they use in *Serratia* to cross the outer membrane.

Protein traffic is one of the major areas of research in bacterial molecular biology. Studies on the bacterial Sec system have long been at the forefront of studies of protein translocation and provided the ground rules for understanding the related Sec pathway in eukaryotes. Studies of the more exotic systems revealed ways in which folded proteins could be translocated across lipid bilayer membranes and are of particular interest in relation to their requirement for the secretion and/or assembly of important virulence factors such as toxins and pili. The recent discovery of apparently cryptic secreton and type IV piliation genes in *E. coli* K-12 raises the possibility that these secretion systems are even more wide-spread than hitherto appreciated. Added impetus for further research into these secretion pathways comes from the possibility that they might be put to good use (e.g., using the CDSP to inject hybrid proteins into epithelial cells; assembly of immunogenic peptides into pili) or that they might provide targets for inhibitors of secretion or piliation that would reduce the virulence of pathogenic bacteria. One important area that has not been discussed at all here is how proteins cross the periplasm and how contacts are established between the inner and outer membranes. Regrettably this review has only skimmed the surface of a fascinating area of research and cannot do anything like justice to the many groups who have contributed to our current knowledge of these sometimes extremely complex processes.

Altman E, Kumamoto CA, Emr SD (1991) Heat-shock proteins can substitute for SecB function during protein export in *Escherichia coli.* EMBO J 10:239-445.

Anderson H, von Heijne G (1993) *Sec* dependent and *sec* independent assembly of *E.coli* inner membrane proteins: the topological rules depend on chain length. EMBO J 12:683-691.

Bassilana M, Gwizdek C (1996) *In vivo* membrane assembly of the *E. coli* polytopic protein, meilibiose permease, occurs via a Sec-independent process which requires the protonmotive force. EMBO J 15:5202-5208.

Boland A, Sory M-P, Iriate M, Kerbourch C, Wattiau P, Cornelis GR (1996) Status of YopM and YopN in the *Yersinia* Yop virulon: YopM of *Y. entyerocolitica* is internalized inside the cytosol of PU5-1.8 marcrophages by the YopB, D, N delivery system. EMBO J 15:5191-5201.

Brundage L, Hendrick JP, Schiebel E, Driessen AJM, Wickner W (1990) The purified E. coli integral membrane protein SecY/E is sufficient for reconstitutuion of SecA-dependent precursor protein translocation. Cell 62:649-657.

Cao G, Kuhn A, Dalbey RE (1995) The translocation of negatively charged residues across the membrane is driven by the electrochemical potential: evidence for an electrophoresis-like membrane transfer mechanism. EMBO J 14:866-875.

Cornelis GR, Wolf-Watz H (1997) The *Yersinia* Yop virulon: a bacterial system for subverting eukaryotic cells. Mol Microbiol 23:861-867.

Cowan SW, Schrimer T, Rummer G, Steiert M, Ghosh R, Pauptit RA, Jansonius JN, Rosenbush JP (1992) Crystal structures explain functional properties of two *E. coli* porins. Nature 358:727-733

de Cock H, Overeem W, Tommassen J (1992) Biogenesis of outer membrane PhoE of *Escherichia coli*. Evidence for multiple SecB-binding sites in the mature portion of the PhoE protein. J Mol Biol 224:369-379.

Economou A, Polliano JA, Beckwith J, Oliver DB, Wickner W (1995) SecA membrane cycling at SecYEG is driven by distinct ATP binding and hydrolysis events and is regulated by SecD and SecF. Cell 83:1171-1181.

Finke K, Plath K, Panzer S, Prehn S, Rapoport TA, Hartmann E, Sommer T (1996) A second trimeric complex containing homologs of the Sec61p complex functions in protein translocation across the ER membrane of *S. cerevisiae*. EMBO J 15:1482-1494.

Hakansson S, Schesser K, Persson C, Galyov EE, Rosquivst R, Homblé F, Wolf-Watz H (1996) The YopB protein of *Yersinia pseudotuberculosis* is essential for the translocation of Yop effector proteins across the target cell plasma membrane and displays contact-dependent membrane disrupting activity. EMBO J 15:5812-5823.

Hammar M, Arnqvist A, Bian Z, Olsén A, Normark S (1995) Expression of two csg operons is required for production of fibrobectin- and congo red-binding curli polymers in *Escherichia coli* K-12. Mol Microbiol 18:661-670.

Hammar M, Bian Z, Normark S (1996) Nucleator-dependent intracellular assembly of adhesive curli organelles in *Escherichia coli*. Proc Natl Acad Sci USA 93:6562-6566.

Hanada M, Nishiyama K-i, Mizushima S, Tokuda H (1994) Reconstitution of an efficient protein translocation machinery comprising SecA and the three membrane proteins SecY, SecE and SecG (p12). J Biol Chem 269:23625-23631.

Hardie KR, Lory S, Pugsley AP (1996a) Insertion of an outer membrane protein in *Escherichia coli* requires a chaperone-like protein. EMBO J 15:978-988.

Hardie KR, Schulze A, Parker MW, Buckley JT (1995) *Vibrio spp.* secrete proaerolysin as a folded dimer without the need for disulphide bond formation. Mol Microbiol 17:1035-1044

Hardie KR, Seydel A, Guilvout I, Pugsley AP (1996b) The secretin-specific, chaperone-like protein of the general secretory pathway: separation of proteolytic protection and piloting functions. Mol Microbiol 22:967-976

Hartl F-U, Lecker S, Schiebel E, Hendrick JP, Wickner W (1990) The binding cascade of SecB to SecA to SecY/E mediates preprotein targeting to the E. coli plasma membrane. Cell 63:269-279.

Hirst TR, Holmgren J (1987) Conformation of protein secreted across bacterial outer membanes: a study of enterotoxin translocation from *Vibrio cholerae*. Proc Natl Acad Sci USA 84:7418-7422

Hoyt DW, Gierasch LM (1991) A peptide corresponding to an export-defective mutant OmpA signal sequence with asparagine in the hydrophobic core is unable to insert into model membranes. J Biol Chem 266:14406-14412.

Hultgren SJ, Normak S, Abraham SN (1991) Chaperone-assisted assembly and moleular architecture of adhesive pili. Ann. Rev. Microbiol. 45:383-415

Hung DL, Knight SD, Woods RM, Pinkner JS, Hultgren SJ (1996) Molecular basis of two subfamilies of immunoglobulin-like chaperones. EMBO J 15:3792-3805.

Ito K (1992) SecY and integral membrane components of the *Escherichia coli* protein translocation system. Mol Microbiol 6:2423-2428.

Jacob-Dubuisson F, Heuser J, Dodson K, Normark S, Hultgren S (1993) Initiation of assembly and association of the structural elements of a bacterial pilus depend on two specialized tip proteins. EMBO J 12:837-847.

Joly JC, Wickner W (1993) The SecA and SecY subunits of translocase are the nearest neighbors of a translocating preprotein, shielding it from phospholipids. EMBO J 12:255-263.

Klauser T, Pohler J, Meyer TF (1992) Selective extracellular release of cholera toxin B subunit by *Escherichia coli*: dissection of *Neisseria* Igaß-mediated outer membrane transport. EMBO J 11:2327-2335.

Koronakis V, Hughes C, Koronakis E (1991) Energetically distinct early and late stages of HlyB/HlyD-dependent secretion across both *Escherichia coli* membranes. EMBO J 10:3263-3272.

Lazar SW, Kolter R (1996) SurA assists the folding of *Escherichia coli* outer membrane proteins. J. Bacteriol. 178:1770-1773.

Lecker SH, Driessen AJM, Wickner W (1990) ProOmpA contains secondary and tertiary structure prior to translocation and is shielded from aggregation by association with SecB protein. EMBO J 9:2309-2314.

Letellier L, Howard SP, Buckley TJ (1997) Studies on the energetics of proaerolysin secretion across of the outer membrane of *Aeromonas* spp: evidence for requirement for both the protonmotive force and ATP. J Biol Chem :in press

Létoffé S, Delepelaire P, Wandersman C (1996) Protein secretion in Gram-negative bacteria: assembly of the three components of ABC protein-mediated exporters is ordered and promoted by substrate binding. EMBO J 15:5804-5811.

Matsuyama S, Fujita Y, Mizushima S (1993) SecD is involved in the release of translocated secretory proteins from the cytoplasmic membrane of *Escherichia coli*. EMBO J 12:265-270.

Ménard R, Sansonetti PJ, Parsot C (1994) The secretion of the *Shigella flexneri* Ipa invasins is activated by epithelial cells and controlled by IpaB and IpaD. EMBO J 13:5293-5302.

Missiakis D, Betton J-M, Raina S (1996) New components of protein folding in extracytoplasmic compartments of *Escherichia coli*: SurA, FkpA and Skp/OmpH. Mol Microbiol 21:871-884

Nishiyama K-i, Suzuki T, Tokuda H (1996) Inversion of the membrane topology of SecG coupled with SecA-dependent preprotein translocation. Cell 85:71-81.

Parsot C, Ménard R, Gounon P, Sansonetti PJ (1995) Enhanced secretion through the *Shigella flexneri* Mxi-Spa translocon leads to assembly of extracellular proteins into macromolecular structures. Mol Microbiol 16:291-300

Phillips GJ, Silhavy TJ (1992) The *E. coli ffh* gene is necessary for viability and efficient protein export. Nature 359:744-746.

Possot O, Letellier L, Pugsley AP (1997) Energy requirement for pullulanase secretion by the main terminal branch of the general secretory pathway. Mol Microbiol :in press.

Pugsley AP (1989) Protein targeting. Academic Press, San Diego, USA

Pugsley AP (1992) Translocation of a folded protein across the outer membrane via the general secretory pathway in *Escherichia coli.* Proc. Natl. Acad. Sci. USA 89:12058-12062

Pugsley AP (1993) The complete general secretory pathway in gram-negative bacteria. Microbiol Rev 57:50-108

Pugsley AP, d'Enfert C, Reyss I, Kornacker MG (1990) Genetics of extracellular protein secretion by Gram-negative bacteria. Ann. Rev. Genet. 24:67-90

Randall LL, Topping TB, Hardy SJS (1990) No specific recognition of leader peptide by SecB, a chaperone involved in protein export. Science 248:860-863.

Rosqvist R, Magnusson K-E, Wolf-Watz H (1994) Target cell contact triggers expression and polarized transfer of *Yersinia* YopE cytotoxin into mammalian cells. EMBO J 13:964-972.

Schiebel E, Driessen AJM, Hartl F-U, Wickner W (1991) $\Delta\mu H^+$ and ATP function at different steps of the catalytic cycle of preprotein translocase. Cell 64:927-939.

Schiebel E, Schwartz H, Braun V (1989) Subcellular location and unique secretion of the hemolysin of *Serratia marcescens*. J Biol Chem 264:16311-16320.

Simonen M, Palva I (1993) Protein secretion in *Bacillus* species. Microbiol Rev 57:109-137

Sory M-P, Boland A, Lambermont I, Cornelis GR (1995) Identification of the YopE and YopH domains required for secretion and internalization into the cytosol of macrophages using the *cyaA* gene fusion approach. Proc Natl Acad Sci USA 92:11998-12002

Strom MS, Nunn DN, Lory S (1993) A single bifunctional enzyme, PilD, catalyzes cleavage and N-methylation of proteins belonging to the type IV pilin family. Proc. Natl. Acad. Sci. USA 90:2404-2408

Valent QA, Kendall DA, High S, Kusters R, Oudega B, Luirink J (1995) Early events in preprotein recognition in *E.coli*: interaction of SRP and trigger factor with nascent polypeptides. EMBO J 14:5494-5505

van Belkum MJ, Worobo RW, Stiles ME (1997) Double-glycine-type leader peptides direct secretion of bacteriocins by ABC transporters: colicin V secretion in *Lactococcus lactis*. Mol Microbiol 23:1293-1301

von Heijne G (1986) A new method for predicting signal sequence cleavage sites. Nucleic Acids Res 14:4683-4690

Wandersman C (1996) Secretion across the bacterial outer membrane. In: Neidhardt FC, Curtiss III R, Ingraham JL, Lin ECC, Low KB, Magasanik B, Reznikoff WS, Riley M, Schaechter M, Umbarger HE (eds) *Escherichia coli* and *Salmonella*. Cellular and Molecular Biology. ASM Press, Washington D.C., pp 955-966.

Watanabe M, Blobel G (1989) SecB functions as a cytosolic signal recognition factor for protein export in E. coli. Cell 58:695-705.

Wattiau P, Woestyn S, Cornelis GR (1996) Customized secretion chaperones in pathogenic bacteria. Mol Microbiol 20:255-262.

Weiss A, Johnson FD, Burns DL (1993) Molecular characterization of an operon required for pertussis toxin secretion. Proc Natl Acad Sci USA 90:2970-2974.

Wu SS, Wu J, Kaiser D (1997) The *Myxococcus xanthus pilT* locus is required for gliding motility although pili are still produced. Mol Microbiol 23:109-121.

OXYGEN TOXICITY, OXYGEN STARVATION AND THE ASSEMBLY OF CYTOCHROME c-DEPENDENT ELECTRON TRANSFER CHAINS IN *ESCHERICHIA COLI*

Jeff Cole and Helen Crooke

School of Biochemistry, University of Birmingham, Birmingham B15 2TT, UK.

1. Introduction

One of the major challenges faced by any microbial population is the regulated response to changes in the availability of oxygen and alternative electron acceptors that influence the redox potential of the growth environment. Organisms found only in a constant environment can ignore such a challenge, but at their peril. The price payed for simplicity is condemnation to live as obligate aerobes, obligate anaerobes, or obligately fermentative organisms. As detailed knowledge of molecular microbiology accumulates, the range of organisms falling into these stereotypes is rapidly decreasing: perhaps all prokaryotes have at least some capacity to respond to changes in the availability of different growth-supporting electron donors and energy-releasing electron acceptors. Certainly the most versatile bacteria include the rhizobia, some of which survive as free-living aerobes, as anaerobic denitrifiers and in symbiosis with a host plant in legumes where the dissolved oxygen concentration is in the nanomolar range; purple non-sulfur photosynthetic bacteria which can generate energy by aerobic respiration, anaerobic denitrification, photosynthesis, and by fermentation; and the enteric bacteria, most of which are facultative anaerobes that can also survive excessive oxygen. This article will first summarise how this last group adapts to extreme changes in their redox environment, the regulatory circuits that mediate the response to oxygen stress and then focus on the biosynthesis of periplasmic components, especially c-type cytochromes, of respiratory chains that enable enteric bacteria to adapt to anaerobic growth.

NATO ASI Series, Vol. H 103
Molecular Microbiology
Edited by Stephen J. W. Busby,
Christopher M. Thomas and Nigel L. Brown

2. Response of *E. coli* to oxygen limitation or excess.

Biochemical engineers make considerable efforts to supply sufficient oxygen to aerated microbial cultures to ensure optimal growth. However, excess oxygen causes just as much stress to a culture as insufficient aeration. Both result in different types of stress response. Both can drastically decrease the growth rate and yield of biomass; and excess oxygen induces genetic instability in the form of abnormally high mutation frequencies. At least seven genetic regulatory circuits control the growth rate and metabolism of enteric bacteria in response to changes in the availability of oxygen (Table 1). The inclusion of some of these circuits such as those controlled by NarX-NarL, NarQ-NarP and FhlA or omission of others such as Fur and RpoS are slightly arbitrary. This is because the former regulatory proteins do not respond directly to oxygen, but to consequences of oxygen starvation, while the latter are directly involved in regulating the synthesis of cellular components required to cope with excess oxygen, for example, during iron-limited growth or in the stationary phase.

Bacteria generate hydrogen peroxide as a by-product of respiration during aerobic exponential growth (Gonzalez-Flecha and Demple, 1995). Enteric bacteria also encounter hydrogen peroxide generated by macrophages during engulfment. If they are first exposed to a sub-lethal dose of peroxide, they synthesize 30 new proteins, 12 of them immediately after exposure and the rest of them after 10 to 30 minutes (Christman *et al.* 1985; Hidalgo and Demple, 1996). These proteins can protect the bacteria against further exposure to higher, lethal concentrations of hydrogen peroxide (Demple and Harrison, 1994). As part of this adaptive response, they synthesize two types of catalase, HP-I and HP-II, to degrade hydrogen peroxide to oxygen and water. The gene for HP-I is *katG* (Claiborne and Fridovich, 1979); the gene for HP-II is *katE* (Loewen and Triggs, 1984). The *katG* gene is regulated primarily by the transcription factor, OxyR, which is a transcription activator when a key cysteine residue in the central domain of the protein becomes oxidised. The other gene, *katE* encoding HP-II, is regulated by the stationary phase sigma factor,

Table 1. Regulatory circuits which enable *E. coli* to respond to oxygen limitation or excess.

Regulator	Stress response
OxyR	Oxidative stress: excess O_2 and H_2O_2
SoxR-SoxS	Oxidative stress: superoxide
ArcB-ArcA	Anaerobic repression of TCA cycle
Fnr	ON switch for anaerobic respiration
NarX-NarL/NarQ-NarP	Response to nitrate and nitrite
FhlA	Response to excess formate
Plus at least one other	Anaerobic control of TMAO reduction.

RpoS. OxyR also regulates the synthesis of alkyl hydroperoxide reductase, an enzyme involved in the removal of organic peroxides; and glutathione reductase.

When redox cycling components in the bacterial electron transfer chain fail to function optimally, considerable quantities of superoxide are generated. Although superoxide is not particularly toxic, it is the source of the far more toxic hydroxyl radical. About 40 new proteins are formed in response to exposure to superoxide radicals (Greenberg and Demple, 1989; Hidalgo and Demple 1996). At least nine of these proteins are regulated by the two-component system, SoxR-SoxS. One of these nine proteins is the Mn^{++}-dependent superoxide dismutase, SodA. Others include two TCA cycle enzymes and a hydroperoxidase which degrades organic peroxides generated, for example, as a result of lipid damage.

Decreases in the oxygen concentration, caused for example by the increased rate of reduction as the bacterial population increases during growth, lead to parallel responses in cell physiology. During oxygen-limited growth, many bacteria synthesize a high affinity cytochrome oxidase which is not found in oxygen-sufficient cultures. The major cytochrome oxidase of *E. coli* during oxygen-sufficient growth is cytochrome *o*, but as oxygen becomes limiting, cytochrome *d* becomes the dominant oxidase (Gennis and Stewart, 1996). Many other changes in cell physiology have to be controlled as oxygen becomes limiting. Genes for the citric acid cycle and lipid oxidation are down-regulated during oxygen starvation, but genes for anaerobic electron transfer reactions are switched on. The OFF switch for the TCA cycle and lipid oxidation is the two-component regulatory system, ArcB-ArcA (Lynch and Lin, 1996a,b): the FNR protein provides the critical ON switch for anaerobic electron transfer pathways and also for many other enzymes as well (Spiro & Guest, 1990; Green & Guest, 1997). The best alternative electron acceptor to oxygen during anaerobic growth is nitrate: there are two, two-component regulatory systems, NarX NarL and NarQ NarP, which detect the presence of nitrate and nitrite in the environment. The two DNA-binding proteins, NarL and NarP, regulate gene expression in response to changes in the availability of nitrate and nitrite (Stewart, 1993; 1997).

During anaerobic growth in the absence of nitrate, nitrite or fumarate, many bacteria survive by generating ATP by substrate level phosphorylation as a result of fermentation reactions. Formate is perhaps the single most important fermentation product of enteric bacteria Its metabolism is regulated by both FNR and by a novel regulatory protein, FHLA (Rossman *et al*, 1991). The FHLA protein is not phosphorylated like the response regulators of two-component systems: instead, it binds formate directly (Korsa & Böck, 1997). In the presence of formate, it turns ON genes for formate metabolism to generate hydrogen gas and carbon dioxide.

Finally, although Table 1 illustrates the complexity of how at least one group of bacteria respond to oxygen excess or oxygen starvation, we know the list is incomplete. For example, trimethylamine-N-oxide (TMAO) is an excellent electron acceptor for anaerobic growth of many bacteria. In *E. coli*, synthesis of reductase is regulated by two factors, the presence of TMAO and the absence of oxygen. A two-component regulatory system responds to the presence of TMAO in the environment (Simon *et al.*, 1995); another unidentified regulatory circuit enables the *tor* genes to respond to anaerobiosis (Méjean *et al* 1994).

3. Duplication of anaerobically-induced enzymes: are they mutually redundant?

Survival in the absence of oxygen requires the synthesis of alternative mechanisms for generating ATP. This can be achieved by photosynthesis, by substrate level phosphorylation during fermentation reactions, or by anaerobic respiration. Even within the same family, different bacterial genera synthesize a different range of electron transfer chains during anaerobic growth. Publication of the complete DNA sequence of the *E. coli* genome has defined the upper limits of the possible range of electron transfer chains in that organism, but it is still uncertain whether all of the putative genes are expressed. With this caveat, the total range of anaerobically-induced electron transfer pathways synthesized by *E. coli* during anaerobic growth likely includes three nitrate reductases, two nitrite reductases, fumarate reductase, two S-oxide reductases currently known as dimethylsulphoxide reductase and biotin sulphoxide reductase, two N-oxide reductases, one of which is specific for TMAO and the other of much lower substrate specificity; and four hydrogenases (Cole 1996). A key question is: Why does *E. coli* synthesize multiple enzymes to catalyse the same reaction? Why, for example, are three different nitrate reductases synthesized during anaerobic growth in the presence of nitrate, all of which result in energy conservation?

The three nitrate reductases are NarG, NarZ and Nap. NarG is a membrane-bound enzyme which transfers electrons from either benzoquinone or naphthoquinones via a *b*-type cytochrome (NarI), an iron-sulphur protein (NarH) and a molybdoprotein (NarG) to nitrate (Chaudry and MacGregor, 1982; Sondergen and DeMoss, 1988).

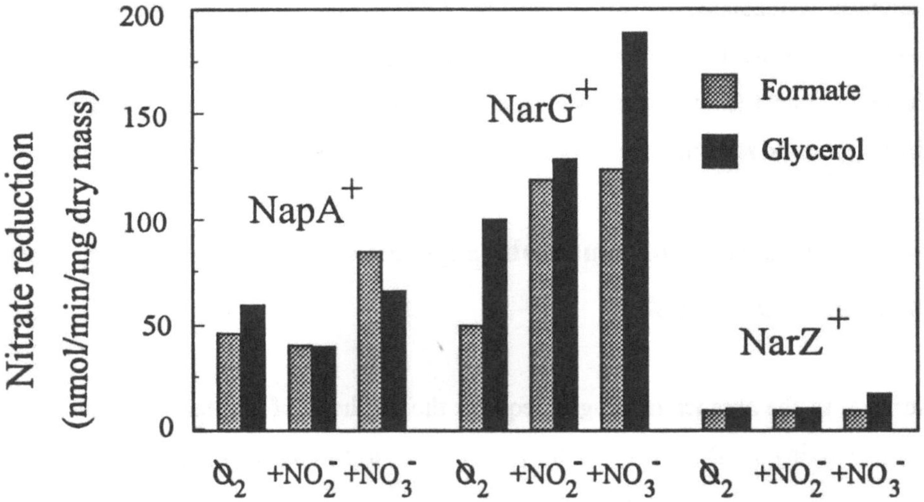

Fig 1 Activity of the three nitrate reductases of *Escherichia coli* Double mutants in which only one of the three nitrate reductases synthesised by *Escherichis coli* remains active were grown anaerobically in the presence of glycerol, fumarate and, where indicated, either 2.5 mM nitrite or 20 mM nitrate. The ratess of nitrate reduction by bacteria incubated with either formate or gylcerol were measured using a nitrate-specific electrode.

An important point is that nitrate is reduced to nitrite on the cytoplasmic side of the membrane. NarZ is almost a carbon copy of NarG with three homologous sub-units, a b-type cytochrome NarV, an iron-sulphur protein, NarY and a molybdoprotein, NarZ: clearly the *narG* and *narZ* operons have evolved by gene duplication. They are so closely related that the protein products are inter-changeable (Blasco *et al.* 1990, 1992). The individual sub-units have been purified and shown that NarG, for example, can accept electron from NarY and NarV to reduce nitrate to nitrite (Blasco *et al.* 1992). The third enzyme is called Nap (nitrate reductase located in the periplasm).

Double mutants in which two of the three nitrate reductases are absent have been used to determine whether each nitrate reductase on its own is sufficiently active to reduce nitrate rapidly and can support an increased growth rate or yield of biomass during anaerobic growth in the presence of nitrate (Figure 1). During growth in the presence of nitrate, NarG is by far the most active enzyme. The assumption that this is therefore physiologically the most important enzyme might not be correct. In contrast, although NarZ appears to be so inactive that it must be useless, NarZ is the only nitrate reductase that is also present during aerobic growth, so perhaps NarZ helps bacteria to adapt rapidly as the culture becomes anaerobic. The real puzzle is the third nitrate reductase, Nap, located in the periplasm which, using a double mutant defective in the other two nitrate reductases, was recently shown to be almost equally as effective as NarG in stimulating the rate of anaerobic growth. The question remains: what are the physiological benefits to the bacteria of retaining both of them? A possible answer comes from the realisation that broth cultures of single strains of bacteria used for most laboratory experiments are supplemented with 20 or 100 mM nitrate. However, the real world is not full of Lennox broth: In any natural environment, especially the human body, the supply of nitrate, nitrite and oxygen are all limited. There are several reasons for proposing that, under these conditions, Nap might be a significant enzyme for nitrate reduction. First, enteric bacteria encounter nitrate outside the cell: they do not synthesise it internally. Consequently, Nap in the periplasm is better placed than NarG, with its catalytic site in the cytoplasm, to reduce a limited supply of nitrate. Secondly, Nap is sufficiently active to compete with NarG for a limited supply of nitrate (Figure 1). Perhaps NarG is only significant when a large excess of nitrate is available?

4. Duplicated enzymes for nitrite reduction to ammonia

The product of nitrate reduction is nitrite which is toxic to all cell types. Enteric bacteria - indeed, bacteria from virtually all carbon-rich but nitrogen poor environments - reduce nitrite rapidly to ammonia (Cole and Brown, 1980). In *E. coli* there are two different nitrite reductases to catalyse exactly the same chemical

Table 2. Control of transcription of genes for two *E. coli* nitrite reductases, both of which catalyse nitrite reduction to ammonia during anaerobic growth.

Environmental signal	Effect on Nir*	Effect on Nrf*
Lack of oxygen	Induced by FNR protein	Induced by FNR protein
Presence of nitrite	Induced by NarL / NarP	Induced by NarL / NarP
Presence of nitrate	Induced by NarL / NarP	Repressed by NarL but induced by NarP
Presence of glucose	Unaffected	Repressed

* Nir is the soluble, cytoplasmic nitrite reductase involved in reducing toxic nitrite which accumulates in the cytoplasm; Nrf is the electrogenic, formate-dependent niterite reductase which includes the periplasmic nitrite reductase, cytochrome c_{552}.

reaction. Both of these enzymes are regulated by the FNR protein (Page *et al.* 1990 Tyson *et al.* 1994): the enzymes are formed only during anaerobic growth.

One of the nitrite reductases uses formate as the electron donor in a membrane-bound electron transfer pathway. Formate reduces naphthoquinones: electrons are then transferred from the naphthoquinones via the Nrf proteins to nitrite in the periplasm (Darwin *et al.* 1993a,b; Tyson *et al.* 1997). The formate pathway is most active when nitrite is present but nitrate is absent (Table 2). As formate is also an excellent electron donor for nitrate as well as for nitrite reduction, nitrate inhibits rather than induces the synthesis of formate-dependent nitrite reductase (Page *et al.* 1990). The reason is simple: more energy is released during nitrate reduction than during nitrite reduction, so this complex series of control mechanisms enable the bacteria to select and use the most favourable source of energy first.

The second enzyme is the product of the *nir* operon (Jayaramman *et al.*, 1987; Tyson *et al.*, 1993). It uses NADH in the cytoplasm to reduce nitrite to ammonia and its physiological role is to remove the toxic nitrite which accumulates during nitrate reduction. No energy is conserved by this pathway, but it is most useful when nitrate is available in excess and is being converted to nitrite which accumulates in the cytoplasm. Consequently, synthesis of the NADH-dependent nitrite reductase is induced during growth in the presence of either nitrate or nitrite. Thus we can explain why two biochemically and genetically independent nitrite reductases are required: the two enzymes are useful in different ways, but they must be regulated independently so one can be used when nitrate is present; the other is most useful when nitrate is absent.

5. Assembly of c-type cytochromes in the periplasm.

In most organisms including mammals, cytochrome *c* is an essential component of the aerobic respiratory chain. In the mammalian respiratory chain there are two *c*-type cytochromes. One is attached to the membrane as part of the bc_1 complex; the other transfers electrons from the bc_1 complex to the copper-containing cytochrome oxidase, cytochrome aa_3. Furthermore, there is essentially only a single electron transfer pathway in mitochondria. In contrast, electron transfer pathways in bacteria are usually branched. The surprise is that neither of the two branches of the *E. coli* electron transfer chain to oxygen contain *c*-type cytochromes. However, during anaerobic growth, three of the anaerobically-induced electron transfer chains include *c*-type cytochromes that are not synthesised during aerobic growth. All of these cytochromes are located in or face the periplasm. These three electron transfer chains include five different types of cytochrome c. One of them is essential for the reduction of TMAO; two are part of the periplasmic nitrate reductase and two more are components of the formate-dependent nitrite reductase pathway (Grove *et al.* 1996a,b). Bacterial c-type cytochromes are rigid molecules with planar haem groups attached covalently by thioether bonds to pairs of cysteine residues organised as a -C-X-X-C-H- motif in the apoproteins. Attempting to explain how such a structure is either assembled in the cytoplasm and exported to the periplasm, or assembled in the periplasm, represents a significant problem in prokaryote cell biology.

Only unfolded proteins can be transported across the bacterial membrane by the conventional secretion (Sec) pathway (Pugsley, 1997), so a novel export mechanism would be required if haem is attached to the apoprotein in the cytoplasm. The alternative possibility is that c-type cytochromes are assembled outside the cell membrane in the periplasm. If so, cytochrome c pre-apoproteins should include normal N-terminal signal peptide sequences and be secreted by the Sec pathway. A separate transport mechanism, possibly an ABC-type traffic ATPase, might be used to transfer haem to the periplasm, but a haem lyase would also be required to catalyse the formation of the thioether bonds that link the pairs of cysteine residues to the haem c prosthetic group. Furthermore, the putative haem lyase must either be a membrane-anchored protein which faces the periplasm; or it must be located in the periplasm.

A considerable body of evidence now persuasively argues that the periplasmic assembly model is correct. First, genes coding for c-type cytochromes that are found in the periplasm include signal sequences that make the apoproteins ideal substrates for the conventional Sec pathway. Secondly, cytochrome *c* apoproteins are secreted by mutants which cannot make haem (Page and Ferguson, 1990). Furthermore, these apoproteins lack the signal peptide, suggesting that they had been cleaved by the "leader peptidase", Lep. Direct evidence that the Sec pathway and the Lep peptidase are essential for cytochrome c assembly has recently been obtained (Thöny-Meyer and Künzler, 1997). The apoproteins accumulate in the cytoplasm of the Sec mutants; in a Lep^- mutant, cytochrome pre-apoproteins are secreted and haem is attached in the absence of signal peptide cleavage.

The seven genes for the periplasmic nitrate reductase were discovered from results from the *E. coli* Genome Sequencing Project which revealed that they are part of a large operon, currently reported to include 15 genes, in the 47 minute region of the chromosome. The periplasmic nitrate reductase molybdoprotein is NapA. The other *nap* genes encode two c-type cytochromes, NapB and C, three iron-sulphur proteins and one protein of unknown function. Downstream of the seven *nap* genes are eight genes (epithet *ccm* for cytochrome *c* maturation) which are again similar to genes essential for cytochrome c assembly in Rhodobacter, Bradyrhizobium and Paracoccus (Beckman *et al.* 1993; Thöny-Meyer *et al.* 1994). The first four genes are predicted to code for a traffic ATPase, possibly to transfer haem from the cytoplasm to the periplasm. The products of the last four genes are predicted to include three haem binding proteins, which might possibly form a haem lyase complex, and a thioredoxin-like protein, CcmG. Parallel but independent work in Birmingham and Zurich established that these eight downstream genes are essential for the synthesis of all c-type cytochromes in *E. coli* (Thöny-Meyer *et al.*, 1995; Grove *et al.*, 1996a). Subsequently, each individual gene has been shown to be essential for the synthesis of all five well-characterised c-type cytochromes in *E. coli*.

Two further genes, *cydC* and *cydD*, have also been considered to be candidates for encoding the haem transporter (Poole *et al.*, 1994; Goldman *et al.*, 1996). When grown under optimal conditions, mutants defective in these genes accumulate detectable but less cytochrome c than the parental strain (Crooke & Cole, unpublished data). An auxilliary role for CydD and CydC in haem transport or metabolism therefore seems likely. The primary candidate for the haem transporter remains the products of the *ccmA, B, C* and *D* genes.

A second group of genes have also been proposed to encode a cytochrome *c* haem lyase. The *nrf* operon coding for the formate-dependent nitrite reductase consists of seven genes (Hussain *et al.*, 1994). The proposed functions of the gene products, all except one of which are essential for Nrf activity (Grove et al., 1996b), are summarised in Table 3. The first two genes encode c-type cytochromes: *nrfA* encodes cytochrome c_{552}, the terminal nitrite reductase located in the periplasm, and NrfB is a membrane-anchored, 18 kDa cytochrome c believed to be the immediate electron donor to cytochrome c_{552}. The next two genes, C and D, encode a quinol

Table 3. Function of the products of the genes of the *nrf* operon

Gene	Proposed function of product	Features of product
nrfA	Nitrite reductase	Cytochrome c_{552} with 4 C-X-X-C-H motifs
nrfB	Electron transfer	Pentahaem cytochrome c
nrfC	Quinol oxidase	Similar to NADH-quinone oxidoreductase
nrfD	Quinol oxidase	Includes four x 4Fe-4S iron sulphur centres
nrfE	Heam lyase component	Putative haem binding site, like CcmF , Ccl1
nrfF	Haem lyase component	Like N-terminal domain of CcmH and Rc-Ccl2
nrfG	Haem lyase component	Like C-terminal domain of CcmH and Bj-CycH

oxidase. The final three genes encode proteins similar to the proteins in *Rhodobacter* which were suggested by Beckman *et al.* (1992) to be a haem lyase. No biochemical assay for a bacterial haem lyase activity is currently available, so direct evidence for or against CcmFH or NrfEFG being able to catalyse haem ligation remains elusive. The following questions therefore remain unanswered. First, why are genes proposed to be implicated in haem attachment duplicated in *E. coli*? Secondly, how can both sets of haem lyase genes be essential for formate-dependent nitrite reductase activity? Thirdly, why is one set of the putative haem lyase genes tacked onto the end of the *nrf* operon? Finally, how can the *ccm* genes be controlled by FNR and NarL and yet the protein products still be made independently of FNR and NarL during TMAO reduction?

6. Chaperones for thiols in the periplasm?

The substrate requirements of haem lyases present a fascinating problem in prokaryote physiology. Haem lyases catalyse the attachment of haem groups to pairs of reduced cysteine residues: but the periplasm is an oxidizing environment, even during anaerobic growth in the presence of nitrate or nitrite. So how can the cysteines to which haem will be attached be kept reduced in an oxidizing environment?

The first clue came from the discovery of a disulphide isomerase-like protein, DipZ, which is essential for the synthesis, assembly or stability of all c-type cytochromes in *E. coli* (Crooke and Cole, 1995). Although DipZ is very hydrophobic and therefore firmly attached to the cytoplasmic membrane, the predicted active centre of DipZ is a C-X-X-C motif located in the periplasm (Metheringham *et al.* 1996). Notice that this is similar to the motif to which haem is attached in cytochrome c. This suggested that DipZ might be required to reduce any disulphide bonds that form spontaneously in the oxidizing environment of the periplasm. The next result was far more surprising, however: a mutation resulting in total loss of cytochrome *c* accumulation was mapped to *dsbA* which codes for a protein known to be essential for disulphide bond formation in the periplasm (Metheringham *et al.*, 1995). This provided the first evidence that there might be a cycle of oxidation-reduction reactions in which one or more substrate - apoprotein, or haem, or both - are presented to a periplasmic assembly complex in a usable redox state. Indeed, the simplest explanation for the observations that DsbA and DipZ are both required for cytochrome c assembly is that they are both components of an assembly pathway in the periplasm.

Electrons transferred to DsbA during disulphide bond formation are passed into the electron transfer chain via a membrane-bound partner protein, DsbB (Missiakas and Raina, 1997). It has recently been shown that DsbB is also essential for cytochrome c synthesis (Metheringham *et al.*, 1996). So how are electrons transferred to or from DipZ to reduce the disulphide bond in preparation for haem attachment? The thioredoxin-like protein, CcmG, which is predicted to be periplasmic and a disulphide reductase rather than a thiol oxidase, is an obvious candidate for such a role.

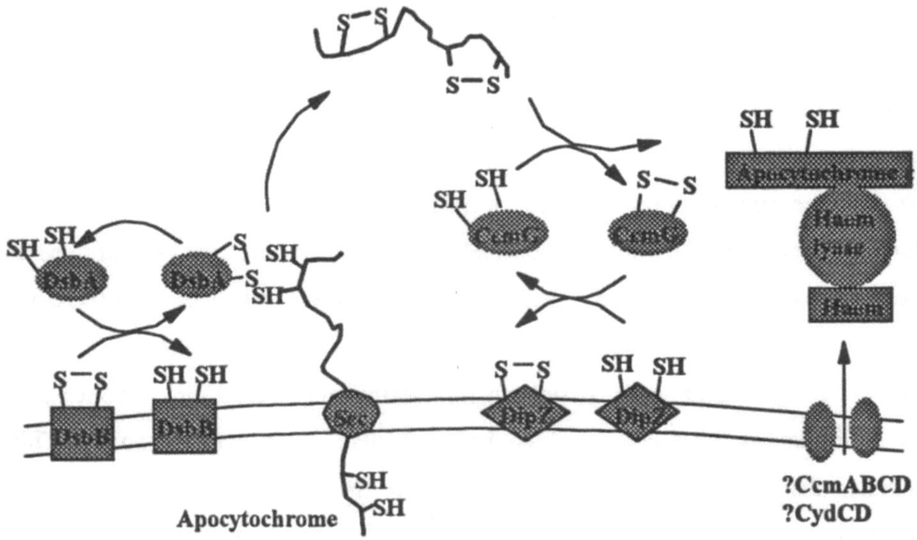

Fig. 2 Proposed role for protein disulphide oxidoreductases in cytochrome c biogenesis. The model proposes that reduced cysteine residues of the newly translocated apocytochromes are oxidised in the periplasm to a disulphide by DsbA which is recycled to its active oxidised form by DsbB. This would prevent unwanted side reactions such as the formation of random. unproductive disulphides and help fold the apocytochrome into the correct conformation. These disulphides must be re-reduced prior to haem attachment, possibly by the periplasmic, thioredoxin-like CcmG working in partnership with the membrane-anchored disulphide isomerase-like protein DipZ.

If haem is attached to reduced cysteine residues, why is it necessary to form disulphide bonds in apocytochrome c in the first place? In the oxidizing environment of the periplasm, *c*-type cytochromes are at risk of forming unproductive disulphide bonds at random. These bonds could be inter-molecular or intra-molecular. By locking each pair of cysteines into a disulphide bond as they emerge in pairs from the membrane-bound Sec complex into the periplasm, DsbA would effectively prevent the spontaneous formation of incorrect disulphide bonds either within a single polypeptide or even between different polypeptides. Having locked the cysteines in their correct pairs, CcmG and its partner, DipZ, would then be required to re-reduce each disulphide bond and present the correct pairs of cysteines to the haem lyase for haem attachment. According to this model (Figure 2), CcmG and DsbA are functioning together as periplasmic chaperones. If so, CcmG, like DsbA, DsbB DipZ, should be essential for cytochrome c synthesis and hence for both Nrf and Nap activity.

7. An alternative export pathway for complex redox proteins?

Finally, how are other types of electron transfer proteins with novel co-factors assembled into periplasmic electron transfer chains? All three nitrate reductases, including the periplasmic nitrate reductase, Nap, are molybdoproteins. In bacteria, almost twenty genes are required for the synthesis of the co-factor molybdopterin guanine dinucleotide which is synthesised in the cytoplasm. This raises the same series of awkward questions about how the periplasmic nitrate reductase molybdoprotein gets out of the cytoplasm: is the Sec pathway involved and, if not, is there a second, so far undiscovered export pathway? If a second export pathway exists, do any other proteins - for example - non-heam iron sulphur proteins, use the same pathway?

Berks (1996) compared the N-terminal sequences of over 150 periplasmic pre-apoproteins known or believed to be associated with such cofactors as molybdopterin, iron-sulphur centres, polynuclear copper sites, flavin adenine dinucleotide and tryptophan tryptophylquinone. Such polypeptides lack conventional leader peptides typical of those exported by the Sec pathway. Instead, a well-conserved motif with the consensus S/TRRXFLK was found close to the amino terminus of all of these pre-apoproteins, and there are also noticeable similiarities in the sequences before and after the common "double arginine" motif. This prompted Berks (1996) to propose that there might be an alternative export pathway which is specific for partially folded substrates. Consistent with this suggestion is the demonstration that signal peptidase cleavage sites of such polypeptides are located at much more variable positions than those of substrates for the Sec pathway and do not always follow von Heijne's (1987) rules. Biochemical and genetic evidence for such a pathway and direct proof that the double arginine motif is essential for this pathway to function are now eagerly awaited.

Acknowledgements

This research was supported by Projects Grants G01944 and GRJ26427 from the UK Biotechnology and Biological Sciences Research Council

References

Beckman, D.L., Trawick, D.R. and Kranz, R.G. (1992) Bacterial cytochromes *c* biogenesis. *Genes Devel* 6: 268-283.

Berks BC (1996) A common export pathway for proteins binding complex redox cofactors? Mol Microbiol 22:393-404

Blasco F, Iobbi C, Ratouchniak J, Bonnefoy V, Chippaux M (1990) Nitrate reductases of *Escherichia coli*: Sequence of the second nitrate reductase and comparison with that encoded by the *narGHJI* operon. Mol Gen Genet 222: 104-111

Blasco F, Nunzi F, Pommier J, Brasseur R, Chippaux M, Giordano G (1992) Formation of active heterologous nitrate reductases between nitrate reductases A and Z of *Escherichia coli*. Mol Microbiol 6: 209-219

Christman MF, Morgan RW, Jacobson FS, Ames BN (1985) Positive control of a regulon for defenses against oxidative stress and some heat shock proteins in *Salmonella typhimurium*. Cell 41:753-762

Chaudhry, G.R. and MacGregor, C.H. (1982) Cytochrome b from *Escherichia coli* nitrate reductase: its properties and association with enzyme complex. J. Biol. Chem. 258, 5819-5827.

Claiborne A, Fridovich I (1979) Purification of the *o*-dianisidine peroxidase from *Escherichia coli* B. Physicochemical characterisation and analysis of its dual catalytic and peroxidatic activities. J Biol Chem 254:4245-4252

Cole J (1996) Nitrate reduction to ammonia by enteric bacteria: redundancy, or a strategy for survival during oxygen starvation? FEMS Microbiology Letters 136:1-11

Cole JA, Brown CM (1980) Nitrite reduction to ammonia by fermentative bacteria: a short circuit in the biological nitrogen cycle. FEMS Microbiol Lett 7:65-72

Crooke H, Cole J (1995) The biogenesis of *c*-type cytochromes in *Escherichia coli* requires a membrane-bound protein, DipZ, with a protein disulphide isomerase-like domain. Mol Microbiol 15: 1139-1150

Darwin A, Hussain H, Griffiths L, Grove J, Sambongi Y, Busby S, Cole J (1993a) Regulation and sequence of the structural gene for cytochrome c_{552} from *Escherichia coli*, not a hexahaem but a 50kDa tetrahaem nitrite reductase. Mol Microbiol 9: 1255-1265

Darwin A, Tormay P, Page L, Griffiths L, Cole J (1993b) Identification of the formate dehydrogenases and genetic determinants of formate-dependent nitrite reduction by *Escherichia coli* K12. J Gen Microbiol 139: 1829-1840

Demple B, Harrison L (1994) Role of oxidative damage to DNA: enzymology and biology. Annu Rev Biochem 63:915-948

Goldman BS, Gabbert KK, Kranz RG (1996) Use of heme reporters for studies of cytochrome biosynthesis and heme transport. J Bacteriol 178:6338-6347

Gennis RB, Stewart V (1996) Respiration. 217-261. *Escherichia coli* and *Salmonella*. Cellular and molecular biology. Neidhardt FC, Curtiss R, III, Ingraham JL et al., Ed. Washington D. C., ASM Press.

Gonzalez-Flecha B, Demple B (1995) Metabolic sources of hydrogen peroxide in aerobically growing *Escherichia coli*. J Biol Chem 270:13681-13687

Green J, Guest JR (1997) The citric acid cycle and oxygen-regulated gene expression in *Escherichia coli*. This volume.

Greenberg G, Demple B (1988) Overproduction of superoxide scavenging enzymes in *Escherichia coli* suppresses spontaneous mutagenesis and sensitivity to redox-cycling agents in *oxyR*⁻ mutants. EMBO J 7:2611-2617

Greenberg G, Demple B (1989) A global response in *Escherichia coli* induced by redox-cycling agents overlaps with that induced by peroxide stress. J Bacteriol 171:3933-3939

Grove J, Tanapongpipat, S, Thomas G, Griffiths L, Crooke H, Cole J (1996a) *Escherichia coli* K-12 genes essential for the synthesis of *c*-type cytochromes and a third nitrate reductase located in the periplasm. Mol Microbiol 19:467-481

Grove J, Busby S, Cole J (1996b) The role of the genes *nrfEFG* and *ccmFH* in cytochrome *c* biosynthesis in *Escherichia coli*. Mol Gen Genet 252:332-341

von Heijne G (1987) In *Sequence Analysis in Molecular Biology: Treasure Trove or Trivial Persuit*. New York. Academic press. pp.113-117

Hidalgo E, Demple B (1996) Adaptive response to oxidative stress: the *soxRS* and *oxyR* regulons. 435-452. Regulation of gene expression in *Escherichia coli*. LynchAS, Lin ECC. Ed. Austin, Texas. RG Landes Co.

Hussain H, Grove J, Griffiths L, Busby S, Cole J (1994) A seven-gene operon essential for formate-dependent nitrite reduction to ammonia by enteric bacteria. Mol Microbiol 12: 153-163

Jayaraman P-S, Peakman T, Busby S, Quincey R, Cole J (1987) Location and sequence of the promoter of the gene for the NADH-dependent nitrite reductase of *Escherichia coli* and its regulation by oxygen, the FNR protein and nitrite. J Mol Biol 196: 781-788

Korsa I, Böck A (1997) Characterization of *fhlA* mutations resulting in ligand-independent transcriptional activation and ATP hydrolysis. J Bacteriol 179:41-45

Loewen PC, Triggs BL (1984) Genetic mapping of *katF*, a locus that with *katE* affects the synthesis of a second catalase species in *Escherichia coli*. J Bacteriol 160:668-675

Lynch AS, Lin ECC (1996a) Regulation of aerobic and anaerobic metabolism by the Arc system. 361-381. Regulation of gene expression in *Escherichia coli*. Lynch AS, Lin ECC. Ed. Austin, Texas. RG Landes Co.

Lynch AS, Lin ECC (1996b) Responses to molecular oxygen. 1526-1538. *Escherichia coli* and *Salmonella*. Cellular and molecular biology. Neidhardt FC, Curtiss R, III, Ingraham JL et al., Ed. Washingto D. C., ASM Press.

Méjean V, Iobbi-Nivol C, Lepelletier M, Giordano G, Chippaux M, Pascal MC (1994) TMAO anaerobic respiration in *Escherichia coli*: involvement of the *tor* operon. Mol Microbiol 11:1169-1179

Metheringham R, Griffiths L, Crooke H, Forsythe S, Cole J (1995) An essential role for DsbA in cytochrome *c* synthesis and formate-dependent nitrite reduction by *Escherichia coli* K-12. Arch Microbiol 164: 301-307

Metheringham R,Tyson K, Crooke H, Missiakas D, Raina S, Cole J (1996) Effects of mutations in genes for proteins involved in disulfide bond formation in the periplasm on the activities of anaerobically induced electron transfer chains in *Escherichia coli* K-12. Mol Gen Genet 253:95-102

Missiakas D, Raina, S (1997) Protein folding in the bacterial periplasm. J Bacteriol 179:2465-2471

Page MD, Ferguson SJ (1990) Apo forms of cytochrome c_{550} and cytochrome cd_1 are translocated to the periplasm of *Paracoccus denitrificans* in the absence of haem incorporation caused either by mutation or inhibition of haem synthesis. Mol Microbiol 4:1181-1192

Page L, Griffiths L, Cole JA (1990) Different physiological roles for two independent pathways for nitrite reduction to ammonia by enteric bacteria. Arch Microbiol 154:349-354

Poole RK, Gibson F, Wu G (1994) The *cydD* gene product, component of a heterodimeric ABC transporter, is required for the assembly of periplasmic cytochrome c and of cytochrome *bd* in *Escherichia coli*. FEMS Microbiol Lett 117:217-223.

Pugsley AP (1997) Protein traffic in bacteria. This volume

Rossman R, Sawers G, Böck (1991) Mechanism of regulation of the formate-hydrogenlyase pathway by oxygen, nitrate and pH: definition of the formate regulon. Mol Microbiol 5:2807-2814

Simon G, Jourlin C, Ansaldi M, Pascal MC, Chippaux M, Méjean V (1995) Binding of the TorR regulator to *cis*-acting direct repeats activates *tor* operon expression. Mol Microbiol 17:971-980

Sondergen, E.J. and DeMoss, J.A. (1988) *narI* region of the *Escherichia coli* nitrate reductase (*nar*) operon contains two genes. J. Bacteriol. 170, 1721-1729.

Spiro S, Guest J (1990) FNR and its role in oxygen regulated gene expression in *Escherichia coli*. FEMS Microbiol Rev 75:399-428

Stewart V (1997) Bacterial two-component regulatory systems. This volume

Stewart (1993) Nitrate regulation of anaerobic respiratory gene expression in *Escherichia coli*. Mol Microbiol 9:425-434

Thöny-Meyer L, Fischer F, Künzler P, Ritz D, Hennecke H (1995) *Escherichia coli* genes required for cytochrome c maturation. *J Bacteriol* **177**: 4321-4326.

Thöny-Meyer L, Künzler P (1997) transduction to the periplasm and signal sequence cleavage of preapoprotein *c* depend on *sec* and *lep*, but not on the *ccm* gene products. Eur J Biochem. In press.

Tyson KL, Bell AI, Cole JA, Busby S (1993). Definition of the nitrite regulation at the promoters of two *Escherichia coli* operons encoding nitrite reductase. Mol Microbiol 13:1045-1055.

Tyson KL, J A Cole JA, Busby SJW (1994). Nitrite and nitrate response elements at the anaerobically inducible *Escherichia coli nirB* promoter: interactions between FNR and NarL. Mol Microbiol 7:151-157.

Aspects of the Molecular Genetics of Antibiotics

Julian Davies

Dept of Microbiology and Immunology, University of British Columbia
300 - 6174 University Blvd., Vancouver, B.C. V6T 1Z3, Canada

Antibiotics have been in use for more than half a century, and their clinical applications have severely restricted the spread of diseases resulting from infections by bacteria, yeasts, and fungi. Although most of the common antibiotics have their structural roots in natural products of bacteria and fungi, many of them have been chemically modified in attempts to improve their pharmacologic properties and spectrum of action. A number of antibiotics (perhaps more correctly antimicrobial agents) have no natural counterpart, being purely synthetic compounds. A list of some of the more common antibiotics is given in Table 1. The objective of this brief review is to summarise current knowledge of antibiotic biosynthesis, modes of action, and mechanisms of resistance.

Antibiotics are a subgroup of the so-called secondary metabolic products of microbes. These constitute an enormous range of molecular diversity, and thousands of new compounds are isolated annually, with many more still to be identified. While the secondary metabolites may well perform roles as antibiotics in nature, it is probable that these molecules have diverse biological functions, including activities as microbial pheromones, inter-cellular signalling agents, and cell regulators (Davies 1990).

Not surprisingly, the molecular complexity of secondary metabolites is paralleled by their biosynthetic complexity, and to date relatively few of the biosynthetic pathways of antibiotics have been characterised in detail. However, for some classes, the general rules of structural assembly have been worked out, and a few examples are given here. It has been shown that there are (at least) seven biosynthetic families (with numerous sub-groups) (Table 2) of secondary metabolites (Vining 1995).

NATO ASI Series, Vol. H 103
Molecular Microbiology
Edited by Stephen J. W. Busby,
Christopher M. Thomas and Nigel L. Brown

Table 1. Common antimicrobial agents

Antibiotic	Example	Spectrum of use	Target	Mode of action
β-lactam	ampicillin cephalexin	broad	cell wall synthesis	attach to penicillin-binding proteins
aminoglycoside	gentamicin	broad	translation	bind to 30S ribosomal subunit (16S RNA), interfere with initiation, decoding
tetracycline	oxytetracycline minocycline	broad	translation	bind to 30S ribosomal subunit, interfere with aminoacyl tRNA binding into the A site
chloramphenicol	chloramphenicol	broad	translation	bind to 50S ribosomal subunit, interfere with peptide bond formation
fluoroquinolone	ciprofloxacin	broad	DNA synthesis	interfere with topoisomerase function by binding to GyrA (gram-) and/or topoisomerase IV (gram+)
macrolide	erythromycin clarithromycin	gram +	translation	bind to 50S ribosomal subunit (23S RNA), interfere with peptidyl transferase and possibly release of peptide chain
glycopeptide	vancomycin teicoplanin	gram +	cell wall synthesis	bind to D-Ala-D-Ala component of growing peptide chain, prevent cross-linking in cell wall
sulphonamide	--	--	folic acid synthesis	inhibit dihydropteroate synthase
trimethoprim	trimethoprim	--	folic acid synthesis	inhibit dihydrofolate reductase
rifamycin	rifamycin	TB	RNA synthesis	bind to RNA polymerase, prevent mRNA chain elongation
isoniazid	INH	TB	?	pro-drug needs activation in cell; several biochemical consequences

Table 2. Biosynthetic classes of antibiotics

Family	Subgroup	Antibiotic	Producing organism
Chorismic acid	phenylpropanoid	chloramphenicol	*S. venezuelae*
Amino acid	serine	cycloserine	*S. lavendulae*
	tyrosine	lincomycin	*S. lincolnensis*
Peptide	polypeptide	gramicidin	*B. brevis*
	glycopeptide	vancomycin	*A. orientalis*
	ß-lactam	penicillin	*P. chrysogenum*
		cephamycin C	*S. clavuligerus*
	lantibiotic	nisin	*L. lactis*
Polyketide	tetracycline	oxytetracycline	*S. rimosus*
	macrolide	erythromycin	*Sac. erythraea*
	ansamycin	rifamycin	*A. mediterranei*
Isoprenoid	triterpene	fusidic acid	*Fd. coccineum*
Saccharide	aminocyclitol	spectinomycin	*S. spectabilis*
	aminoglycoside-aminocyclitol	gentamicin	*M. purpurea*
		streptomycin	*S. griseus*
Nucleoside	purine	puromycin	*S. alboniger*

A, *Amycolatopsis*; B, *Bacillus*; Fd, *Fusidium*; L, *Lactococcus*; M, *Micromonospora*; P, *Penicillium*; Sac, *Saccharopolyspora*; S, *Streptomyces*.

The **polyketides** constitute a large group of secondary metabolites formed by the condensation of C2 or C3 units which undergo ring formation, reduction, and other enzymic modifications (Hutchinson and Fujii 1995). There are two classes of polyketide molecule, type I and II. The type I polyketides produced by bacteria are synthesized in a non-reiterative manner by very high-molecular-weight modular polyketide synthases with sequential catalytic sites which are encoded by modular genetic clusters. The archetypical example is erythromycin, for which the biosynthetic gene clusters are shown in Figure 1 (Donadio et al. 1991). The type I polyketides of fungi are produced by combined reiterative and non-reiterative processes on single high-molecular-weight polyketide synthetases. As with the bacterial type I clusters, a number of accessory genes that are required for modifications of the core structure, including glycosylation and methylation, are found flanking the PKS genes. These structural modifications frequently modulate the biological activity of the polyketides; the type I polyketides are usually fully reduced molecules.

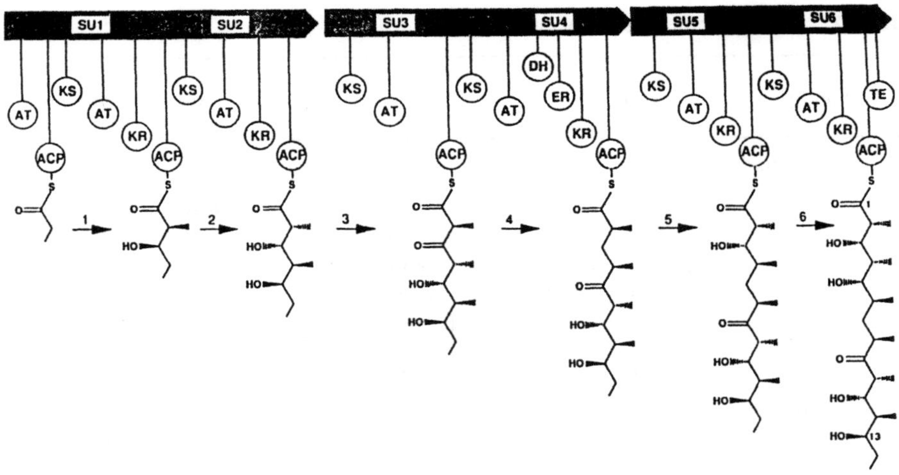

Figure 1. The sequential formation of 6-deoxyerythronolide B: the macrocyclic core of erythromycin. Note the organisation of enzymatic activities belonging to adjacent synthase units (SUs). ACP, acyl carrier protein; AT, acyltransferase; DH, dehydratase; ER, enoylreductase; KR, ß-ketoreductase; KS, ß-ketoacyl-ACP synthase; TE, thioesterase.

On the other hand, the synthesis of type II polyketides, widely distributed among bacterial genera, involves clusters of independent genes. Many such biosynthetic clusters have been identified, but only a few have been studied in some detail. Mechanistically, the type II PKS enzymes can be distinguished from those of type I since they lead to primarily unsaturated compounds. Because of their multicomponent genetic structure, type II PKS pathways can be readily manipulated by recombinant DNA methods to mix pathways and generate hybrid molecules.

Another good example of an antibiotic biosynthetic pathway which involves non-reiterative process is provided by the **peptide antibiotics**; they are produced by a 'non-ribosomal' process on very large peptide synthase enzymes. The peptides are a numerous and structurally diverse group of molecules that contain D- or L-amino acids in mixed configurations and with various modifications of the individual residues. The peptide antibiotics, like the polyketides, have a very wide range of biological activities (Stachelhaus and Marahiel 1995).

The **nucleoside** (Lacalle et al. 1992) and **sugar/cyclitol antibiotics** (Rinehart and Suami 1980) are two other well-known classes of secondary metabolites. The aminoglycoside antibiotics contain sugars and cyclitols in combination as pseudosaccharides; they frequently include amino-sugars. The biosynthetic pathway to the aminoglycoside streptomycin is the most studied; the formation of this molecule from glucose requires more than thirty different genes which encode the enzymes required for assembling the three components of streptomycin by three separate pathways (Distler et al. 1992).

From the point of view of **genetic regulation** of antibiotic biosynthesis, streptomycin is a good example. A series of studies by Ohnuki (Ohnuki et al. 1985) and by Piepersberg, (Ahlert et al. in press) and their co-workers have characterised the organisation and expression of the genetic units of the streptomycin biosynthetic pathway; a number of divergent transcripts have been identified and initiation occurs from adjacent promoters whose functions appear to be co-regulated. For all known secondary metabolites a variety of factors influence the pathways of biosynthesis, and both internal and external signals play important roles in the process. Alterations in growth patterns of producing organisms by changing C, N, and P sources have played an important role in enhancing production yields, and extensive (but little characterised!) mutagenic studies have been carried out. However, much remains to be learned of the biosynthetic pathways of secondary metabolite production and the ways in which they can be regulated by environmental factors. Suffice it to say that this is a very poorly studied aspect of microbial metabolism, especially considering that a single producing organism can make not one, but often dozens of different secondary metabolites.

The **biochemical mode of action** of antibiotics has been a subject of great interest, and antibiotics have proved to be important tools in the analysis of biochemical pathways and processes (Gale et al. 1981). Table 1 indicates the known targets and modes of action for the commonly used antibiotics; much information about the processes of macromolecular synthesis (transcription and translation, in particular) has been revealed from studies of the mode of action of antibiotic inhibitors. Genetic and biochemical studies of antibiotic-resistant mutants have been extremely valuable in the functional analysis of numerous prokaryote and eukaryote cellular processes. For example, detailed studies of the structural aspects of antibiotic interactions with DNA, ribosomal RNA, or enzymes such as DNA topoisomerases, have been carried out in recent years and have led to high-resolution definitions of the atomic interactions between macromolecules and low-molecular-weight inhibitors. In particular, significant advances have been made in the identification of the receptors for the aminoglycosides, which have been shown to be domains within ribosomal RNA. A single base alteration in these domains leads to an antibiotic-resistant phenotype (Cundliffe 1990; Fourmy et al. 1996). One benefit of these studies will likely be the rational design of more effective inhibitors and the development of potent new agents for the treatment of a variety of diseases.

The introduction of antibiotics into clinical practice in the late 1940's created an intense survival pressure on the microbial population, with the result that there has been an explosion of antibiotic-resistant strains that have compromised the use of essentially all antimicrobial agents in hospitals and in the community. This has led to serious problems, in many instances on a world-wide basis; for example, there have been frequent epidemics of nosocomial infections by methicillin-resistant *Staphylococcus aureus*; in some hospitals more than 50% of the isolates may be multidrug resistant (de Lencastre et al. 1994). Table 3 lists the **biochemical mechanisms of resistance** of the commonly used antibiotics. It should be noted that the development of antibiotic resistance may occur as the result of chromosomal mutations in the host or by the acquisition of resistance determinants from other sources. In some cases, both of these genetic processes play a role; for example, the phenomenon of expanded-spectrum β-lactamases is due to the mutational evolution of plasmid-encoded β-lactamases. A succession of single base changes results in the substitution of amino acids residing in the active site of the enzyme, forming modified enzymes that are capable of hydrolysing a wide structural range of β-lactam antibiotics (Bush et al. 1995).

Since current studies of antibiotic resistance development are retrospective exercises, it is difficult to assess the role of mutation in the early development of resistance. However, it is likely that mutations leading to partial (low-level) resistance, for example by increasing the activity of the endogenous efflux pumps of the bacterial host, provided the initial steps prior to the development of high-level plasmid-mediated resistance in bacteria. In general, chromosomal mutation will generate resistance alleles that are recessive to sensitivity; exceptions are the cases where mutations generate increased plasmid-copy number or up-promoter changes. It should be emphasized that resistance occurring by gene acquisition is dominant, since in most cases the acquired resistance determinant is novel with no corresponding function (recessive allele) in the host.

Table 3. Biochemical mechanisms of antibiotic resistance and their genetic determinants

Mechanism	Examples	Genetic determinants Mutation	Gene acquisition
Reduced permeability	aminoglycosides	+	+
Pro-drug not activated	isoniazid	+	
Active efflux	tetracycline		+
	fluoroquinolones	+	
Alteration of drug target	erythromycin		+
	fluoroquinolones	+	
	rifampicin	+	
	tetracycline		+
Inactivation of drug	aminoglycosides		+
	chloramphenicol		+
	β-lactams	+	+
'By-pass' inhibited step	sulphonamides		+
	trimethoprim		+
Immunity protein	bleomycin		+
Amplification of target	trimethoprim	+	
	sulphonamides	+	
Sequestration of drug	β-lactams	+	+

The rapid development of resistance to antibiotics in the bacterial population owes much to the transferability of resistance genes (usually on plasmids) between bacterial genera and species; wide-host-range transfer is common. This aspect will not be discussed here; plasmid gene transfer and its consequences have been adequately discussed in other papers in this book. However, it is worth noting that the inter- and intra-specific exchanges that take place in natural populations are a complex and chaotic process involving many hosts and a variety of different mechanisms of gene exchange. Demonstrations of gene transfer between different bacterial species in the laboratory can be used to define mechanisms, but cannot provide much information about the actual processes of the multiplicity of interspecies transfers which occur naturally.

Returning to the mechanisms of antibiotic resistance as shown in Table 3, there are at least ten different biochemical processes that may be involved in clinical isolates. These biochemical mechanisms will not be discussed in detail since the topic has been reviewed extensively in the literature; only a few of the more relevant examples are presented to illustrate some points of interest (Davies 1997).

Vancomycin is a glycopeptide antibiotic that has been available for more than thirty years; however, it is only recently that it has come into prominence as a front-line therapeutic agent. Firstly, improvements in production provided a more consistent product, which made vancomycin the drug of choice (the only one in fact) for the treatment of infections by methicillin-resistant staphylococci (MRSA). Secondly, the widespread development of vancomycin-resistant enterococci has raised concerns that gene exchange between enterococci and staphylococci, which is known to take place in the laboratory (Arthur et al. 1996), may result in vancomycin-resistant MRSA which would be refractory to currently available antibiotics.

It was mentioned earlier that most acquired resistance genes are dominant; in the case of vancomycin resistance, a mechanism to establish dominance is ancillary to the process of vancomycin resistance. The most thoroughly studied form of vancomycin resistance is that of *vanA*, which consists of a cluster of seven genes encoded on a conjugative transposon Tn*1546* (Fig. 2) (Arthur et al. 1996). Of these seven genes, two (*vanR* and *vanS*) have been implicated in the inducible expression of the resistance genes. The exact nature of the inducer is unknown, but it is thought that cell wall peptide intermediates produced by the action of the inhibitor are responsible; *vanR* and *vanS* are proposed to encode a typical two-component regulatory system that responds to the presence of the inducer peptides in the environment. Three genes, *vanA, H* and *X* determine the actual resistance mechanism. The mode of action of vancomycin is unusual for an antibiotic, since it binds to a dipeptide target D-Ala-D-Ala which is part of the pentapeptide chain linking glycan polymers in the gram-positive cell wall; most antibiotic inhibitors recognise a macromolecule such as a protein or nucleic acid target in susceptible hosts. The *vanA* gene encodes a depsipeptide ligase synthesizing D-Ala-D-Lac that is inserted into the cell wall pentapeptide in place of D-Ala-D-Ala; the replacement of D-Ala by D-Lac significantly reduces the binding of vancomycin to the target, and lethal inhibition of cell wall synthesis is averted. The production of D-lactate (not normally found in bacterial cells) requires a specific dehydrogenase *vanH* that reduces pyruvate to D-lactate. Thus *vanA* and *vanH* ensure the production of a cell wall component that is unable to form an inhibitory complex with vancomycin.

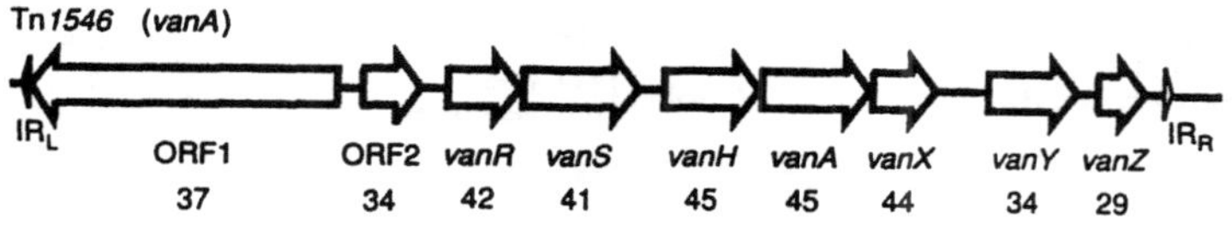

Figure 2. The vancomycin resistance gene cluster of TN*1546*. This gene cluster is responsible for glycopeptide resistance in *Enterococcus faecalis*. The functions of the various genes are as follows: *vanR* and *vanS*, regulator and sensor of two component regulatory systems; *vanH*, dehydrogenase to convert pyruvate to lactate; *vanA*, ligase that inserts lactate into cell wall peptide through ester bond formation; *vanX* and *vanY*, hydrolases to remove the normal vancomycin-binding fragment. ORF1 and ORF2 are the transposase and resolvase, and IR the inverted repeats. (For more complete information, see Arthur et al, 1996.)

The perceptive reader will note that in order for vancomycin resistance to be fully expressed, all of the cell wall pentapeptide must be replaced by molecules containing the D-Ala-D-Lac depsipeptide. Enter *vanX*, a D,D-dipeptidase that specifically degrades the natural peptidoglycan precursors that contain D-Ala-D-Ala. The *vanY* product (a D,D-carboxypeptidase) appears to be a 'back-up' process to ensure that the cell wall contains only the 'resistant' D-Ala-D-Lac. Variants on this theme have been identified in other species of enterococci that are resistant to glycopeptide antibiotics (Arthur et al. 1996). The novelty of this multi-component process and the process of its evolution is the topic of much speculation; the *vanA* ligase is similar in amino acid sequence to known bacterial cell wall ligases but has no clear relationship to any other members of the phylogenetic group (Evers et al. 1996).

The principal mechanism of resistance to aminoglycoside antibiotics is detoxification by enzymic modification of amino- or hydroxyl-groups on the antibiotic; acetyl CoA and ATP are the co-substrates in these reactions. What is extraordinary about aminoglycoside resistance is the large number of independent isoenzymes that have been identified in different resistant strains. For some of the target positions on the aminoglycoside molecule, as many as ten different isozymes have been identified, each encoded by an independent gene (Fig. 3) (Davies and Wright 1997).

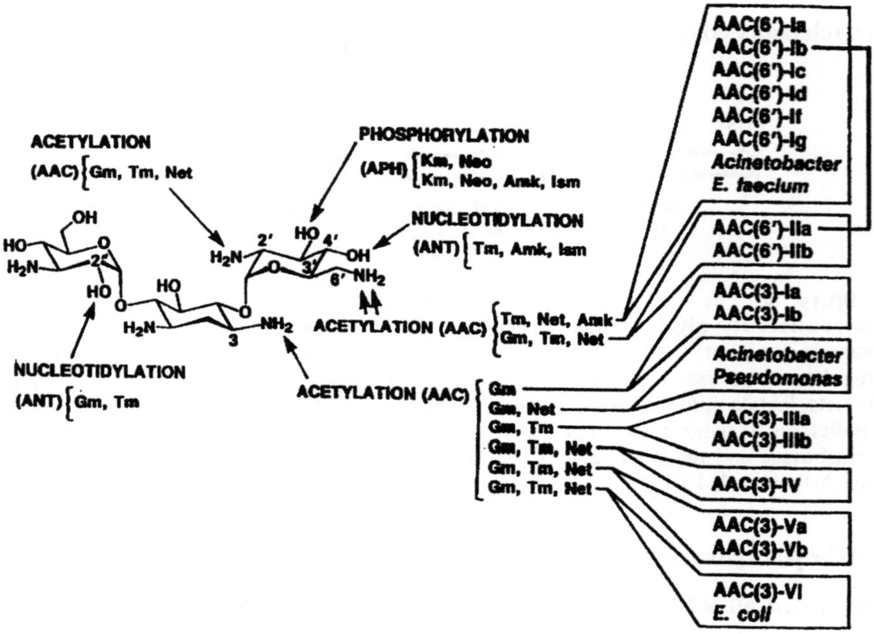

Figure 3. The enzymic modification of the aminoglycoside antibiotics of the gentamicin, kanamycin, and tobramycin groups in gram-negative bacteria. The structure shown is kanamycin B. The arrows indicate the sites of modification by *N*-acetyltransferases (AAC), *O*-phosphotransferases (APH), and *O*-nucleotidyltransferases (ANT). The substrate ranges of the various enzyme forms are indicated: neomycin (Neo), kanamycin (Km), gentamicin (Gm), tobramycin (Tm), amikacin (Amk), netilmicin (Net), isepamicin (Ism). (For additional information, see Evers et al, 1996.)

Analyses of the nucleotide sequences of the genes for aminoglycoside-modifying enzymes indicate limited sequence relationships, which implies that they must have been derived from independent heterologous sources. The different classes and sub-classes of aminoglycoside-modifying enzymes clearly could not have evolved (recently) as a result of a limited number of mutations (amino acid substitutions) in single precursor genes. This distinguishes the evolutionary route to the aminoglycoside-modifying enzymes from that established for the extended-spectrum **β-lactamases**. The genes for the latter are related by single base changes leading to amino acid substitutions that alter substrate specificity (Bush et al. 1995; Medeiros 1997). By contrast, the genes encoding aminoglycoside-modifying enzymes that have different substrate ranges must have been acquired from independent bacterial sources during the recent period of antibiotic use. The evidence available at the present time suggests strongly that these sources are the microbial species that produce different members of the aminoglycoside antibiotic family (Benveniste and Davies 1973). The genetic and biochemical mechanisms for the process of resistance gene acquisition remain poorly understood.

A similar, multigenic origin has been predicted for the ribosome methyltransferase genes that encode resistance to the **macrolide, lincosamide and streptogramin antibiotics** (MLS). These methylase enzymes render the target ribosomes (50S subunits) unable to bind the antibiotic as a result of specific mono- or di-methylation of A2048 in 23S rRNA (22,23). MLS resistance is quite common within isolates of staphylococci and enterococci in hospital environments and is usually inducible. The induction of the expression of MLS resistance is due to a novel post-translational process; in the presence of low concentrations of the antibiotic, the ribosome stalls on a 5'-leader sequence in the mRNA to liberate the translation initiation signals from a folded mRNA complex, thus permitting translation of the methylase gene. The biochemistry of this process has been analysed extensively by Weisblum (Weisblum 1995a; Weisblum 1995b) and Dubnau (Dubnau and Monod 1986) and their colleagues, who isolated and characterised point mutations and deletions in the leader sequence and showed how changes in the secondary structure of the leader influence the synthesis of the leader-peptide and thereby control translation initiation on the ribosome. It was demonstrated that modifications in the leader sequence may also lead to constitutive expression of MLS resistance. The translational attenuation process regulating the *erm* methylases bears a formal similarity to other types of attenuation regulation, such as that found in the case of chloramphenicol acetyltransferases or more particularly, in the control of amino acid biosynthesis. One cannot help but marvel at the simplicity and metabolic frugality of this form of control of gene expression; all that is required is a few extra bases added to the 5' end of a gene!

Fluoroquinolones are man-made antibiotics; they have no known analogues among natural products. Not surprisingly, bacterial resistance to the fluoroquinolones occurs by mutation, and no plasmid-determined fluoroquinolone resistance has yet been identified; however, the incidence of resistant strains carrying mutations altering the target molecule of the fluoroquinolones is increasing. Such mutants have been isolated in the laboratory and shown to alter the GyrA (or topoisomerase or partition) enzymes that are essential to the replication of DNA. There appears to be generic variation in the types of resistance, in the sense that pseudomonads and other gram-negative bacteria most often have alterations in *gyrA*, while in staphylococci the principal determinant of resistance is mutation altering topoisomerase IV function (Huang 1996; Blanche et al. 1996).

Active efflux of fluoroquinolones from host cells has been identified as another factor contributing to fluoroquinolone resistance in bacteria. This mechanism has been increasingly implicated in resistance to a wide variety of antimicrobial agents. Some examples of efflux systems include the NorA protein from *S. aureus*, which removes fluoroquinolones and chloramphenicol, as well as dyes such as acriflavin and rhodamine-6-G, from the cell; and the AcrAB system from *E. coli*, which pumps out fluoroquinolones, chloramphenicol, acriflavin, tetracycline, erythromycin, and β-lactams. In addition to these multi-substrate pumps, efflux pumps that determine resistance to specific antibiotics, such as the Tet K and Tet L systems, have been described(Lewis 1994; Roberts 1996).

The few examples of antibiotic resistance mechanisms presented here illustrate the extraordinary diversity of antibiotic biology. Firstly, there is the extensive biochemical variety of antibiotic compounds that act as inhibitors of cell function, which emphasises the role(s) of low-molecular-weight compounds in biological systems and the range and specificity of their interactions with major functional macromolecules of cells. Secondly, studies of resistance illustrate the variety of biochemical mechanisms that can be employed by microbial pathogens to resist the inhibitory activities of antibiotics. The interplay between the mechanisms of mutation, gene acquisition, gene exchange/transfer, and transposition provides unlimited ways by which the development of antibiotic resistance and its dissemination may occur in bacteria. Microbes are formidable adversaries that are capable of developing functional resistance to any potential therapeutic agent that may be developed. It is considered that the only practical solution to the increasing problem of antibiotic resistance is to try to delay the process by careful control of antibiotic use; complete avoidance of the problem does not appear to be possible (Cohen 1992).

There is one benefit that derives from the natural development of transmissible antibiotic resistance – the resistance genes found on R plasmids have been used extensively, in combination with their corresponding antibiotics, to develop powerful selective systems in the construction of cloning vectors for genetic engineering studies in all types of organisms (Table 4).

Table 4. Dominant resistance genes employed as selective markers

Antibiotic	Target	Resistance gene	Potential applications
Bialaphos	glutamine synthetase	*bar*	P, E *
Blasticidin	translation	*bac*	E
Bleomycin	DNA degradation	*bbp, bat*	P, E
Chloramphenicol	translation	*cat*	P
Hygromycin	translation	*hph*	P, E
β-lactams	peptidoglycan synthesis	*bla*	P
Neomycin/Kanamycin	translation	*aph, aac*	P, E
Puromycin	translation	*pac*	E
Streptothricin	translation	*sat*	P, E
Tetracycline	translation	*tetA, tetM*	P
Thiostrepton	translation	*tsr*	P

* P, prokaryote; E, eukaryote

In a more general sense, if there had been no plasmid-determined resistance, what would have driven the development of the recombinant DNA era?

Acknowledgments

I wish to thank the Canadian Bacterial Diseases Network and the National Science and Engineering Council of Canada for generous support.

References

Ahlert J, Beyer S, Distler J, Neumann T, Retzlaff L, Verseck S, Piepersberg W (in press) Advances in the molecular biology of streptomycin. *In* Lal R, Lal S (eds) Biotechnological Aspects of Industrial Antibiotics. Springer-Verlag, Heidelberg

Arthur M, Reynolds P, Courvalin P (1996) Glycopeptide resistance in enterococci. Trends Microbiol 4:401-407

Benveniste R, Davies J (1973) Mechanisms of resistance to antibiotics. Annu Rev Biochem 42:471-506

Blanche F, Cameron B, Bernard F-X, Maton L, Manse B, Ferrero L, Ratet N, Lecoq C, Goniot A, Bisch D, Crouzet J (1996) Differential behaviors of *Staphylococcus aureus* and *Escherichia coli* type II DNA topoisomerases. Antimicrob Agents Chemother 40:2714-2720

Bush K, Jacoby GA, Medeiros AA (1995) A functional classification scheme for β-lactamases and its correlation with molecular structure. Antimicrob Agents Chemother 39:1211-1233

Cohen ML (1992) Epidemiology of drug resistance: implications for a post-antimicrobial era. Science 257:1050-1055

Cundliffe E (1990) Recognition sites for antibiotics within rRNA. *In* Hill WE, Dahlberg AE, Garrett RA, Moore PB, Warner JR (eds) The Ribosome: Structure, Function and Evolution. American Society for Microbiology, Washington, D.C., pp 479-490

Davies J (1990) What are antibiotics? Archaic functions for modern activities. Mol Microbiol 4:1227-1232

Davies J (1997) Origins, acquisition and dissemination of antibiotic resistance determinants. *In* CIBA Foundation Symposium No. 207: Antibiotic Resistance: Origins, Evolution, Selection and Spread. Wiley, Chichester, pp 15-27

Davies J, Wright GD (1997) Bacterial resistance to aminoglycoside antibiotics. Trends Microbiol 5:234-240

de Lencastre H, de Jonge BLM, Matthews PR, Tomasz A (1994) Molecular aspects of methicillin resistance in *Staphylococcus aureus*. J Antimicrob Chemother 33:7-24

Distler J, Mansouri K, Mayer G, Stockmann M, Piepersberg W (1992) Streptomycin biosynthesis and its regulation in streptomycetes. Gene 115:105-111

Donadio S, Staver MJ, McAlpine JB, Swanson SJ, Katz L (1991) Modular organization of genes required for complex polyketide biosynthesis. Science 252:675-679

Dubnau D, Monod M (1986) The regulation and evolution of MLS resistance. *In* Banbury Report 24: Antibiotic Resistance Genes: Ecology, Transfer, and Expression. Cold Spring Harbor Laboratory, Cold Spring Harbor, N Y, pp 369-387

Evers S, Casadewall B, Charles M, Dutka-Malen S, Galimand M, Courvalin P (1996) Evolution of structure and substrate specificity in D-alanine:D-alanine ligases and related enzymes. J Mol Evol 42:706-712

Fourmy D, Recht MI, Blanchard SC, Puglisi JD (1996) Structure of the A site of *Escherichia coli* 16S ribosomal RNA complexed with an aminoglycoside antibiotic. Science 274:1367-1371

Gale EF, Cundliffe E, Reynolds PE, Richmond MH, Waring MJ (1981) The Molecular Basis of Antibiotic Action. John Wiley, Chichester, UK

Huang WM (1996) Bacterial diversity based on type II DNA topoisomerase genes. Annu Rev Genet 30:79-107

Hutchinson CR, Fujii I (1995) Polyketide synthase gene manipulation: a structure-function approach in engineering novel antibiotics. Annu Rev Microbiol 49:201-238

Lacalle RA, Tercero JA, Jiménez A (1992) Cloning of the complete biosynthetic gene cluster for an aminonucleoside antibiotic, puromycin, and its regulated expression in heterologous hosts. EMBO J 11:785-792

Lewis K (1994) Multidrug resistance pumps in bacteria: variations on a theme. Trends Biochem Sci 19:119-123

Medeiros AA (1997) Evolution and dissemination of β-lactamases accelerated by generations of β-lactam antibiotics. Clin Infect Dis 24:S19-45

Ohnuki T, Imanaka T, Aiba S (1985) Isolation of streptomycin-nonproducing mutants deficient in biosynthesis of the streptidine moiety or linkage between streptidine 6-phosphate and dihydrostreptose. Antimicrob Agents Chemother 27:367-374

Rinehart KW, Jr, Suami T (1980) Aminocyclitol Antibiotics. ACS Symposium Series 125. American Chemical Society, Washington, D.C.

Roberts MC (1996) Tetracycline determinants: mechanisms of action, regulation of expression, genetic mobility, and distribution. FEMS Microbiology Reviews 19:1-24

Stachelhaus T, Marahiel MA (1995) Modular structure of genes encoding multifunctional peptide synthetases required for non-ribosomal peptide synthesis. FEMS Microbiol Lett 125:3-14

Vining LC (1995) Other biosynthetic groups of antibiotics. *In* Vining LC, Stuttard C (eds) Genetics and Biochemistry of Antibiotic Production. Butterworth-Heinemann, Boston, pp 499-504

Weisblum B (1995a) Insights into erythromycin action from studies of its activity as inducer of resistance. Antimicrob Agents Chemother 39:797-805

Weisblum B (1995b) Erythromycin resistance by ribosome modification. Antimicrob Agents Chemother 39:577-585

INTERACTIONS OF THE BACTERIAL PATHOGEN *LISTERIA MONOCYTOGENES* WITH MAMMALIAN CELLS

Pascale Cossart, Unité des Interactions Bactéries Cellules, Institut Pasteur, 28 Rue du Docteur Roux, Paris 75015, France.

Introduction

Despite the extensive use of antibiotics and vaccination programs, infectious diseases, and in particular microbial diseases, continue to be a leading cause of morbidity and mortality worldwide. Recent outbreaks and epidemiologic studies predict that their incidence will increase in the future. Among bacterial pathogens, intracellular bacteria are responsible for a very large number of infectious diseases. Some of them are only present in phagocytic cells, others have evolved the capacity to enter into cells which are normally non-phagocytic. *Listeria monocytogenes* belongs to this latter category. This Gram-positive bacterium has recently been extensively studied and a precise picture of the molecular and cellular basis of the infectious process is emerging (Sheehan et al., 1994). The study of the interactions between *L. monocytogenes* and the mammalian cells provides in addition, tools to address key questions in Cell Biology such as actin-based motility.

The infection by *L. monocytogenes, in vivo*

L. monocytogenes is responsible for severe human food borne infections characterized by meningitis, meningo-encephalitis, septicemias, abortions and also gastroenteritis with a mortality of 30%. It also naturally infects many other animal species including cows and sheep and is therefore of veterinary importance. Most of our knowledge of the human diseases comes from the many studies carried out in mice.

NATO ASI Series, Vol. H 103
Molecular Microbiology
Edited by Stephen J. W. Busby,
Christopher M. Thomas and Nigel L. Brown

Via contaminated food, bacteria reach the gastro-intestinal tract and cross the intestinal barrier. The exact site of entry (the enterocytes or the M cells) is still a matter of debate. Bacteria are subsequently engulfed by macrophages. Then, via the lymph and the blood , they reach the spleen and liver. In this organ, most of the bacteria are killed by the kupffer cells. A fraction of the bacteria reach the hepatocytes where they induce a process of apoptosis with the concommittant release of chemoattractants which will lead to the influx of many neutrophils. These phagocytic cells will ingest bacteria or infected hepatocytes and contribute to the rapid clearing of the infection before the complete sterilisation by the immune response. However, in some cases - the immunocompromised host or in the pregant women - bacteria will multiply unrestrictedly in the hepatocytes from which they will disseminate, via the blood, to the brain and the placenta, resulting in the characteristic clinical features described above. The key step in the establishment of a "successful" infection is the bacterial multiplication step in the liver.

It was shown in the 1960s by Mackaness that complete recovery from infection and protection against secondary infection require the induction of a specific T-cell-dependent immune response, antibodies playing no role in this process of clearing bacteria. Following this pioneering work, *L. monocytogenes* has become a model to study the induction of the immune T-cell response and research in this area has culminated with the discovery of the first bacterial protective CD8+ epitope.

In the last ten years, *L. monocytogenes* has also become one of the best systems to study intracellular parasitism. This is due to several specific properties.

1. It is an opportunnistic bacterium and one is not forced to work with this organism under stringent laboratory security conditions.

2. It is fast growing, which allows one to perform genetics inasmuch as genetic tools are available (transposons, transformation, allelic exchange...)

3. The genus *Listeria* contains, besides the two pathogenic species *L. monocytogenes*, and *L. ivanovii*, four other non pathogenic species. These species, in particular *L. innocua* provide useful tools to identify *L. monocytogenes* virulence specific genes or express *L. monocytogenes* genes in order to evaluate their function.

4. The murine infection is critical to test the relevance of putative virulence factors.
5. Infection of tissue cultured cells provide a very convenient way to study the molecular and cellular basis of infection.

The infectious process by *L. monocytogenes* at the cellular level

Detailed analysis of *Listeria*-infected cell cultures has revealed a complex series of host-pathogen interactions culminating in the direct dissemination of *L. monocytogenes* from one infected cell to another. Host cell infection begins with the internalization of the bacteria either by phagocytosis in the case of macrophages or induced phagocytsosis in the case of non-phagocytic cells. The bacteria are rapidly internalized and reside within membrane-bound vacuoles during a very short period, i.e. 30 min. Bacteria then lyse the vacuoles and reach the cytosol. In this environment, they start to multiply (at about the same rate as in broth medium, i.e. doubling time of one hour). Concommitantly, they become covered with actin filaments which within 2h rearrange in long comet tails left behind in the cytosol while the bacteria are moving ahead at a speed of about 0. 3 μs^{-1}. When moving bacteria contact the plasma membrane, they induce the formation of pseudopod-like protrusions of the membrane. Contact between these protrusions and neighboring cells results in the internalisation of the bacteria containing protrusion. In the newly infected cell, the bacterium is surrounded by two plasma membranes which must be lysed to initiate a new cycle of multiplication and movement. Thus, once *Listeria* has entered the cytoplasm, it can disseminate directly from cell to cell circumventing such host defenses as circulating antibody and complement. This ability to disseminate in tissues by cell-to-cell spreading provides an explanation for the early observation that antibody (although induced and abundant) is not protective and that anti-listeria immunity is T-cell mediated.

These different steps of the infection are schematized in Figure 1:

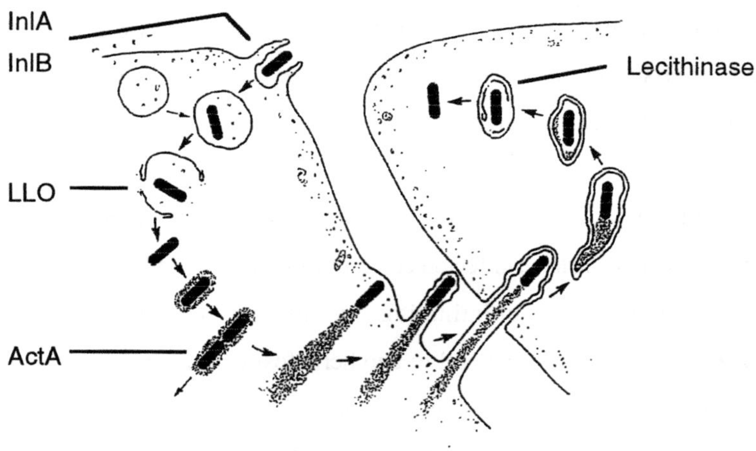

Figure 1: Schematic representation of the cell infectious process by *L. monocytogenes*. The bacterial factors involved are indicated.

In this figure , the bacterial factors involved in the various steps have been indicated. Entry into mammalian cells can be mediated by at least two bacterial factors: internalin and InlB. Escape from the phagocytic vacuole requires expression of listeriolysin O (LLO), a pore forming toxin which in some cells can work synergistically with a phosphatidylinositol-specific phospholipase C (PI-PLC). Intracellular movement requires expression of ActA, and lysis of the two-membrane vacuole is performed by a lecithinase (PlcB), which needs to be preactivated by a metalloprotease, the product of the *mpl* gene.

Most of the genes coding for these virulence factors are clustered on a 15kb region of the chromosome which can be considered as a pathogenicity island since it is absent from the non pathogenic species. The *inlAB* operon is located in another region. All these virulence genes are under the control - either absolute or partial - of a pleiotropic activator protein PrfA (Figure 2).

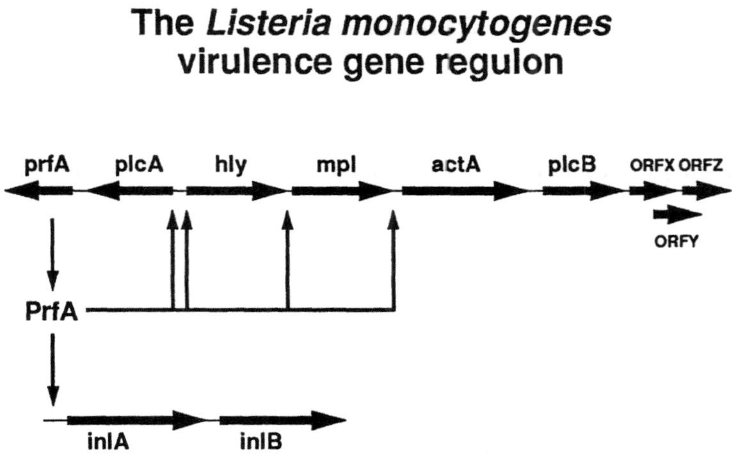

Figure 2. Schematic representation of the *Listeria monocytogenes* virulence gene regulon.

Two aspects of the infectious process will be discussed in details : entry into mammalian cells (invasion of mammalian cells) and the actin-based motility.

1. Invasion of mammalian cells

1. 1 The internalin-Ecadherin pathway

- Identification of the internalin gene locus.

Entry into some epithelial cells is mediated by internalin, a surface protein encoded by the gene *inlA* (Gaillard et al., 1991). Internalin was identified by screening a library of Tn*1545* mutants of *L. monocytogenes* for loss of invasiness into the intestinal epithelial cell line Caco-2. This screening was done using a classical gentamicin survival assay. Three such mutants were obtained. These mutants were unable to adhere to Caco-2 cells and were defective for entry into a variety of epithelial cell lines. In all three mutants, the transposon had inserted into a region upstream from two open reading frames, *inlA* and *inlB*. Transcription of *inlA* and *inlB* was abolished in the non-invasive mutants.

Expression of *inlA* in the non-pathogenic species *L. innocua* confered invasiveness on this non-invasive bacterium. In addition, the *inlA*-transformed *L. innocua* expressed a novel surface protein in agreement with the idea that *inlA* encodes a surface protein which could interact with a receptor located on the mammalian cell surface to mediate entry. The *inlA* gene product was thus named internalin.

- Structure of internalin

Internalin is an 800 amino-acid protein which displays two regions of repeats, the first being a succession of 15 twenty-two amino-acid long leucine rich repeats (LRRs). Internalin has all the features of a protein which is targeted to and exposed on the bacterial surface. i.e. a signal peptide, and a C-terminal region made of a LPXTG peptide followed by a hydrophobic sequence and a few charged residues. This type of C-terminus is now found in more than 50 Gram positive bacterial surface proteins and allows a covalent linkage of the protein to the cell wall peptido-glycan, occurring after cleavage of the T-G link.

- role of the surface associated form of internalin in invasion.

During growth in broth medium, internalin is partially released in the medium and partially present on the bacterial surface. It was thus of interest to address the role of the surface-associated form and that of the released form of internalin in entry. By deleting the region coding for the C-terminus of internalin, evidence was obtained that in these conditions, no internalin can be detected on the surface and that washed bacteria are non invasive. In contrast, washed bacteria expressing wild type internalin or internalin anchored in the bacterial membrane by the ActA C-terminal region can mediate entry. Thus the surface associated form can mediate entry. The role of the released form remains to be determined. This form might well be a laboratory artefact due to cell wall turnover in certain media.

- Identification of the internalin receptor

Taking advantage of the release of internalin in culture supernatants, internalin was purified and shown to be able to bind to mammalian cells. Binding was inhibited by

EDTA. An internalin column was then used to purify the putative receptor from Caco2 extracts. After running the extracts on the column and extensive washings, two protein bands were eluted by EDTA. Their N-terminal sequence was determined. Comparison with protein sequence data banks revealed that these two proteins were E-cadherin and its proteolytic fragment normally produced in the conditions used to prepare the extracts (Mengaud et al., 1996).

E-cadherin is a transmembrane glycoprotein which mediates calcium dependent cell-cell adhesion, through homophilic interactions between extracellular domains of E-cadherin. Cadherins are proteins specifically expressed in different tissues, E-cadherin in epithelial cells, N-cadherin in neuronal cells, etc. Cadherins play a critical role in cell sorting during development and in maintenance of tissue cohesion and architecture during adult life. In polarized epithelial cells, E-cadherin is mainly expressed at the adherens junctions and on the baso-lateral face. Integrity of the intracytoplasmic domain of cadherins is required for adhesion. This domain interacts with proteins named catenins which in turn interact with the cytoskeleton, demonstrating the involvement of the cytoskeleton in maintaining adhesion of adjacent epithelial cells.

To demonstrate that the internalin-E-cadherin interaction promoted specific binding and entry of *L. monocytogenes*, a set of transfected cell lines was used. Entry of both *L. monocytogenes* or *L. innocua* expressing internalin was highly promoted in cells expressing the chicken E-cadherin, in contrast to cells expressing N-cadherin or no cadherin.

Thus E-cadherin is the receptor for internalin and the interaction between these two proteins promotes entry. These results would suggest that *in vivo*, *L. monocytogenes* does not pemetrate the intestinal barrier by the apical pole of enterocytes and favor a mechanism of translocation through M cells as a primary step in infection. Entry in enterocytes would represent a secondary step taking place at the baso-lateral face. This hypothesis is in agreement with previous *in vitro* observations that *Listeria* preferentially invades Caco-2 cell islets at their periphery, i.e., where the baso-lateral surface is accessible.

It is the first time that a cadherin is shown to be able to serve as a receptor for a pathogen but it is not the first time that a cellular molecule involved in adhesion is used as a receptor by a bacterial pathogen. Indeed, *Yersinia pseudotuberculosis* expresses an outer membrane protein named invasin which when expressed in *E. coli* can promote entry of *E. coli* in epithelial cells, after interaction of invasin with a subset of β1 integrins. The morphological events described during *Yersinia* entry into cells are totally different from those described for two other invasive bacteria, *Shigella* and *Salmonella*. In the first case, entry occurs by a "zipper" mechanism while entry of *Salmonella* and *Shigella* triggers dramatic membrane projections and ruffles which engulf the bacteria in a macropinocytic manner. These events are mediated by a number of proteins secreted by the type III secretion system, a cell contact regulated secretion mechanism. Entry of *Listeria* in Caco2 cells or in E-cadherin expressing cells occurs by a mechanism closely related to that of *Yersinia*.

1. 2. The InlB pathway

By creating deletions in *inlA* or *inlB*, and testing the corresponding mutants in various cell lines, it was established that InlB is an invasion protein used for entry in some hepatocyte like cell lines, HeLa cells, Vero cells, CHO cells (Dramsi et al., 1995). InlB is a protein similar to internalin albeit with differences. It has eight tandem leucine rich repeats (LRRs) very similar to those of internalin. It has a signal sequence but does not display any hydrohobic region which would indicate a possible transmembrane region. Nevertheless, this protein is present on the bacterial surface and recent experiments indicate the 231 C-terminal amino-acids of InlB can anchor this protein to the bacterial surface, probably through an interaction with a non-identified compound of the cell wall. Recombinant InlB is able to bind to mammalian cells but its receptor is unknown.

1. 3. Signalling events leading to entry.

Three types of drugs have shed light on the cellular components which may lead to entry. Cytochalasin D totally prevents entry, demonstrating that an intact actin cytoskeleton is required for invasion. Tyrosine kinase inhibitors such as genistein inhibit entry, demonstrating a requirement for at least one tyrosine kinase. (Note that cytochalasin D and genistein do not prevent adhesion). In addition, wortmannin and LY294002 inhibit entry at concentrations sufficiently low to suggest that a phosphoinosistide (PI) 3 kinase is involved in the events leading to internalisation.

The lipid kinase p85/P110 PI3-kinase rapidly appeared as a very good candidate for a protein which could mediate signals from the membrane to the cytoskeleton upon entry of *L. monocytogenes* into mammalian cells. Indeed this signalling protein is activated upon receptor stimulation, to requires tyrosine phosphorylation and leads to cytoskeleton rearrangements. The p85/P110 PI-3 kinase is an heterodimeric lipid kinase which upon receptor stimulation, migrates from the cytosol to a membrane tyrosine phosphorylated tyrosine kinase receptor. Migration to the plasma membrane stimulates activity of PI-3 kinase, by placing the enzyme in a compartment where its substrates are located and also by conformational changes after protein protein interactions. PI 3 kinase phosphorylates the D3 position of PI, PIP and PIP2 giving rise to PI3P, PI3,4P2 and PI3,4,5P3. These last two components are virtually absent in resting cells and dramatically increase upon stimulation. They are not the substrates for phospholipases and may act as second messagers by interacting with kinases such as Akt.

Wortmannin as said above and expression of a dominant negative form of PI-3 kinase inhibit *Listeria* entry, demonstrating the requirement for PI-3 kinase (Ireton et al., 1996). Measurements of the levels of phosphoinositides reveal that bacterial entry stimulates synthesis of both PI3,4P2 and PI3,4,5P3. This synthesis is inhibited by wortmannin, but not cytochalasin D providing evidence that cytoskeleton rearrangements occur after stimulation of PI-3 kinase, during bacterial invasion. In contrast, PI-3 kinase stimulation is inhibited by genistein. In agreement with this observation it was found that shortly after infection, PI-3 kinase is co-immunoprecipitated with at least one tyrosine phosphorylated protein bringing evidence that upon entry PI-3 kinase activity is stimulated after interaction with at least one tyrosine phosphorylated protein.

A ΔinlB mutant still adheres to Vero cells but does not efficiently stimulate PI 3 kinase and does not stimulate association of PI 3 kinase with tyrosine phosphorylated proteins. Thus InlB seems to play a critical role in the stimulation of P85/P110.

How this stimulation of PI3 kinase affects bacterial invasion is not known, but one possible mechanism is by controlling actin polymerisation or reorganization of the actin cytoskeleton. PI3,4P2 and PI3,4,5P3 are able to uncap barbed ends of actin filaments in permeabilized platelets, suggesting a simple means by which stimulation of PI3 kinase activity by adherent bacteria could drive local cytoskeletal changes needed for entry. It is also possible that a high concentration of phosphoinositides may affect curvature of the lipid bilayer facilitating endocytosis. It is also possible that the generation of phospholipids in the membrane may attract to this compartment proteins which specifically bind to these normally absent phospholipids. PI-3 kinase itself can interact with other signalling proteins that regulate organization of the actin cytoskeleton such as pp125FAK or rho GTPases and it is possible that such interactions play a role in entry. PI-3 kinase has also been implicated in endocytic processes and it is possible that some of the components used for endocytosis are also emplyed by bacterial pathogens to gain entry into host cells.

Intriguingly, the internalin E cadherin mediated entry in Caco-2 cells which is affected by PI-3 kinase inhibitors does not stimulate PI-3 kinase activity. In addition, not only the basal levels of PI3,4P2 and PI3,4,5P3 did not increase significantly upon entry but they are already very high in uninfected Caco-2 cells. Thus in Caco-2 cells, the internalin E-cadherin pathway exploits a pre-stimulated PI-3 kinase dependent pathway.

The internalin- and InlB-mediated pathways are schematically represented in Figure 3

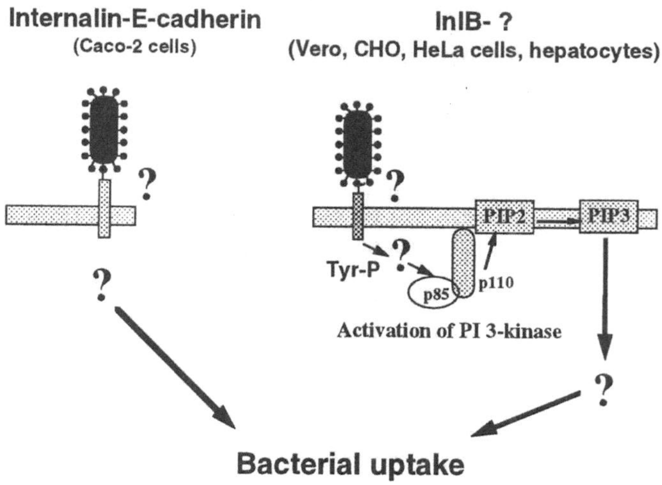

Figure 3: Schematic representation of the the Internalin- and InlB-mediated pathway.

Interestingly, when wortmannin was used to evaluate the role of PI3 kinase in *Salmonella* invasion this drug did not inhibit entry. It even stimulated entry of *Salmonella* reinforcing the view that this invasive bacterium is not using the same cellular functions as *Listeria* for invasion.

2. The actin-based motility.

This spectacular phenomenon which, by coupling actin polymerisation and movement, propels the bacteria inside the cytosol has received a great deal of attention since it is highly reminiscent of cellular events which remain unexplained, in particular the movement of cells such as neutrophils attracted at a site of infection, or the movement of cancer cells. It is believed that similar mechanisms could occur at the plasma membrane and at the rear of the bacteria, hence the enthusiasm for a system which by its relative simplicity can be manipulated and studied more easily. The reader should refer to recent reviews published on this topic (Cossart, 1995; Lasa and Cossart, 1996; Ireton and Cossart, 1997). Well established data will only be summarized here, in order to develop the most recent findings in the field.

2.1. The relationship between the actin tail formation and movement

The early observations of thin cross sections of *Listeria*-infected cells decorated with fragment S1 of myosin revealed that the actin tails are made of cross-linked short filaments with their barbed (fast polymerizing) end oriented towards the bacterium, and suggested that actin polymerisation takes place at the rear of the bacterium. Microinjection of fluorescent actin monomers in live infected cells then demonstrated that the actin polymerisation takes place at the rear of the bacterium and that bacteria move away while the actin tail remains stationary in the cytosol. Video microscopy observations have indicated that there is a strict correlation between the rate of tail formation and speed movement strongly suggesting that the force for propulsion is provided by the actin polymerisation itself.

2. 2 The ActA protein.

ActA was discovered by observing a transposon mutant which inside cells was totally unable to polymerize actin and move. This mutant formed microcolonies inside the cell. It was unable to spread from one cell to the other. This mutant was avirulent when injected to mice. This mutant carried a transposon insertion in a gene named *actA* which encodes a protein of 610 amino-acids.

ActA has all the features of a protein which can be targeted to the bacterial surface. It has a signal sequence and a C-terminal hydrophobic region which was first suspected and further demonstrated to anchor this protein in the bacterial membrane. By use of antibodies raised against a peptide located in the center of the protein, it was established by immunofluorescence labeling and immunogold labeling that ActA has a polar distribution on the bacterial surface, with a higher distribution on one pole of the bacteria. This distribution is established during division but the mechanism underlying this uneven distribution is unexplained. In infected cells, ActA is located at the base of the actin tail suggesting that this polar distribution predetermines the site of actin assembly and the direction of movement.

2.2 ActA is sufficient to induce actin polymerization and movement

To determine whether ActA is sufficient to induce actin polymerisation, the gene *actA* was transfected in mammalian cells where it was able to induce the polymerisation of actin. Three types of experiments were performed. Either *actA* was transfected in its totality. In this case, it was targeted to mitochondria where it induced actin polymerisation around these intracytosolic organelles. When *actA* lacking the 3'end coding for the membrane anchor, was transfected, the resulting soluble protein increased the total amount of cellular F-actin. Finally, when a construction encoding ActA fused to CAAX box, which can target proteins to the inner face of the plasma membrane, was expressed in mammalian cells, ActA was indeed targeted to the plasma membrane where together with actin polymerisation, this ActA-CAAX protein induced the formation of aberrant protrusions, indicating that protrusion formation is an event which can be closely coupled to actin polymerisation.

Two different experimental approaches were used to demonstrate that ActA is sufficient to induce not only actin polymerisation but also actin-based motility: i) by producing ActA in the non-pathogenic species *L. innocua*. This normally non-motile bacterium became converted into an organism capable of actin polymerisation and movement in *Xenopus* cytoplasmic extracts, an *in vitro* system now currently used to observe *listeria* movement. ii) by incubating *Steptococcus pneumoniae* with a recombinant ActA-LytA hybrid protein, which was adsorbed on and covered the streptococcal surface. The decorated bacteria became able to polymerize actin and move. However, these events only occurred after they had divided to generate a polar distribution of ActA on the streptococcal surface. Thus, both ActA production and its polar distribution are prerequisites for actin assembly and movement.

2.3 comparison with other proteins

The ActA protein can be artificially divided into three parts, the N-terminal domain (1-233) which is highly charged, a central proline rich repeat region (234-395) and the C-terminal region (396- 610). When ActA was discovered, the central region which is made of several proline-glutamic acid rich regions was immediately suspected to be a ligand for profilin, a small protein involved in actin polymerisation and which has affinity for proline-rich sequences. Experimental results indicate that it is not the case. This region binds another protein, VASP which itself has proline rich regions and has affinity for profilin (see below).

Recent amino-acid sequence comparisons have revealed that ActA is a composite protein. The proline-rich repeats and the carboxy-terminal domain share significant sequence similarity with zyxin, a protein associated with focal contacts and actin stress fibers. The N-terminal domain of ActA is similar (25% identity) to the C-terminal region (aminoacid 879-1066) of vinculin which is a protein recently shown to be able to bind actin precisely through its C-terminal region. ActA(1-132) is also similar (24%identity) to CAP23 (30-162), a 209 amino-acid chicken cytoskeleton associated protein of unknown function. In summary, ActA seems to have domains similar to eucaryotic proteins involved in the organization of the cytoskeleton.

2. 4. The cellular factors involved.

In order to address the role of ActA, the first question was: Is ActA able to bind actin? All attempts to demonstrate interactions between ActA and actin have failed, suggesting that this protein does not interact directly with actin and that at least one other cellular factor is involved in the actin polymerisation process..

To identify such factors, the main approach has been to use antibdodies against known cytoskeletal proteins and try to detect the corresponding proteins in the actin tails. Proteins identified in this way include α-actinin, tropomysosin, vinculin, villin, talin, ezrin/radixin, fimbrin, profilin and VASP. The two most relevant proteins are profilin and VASP which colocalize only with the beginning of the actin tail. VASP was shown to bind purified ActA *in vitro*, establishing the first direct link between ActA and the cytoskeleton. Since VASP binds profilin, colocalization of the two proteins at the beginning of the actin tail is probably only due to VASP binding to ActA

The role of some of these proteins was assessed in different ways. In the case of profilin , depletion experiments were performed in which polyproline beads were used to deplete cytoplasmic extracts. Bacterial movement was then tested in the depleted extracts. These experiments led to opposite results concerning the role of profilin. Nonetheless, it is clear that most profilin can be depleted without affecting bacterial actin-based motility. Concerning α–actinin, microinjection of a fragment which acted as a dominant negative mutant demonstrated that this crosslinking protein is required for efficient actin tail formation and movement.

A very different approach was reported recently which led to the identification of other cellular factors involved in the actin-based motility. Cytoplasmic extracts isolated from human platelets were fractionated and an eight polypetide complex was purified which was sufficient to initiate ActA-dependent actin polymerisation at the surface of *L. monocytogenes*. Two subunits of this protein complex are actin-related proteins (Arps) belonging to the Arp2 and Arp3 subfamilies. The Arp3 subunit localizes to the surface of stationary bacteria and the tails of motile bacteria in *L. monocytogenes* infected tissue culture cells, consistent with a role for the complex in promoting actin assembly *in vivo*.

2. 5. Genetic analysis of ActA

In order to identify the regions of ActA critical for its function, deletions in *actA* were generated which demonstrated that the N-terminal (ActAN) is absolutely critical, the central region acting as a stimulator. These results were confirmed by expressing in *Listeria* an ActAN-lacZ fusion which was functional for movement. It was also shown that VASP binds to the proline-rich region of ActA. A recent analysis of the N-terminus of ActA has demonstrated that this region contains two critical regions specifically involved in the process. Both are required for the actin polymerisation process but each of them has a specific role to maintain the dynamics of the process. The first region (region T) is critical for filament elongation as shown by absence of a tail in a mutant with a deletion of this region. The second (region C) is more specifically involved in the maintenance of the continuity of the process, probably by F-actin binding and/or prevention of barbed-end capping.

2. 6. Current model

Our current model (Figure 4) is that the central region of ActA is only acting as a stimulator of movement by providing through VASP binding, profilin-actin complexes which will fuel the actin polymerisation process taking place in the N-terminus. Movement is thus generated by three successive steps:

The first is the generation of free barbed ends either by uncapping or severing of actin filaments or by nucleating actin monomers.

The second is then elongation of actin filaments at the free barbed ends. This elongation step provides the driving force for movement .

The actin filaments are then released, capped and cross linked while new free barbed ends are constantly generated.

How the Arp2/Arp3 proteins participate in this process is unknown.

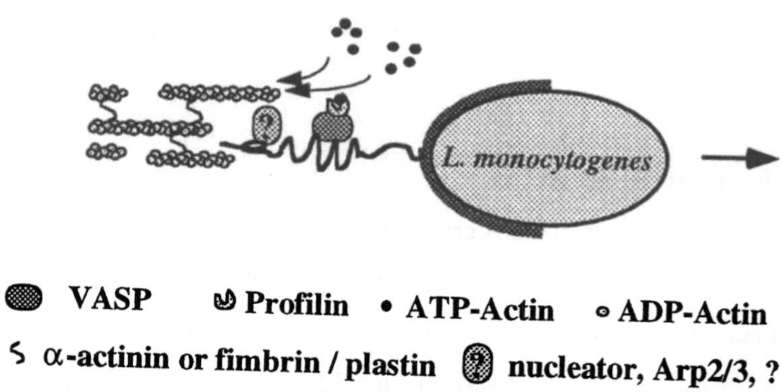

Figure 4: Model of actin assembly by *L. monocytogenes*

CONCLUDING REMARKS

Listeria is, with *Yersinia, Salmonella* and *Shigella* , the intracellular bacterium for which most data have been accumulated (Cossart1997). Entry of *Listeria* appears up to now similar to that of *Yersinia,* However, the cellular infectious process is very similar to that of *Shigella* which, like *Listeria* , invades cells, lyses the vacuole and uses an actin-based motility to spread from one cell to the other. Interestingly, none of the genes used by *Shigella* is similar to the corresponding ones in *Listeria*. The molecular interactions between bacterial pathogens and mammalian cells are far from being understood!

References

1. **Cossart, P.** 1995. Bacterial actin based motility. Curr. Opin. Cell. Biol. **7**: 94-101.

2. **Cossart, P.** 1997. Subversion of the mammalin cell cytoskeleton by invasive bacteria. J. Clin. invest. **99;** 2307-2311.

3. **Dramsi, S., I. Biswas, E. Maguin, L. Braun, P. Mastroeni, and P. Cossart.** 1995. Entry of *L. monocytogenes* into hepatocytes requires expression of InlB, a surface protein of the internalin multigene family. Mol. Microbiol. **16**: 251-261.

4. **Gaillard, J.-L., P. Berche, C. Frehel, E. Gouin, and P. Cossart.** 1991. Entry of *L. monocytogenes* into cells is mediated by internalin, a repeat protein reminiscent of surface antigens from gram-positive cocci. Cell **65**: 1127-1141.

5. **Ireton, K., B. Payrastre, H. Chap, W. Ogawa, H. Sakaue, M. Kasuga, and P. Cossart.** 1996. A role for phosphoinositide 3-kinase in bacterial invasion. Science **274**: 780-782.

6. **Ireton, K. and P. Cossart.**1997 Host pathogen interactions during entry and actin-based movement of *Listeria monocytogenes*.Annu. Rev. Genet. **31**: 113-138.

7. **Lasa, I., and P. Cossart.** 1996. Actin-based motility: towards a definition of the minimal requirements. Trends Cell Biol. **6**: 109-114.

8. **Mengaud, J., H. Ohayon, P. Gounon, R. M. Mège, and P. Cossart.** 1996. E-cadherin is the receptor for internalin, a surface protein required for entry of *Listeria monocytogenes* into epithelial cells. Cell **84**: 923-932.

9. **Sheehan, B., C. Kocks, S. Dramsi, E. Gouin, A. Klarsfeld, J. Mengaud, and P. Cossart.** 1994. Molecular and genetic determinants of the *Listeria monocytogenes* infectious process. Curr. Top. Microbiol. Immun. **192**: 187-216.

THE BEHAVIOUR OF BACTERIAL PATHOGENS *IN VIVO*

Harry Smith

The Medical School, The University of Birmingham, Edgbaston,
Birmingham B15 2TT, UK

1 Introduction

Most of this book deals with general bacterial processes: growth, death, metabolism, genome survival, gene expression and integrated cell biology. The amount and the erudition of this knowlege is impressive. It has been gained mainly from experiments *in vitro* with single bacterial species, notably *Escherichia coli*. Nevertheless, the principles derived from this knowledge probably apply to most bacteria in nature. They should help us to understand the complex ecological interactions that bacteria have, between themselves, with other types of microbes and with animals, plants, air, water and soil. This paper deals with the ecology of a small but important part of the total bacterial world. Pathogenic bacteria cause disease and often death in man, animals and plants. For this reason, although few in number, they attract more attention than the great majority of bacteria which are harmless and sometimes beneficial. Pathogens have unique biological properties. Those that cause disease in man and animals can infect the surfaces of the respiratory, alimentary or urogenital tracts, penetrate these surfaces, grow in the tissues of the host, inhibit host defences and damage the host thus causing sickness and possibly death (Smith, 1995). The molecular bases for these biological properties are called the determinants of pathogenicity (or virulence, a synonymous term). As Pascale Cossart has described in the previous chapter, we know a lot about these determinants. However, this knowledge has also been derived largely from experiments with bacteria grown *in vitro*. Now, there is a rising interest in finding out what happens to bacterial pathogens in the tissues of an infected host. How far do the general bacterial processes discussed in the early chapters of this book and the specific virulence mechanisms of pathogens described in the previous chapter operate *in vivo*? This chapter summarises what is known and unknown about the behaviour of bacterial pathogens *in vivo* and contains suggestions for learning more. Some of the points may apply to ecological interactions of bacteria in environments other than the animal host. To keep the number small, most of the references are either reviews or symposium books. They provide further reading and references to original papers.

First, I will describe the experiments that convinced me many years ago that bacterial behaviour *in vivo* could be different from that *in vitro*. In the 1950's, I worked on anthrax. At that time, a toxin responsible for death from this disease had not been recognised in laboratory cultures of the causative organism, *Bacillus*

NATO ASI Series, Vol. H 103
Molecular Microbiology
Edited by Stephen J. W. Busby,
Christopher M. Thomas and Nigel L. Brown

anthracis. By looking at anthrax bacilli and their extracellular products harvested in quantity from infected guinea pigs, the toxin was demonstrated (Smith and Keppie, 1954). Later it was produced *in vitro* by manipulating the growth conditions. As convincing as the discovery of the toxin was the fact that the anthrax bacilli grown *in vivo* had membrane characteristics different from those grown *in vitro*. Unlike the latter, the bacteria from guinea pigs became swollen when they were suspended in water and they were completely lysed by adding a minute amount of ammonium carbonate (Smith, 1990). These results prompted me to advocate the use of bacteria grown *in vivo* in studies of pathogenicity (Smith, 1958). But this failed to catch on. Only in the past 5-10 years has interest quickened in pathogens grown *in vivo*, encouraged by the immense popularity of studies *in vitro* on regulation of gene expression.

The environment *in vivo* influences bacterial pathogenicity in two ways: by affecting production of virulence determinants and by controlling growth rate. Interest in the first overshadows consideration of the second, although it is equally important in pathogenicity. The neglected subject is considered first.

2 Lack of Knowledge of Growth Rates *In Vivo* and Factors That Control Them

The first three chapters in this book summarise the vast knowledge that has accumulated on growth, metabolism and bioenergetics of bacteria in laboratory cultures. The stark contrast with the situation *in vivo* will become apparent as this section unfolds.

Bacterial growth *in vivo* is essential for pathogenicity. Artificially made, nutritionally deficient mutants of *Salmonella typhimurium* were not lethal to mice unless the nutrients required for growth (purines or aromatic amino acids) were injected with them (Smith, 1990). However, most wild-type pathogens can find sufficient nutrients in the tissues to support some growth. The question is at what rate? Rapid multiplication is needed in acute disease to overwhelm initial defences and cause sickness before the protective immune response can be effective. In chronic disease, a slow growth rate may be more important since it could lead to less stimulation of host defences. In carrier states, by which pathogens like typhoid bacilli persists in a host as a focus for future epidemics, the bacteria may enter a resistant, stationary phase.

Unfortunately, little is known about the growth rates of pathogens during infection. Increase or decrease in bacterial populations in blood, lymph nodes and spleen are recorded frequently, eg. for mutants of salmonellae in mice (Smith, 1990). These population changes are, however, not growth rates but the resultants of bacterial multiplication and destruction by host defences. Rapidly increasing populations means that doubling times are short. The problem arises when populations increase or decrease slowly. True multiplication rates, which may be high but disguised by high death rates due to efficient host defences, are unknown. A stationary population does not necessarily mean that the pathogen is in a stationary state. Bacterial doubling times *in vivo* can be measured (Smith, 1990). One method relies on a genetic marker distributing to only one of the two daughter cells in each successive generation. Examining the proportion of the bacterial population carrying the non-replicating marker at intervals as infection proceeds shows the number of generations. Increase in ratios of wild type organisms (which multiply *in vivo*) to those of temperature sensitive mutants (which should not grow *in vivo*) is another method. Unfortunately, these methods have only been used for *E.coli*, salmonellae and *Pseudomonas aeruginosa*.

Equally frustrating is the position of studies on nutrients and metabolism that underpin bacterial growth *in vivo*. Only one aspect has received adequate attention. Iron is essential for bacterial growth but *in vivo* its availability in body fluids is restricted by chelation to host lactoferrin and transferrin. Pathogens obtain sufficient iron by a number of strategies, some of which have been elucidated in molecular terms, eg siderophore production (Weinberg, 1995). Most bacterial pathogens have been examined in this respect and the erudition of the work highlights the dearth of knowledge on use of other bacterial nurients *in vivo*. For example, the vast knowledge about the nutrition and metabolism of *E.coli in vitro* (see previous chapters) has not yet been applied to the growth of pathogenic strains *in vivo*. In the past, it was shown that local concentrations of preferred nutrients can determine the tissue tropism of pathogens (Smith, 1990). *Proteus mirabilis* causes severe kidney infections in man. One factor contributing to the harmful localisation is that *P.mirabilis* grows extremely well on urea, which is concentrated in the kidney. *Brucella spp*. are bacterial pathogens which cause abortion in cattle, sheep, goats and pigs. This is due to rapid and intense bacterial growth in the placenta and foetal membranes because they contain erythritol, a growth stimulant for brucellae. These promising leads were never extended to other examples of tissue localisation.

The holdup in this area is not due to lack of experimental approaches. As already stated, growth rates *in vivo* can be measured and stationary states could be detected by similar methods. The effects of nutrients such as $PO_4^{3/4-}$, Mg^{2+}, Zn^{2+}, amino acids, sugars and vitamins could be examined by modifications of the methods used for iron. These began with the effects of iron salts on growth in serum-containing media and on virulence in animal models. Observations on auxotrophic mutants could show whether particular nutrients were accessible *in vivo*. Reporter genes, introduced into pathogens could indicate the concentrations of nutrients in various tissues during infection. Then, growth rates *in vivo* could be compared with those under simulant conditions *in vitro*.

The need for animal experiments and the reduced interest in microbial physiology are probably the main reasons for the lack of progress. Animal experiments are more difficult, more expensive and less repeatable than those *in vitro*. The effects of iron limitation *in vivo* have already been demonstrated in animals so that work on this topic can proceed *in vitro* on a sound basis. The initial animal experiments have still to be carried out for the other nutrients. Also, some of the relatively few microbial physiologists have to be persuaded that work on the nutrition and metabolism of pathogens could be rewarding. Overall, the prospect for progress in this important aspect of pathogenicity is not good.

3 Effect of the Environment *In Vivo* on Production of Virulence Determinants

In diseased animals, bacterial pathogens produce all the determinants necessary for the manifestations of virulence. The environmental conditions *in vivo* (osmolarity, pH, Eh and nutrient availability) differ from those in laboratory cultures. They are not only more complex, but alter as infection proceeds due to inflammatory exudates, tissue breakdown and spread from one anatomical site to another. When pathogens are moved from animals to laboratory cultures and vice versa phenotypic change and selection of genotypes occur (Smith, 1990). Similar changes in phenotypes and genotypes will also happen *in vivo* as the environment varies with progress of infection (Dorman, 1994). These facts have three implications for studies of pathogenicity. First, putative determinants of virulence indicated by experiments with organisms grown *in vitro*, may not be formed *in vivo*. Second, one or more of the full armoury of virulence determinants that are formed *in vivo* may not be produced under arbitarily chosen conditions of growth *in vitro*. Third, all members of the armoury are not necessarily produced *in vivo* at the same time or place; the complement could vary with the phase of infection and anatomical site. The first two implications have been heeded in studies of pathogenicity over the last ten years but attention to the third is only just beginning.

3.1 Confirmation that Putative Determinants Detected *In Vitro* are Produced *In Vivo* and Contribute to Virulence

Confirmation of production *in vivo* of putative determinants is now accepted as good practice in studies of pathogenicity. Bacteria obtained from patients or infected animals are examined for the putative determinants by SDS-PAGE and immunoblotting. Convalescent sera are investigated for appropriate antibodies. Table 3.1 includes a few of many examples. However, there are still some gaps. As far as I am aware, antibodies to the product of the *act*A gene of *Listeria monocytogenes* and the invasin of *Yersinia enterocolitica* and *Yersinia pseudotuberculosis* have not been detected in patients.

It is also usual to check that putative determinants contribute to virulence by comparing, in appropriate animal models, the wild type strain with determinant-deficient mutants (Hormaeche *et al.,* 1992; Roth *et al.*, 1995). In most cases, relevance to virulence is proved but sometimes not. The 17 kDa product of the *ail* gene of *Y. enterocolitica* is formed in Peyer's patches of infected mice but observations on an *ail* deficient mutant indicated that the gene is not required for either initial invasion or for establishing systemic infection (Smith, 1996). Type 1 fimbriae of *E.coli* F-18 are produced in abundance in mice but experiments with mutants show they are not essential for virulence (Smith, 1996).

Table 3.1. Confirmation that putative virulence determinants detected *in vitro* are produced *in vivo*

Bacterial species	Host	Site of recognition	Putative determinant demonstrated *in vivo* (and/or its antibody)
Proteus mirabilis	Man	Urine	IgA protease
Bordetella pertussis	Man	Blood	Adenylate cyclase
Pseudomonas aeruginosa	Man	Sputum	Alginate
Listeria monocytogenes	Man	Serum CSF	Listeriolysin O Listeriolysin O
E.coli	Man	Serum	Verotoxin I
Yersinia spp	Man Rabbit Mice	Serum	YadA, YopE, YopH and others

Original references see Smith (1996)

Another healthy sign is that the results of cell-culture tests for virulence determinant *in vitro* are being viewed in relation to the pathology of disease. Studies on intestinal invasion by *Shigella flexneri* are a good example (Hormaeche *et al.*, 1992; Perdomo *et al.*, 1994). Using Hela cells, the role of plasmid gene products IpaB, IpaC, IpaD and IcsB to entry, intracellular movement and transfer between cells was established. However, colonic enterocytes, unlike Hela cells, have a brush border and shigellae cannot penetrate it. The pathology of infections in primates and rabbit intestinal loops suggests that shigellae invade the colonic mucosa initially by a non-specific mechanism. They are ingested by M cells of lymphoid follicles whose normal function is to transfer particles from the gut to antigen processing macrophages in the *lamina propria.* When shigellae enter these macrophages, apoptosis and inflammation follow, leading to disruption of the epithelial cell layer. Shigellae can then invade through the sides and bases of the cells using the determinants recognised by the studies with Hela cells. Other intestinal pathogens such as *L. monocytogenes*, *Yersinia spp.* and *Salmonella spp.* also seem to invade via M cells (Siebers and Findlay, 1996). Hence, the relevance to intestinal penetration *in vivo* of determinants indicated by invasion of cell lines by these pathogens is coming under scrutiny. For example, mutants of *Salm. typhimurium* deficient in the invasive genes *inv*A and *inv*G and unable to invade MDCK cells invaded M cells of Peyer's patches in infected mice (Clarke *et al.*, 1996). This realistic trend in work concerned with intestinal invasion should be encouraged in other areas of pathogenicity.

3.2. Comparison of *In Vivo* and *In Vitro* Grown Bacteria to Reveal Previously Unknown Virulence Determinants

Increasingly, bacteria grown *in vivo* are being compared with those grown *in vitro* by SDS-PAGE and, sometimes, in biological tests. Bacterial components formed *in vivo* and not, or less so, *in vitro*, are frequently demonstrated. Repression *in vivo* of components formed *in vitro* is also detected. Table 3.2 contains a few of many examples. These studies reveal bacterial constituents that may be important for diagnosis and vaccination as well as virulence determinants.

Table 3.2 Differences between bacteria grown *in vivo* and *in vitro* that are possibly important in pathogenicity

Species	Host	Site of bacteria	Different property of *in vivo* organisms
Staph. epidermidis	Guinea pig	Peritoneal chambers	2 Iron-regulated proteins formed and 4 proteins repressed
	Pigs	Peritoneal chambers	2 Proteins not seen *in vitro* and many proteins repressed
Pasteurella haemolytica	Rabbit	Peritoneal chambers	3 Iron-regulated proteins (antibodies to them in bovine serum)
	Cattle	Peritoneal chambers Lung	1 Protein not seen *in vitro*, 3 iron-regulated proteins, several proteins repressed
Campylobacter jejuni	Rabbit	Intestinal loops	Several proteins not seen *in vitro* (antibodies to 2 in human serum)
Yersinia enterocolitica	Mice	Intestinal loops	1 Protein not seen *in vitro*
		Peyer's patches	3 Proteins not seen *in vitro*, several proteins repressed
Bacteriodes fragilis	Mice	Peritoneal chambers	Protein profiles, iron-regulated proteins not induced

Original references see Smith (1996)

The next step is to show that the previously unknown bacterial component is responsible for a biological property related to pathogenicity (eg. interference with phagocytosis) and to contribute to virulence *in vivo*. Comparisons between wild types and determinant-deficient mutants are important. Unfortunately, this essential follow-up is not as popular as making the original revelations by SDS-PAGE. Most of the examples quoted in Table 3.2 have not been followed up. For example, in the intestinal lumen of mice, *Y. enterocolitica* produced a plasmid mediated outer membrane protein (23 kDa) which was not seen in culture. On invasion of Peyer's patches, this protein and two further novel proteins (210 and 240 kDa) were formed

(Smith, 1996). There the matter remains: the possible function of these novel proteins in intestinal invasion has not been investigated. Later in this paper, an investigation will be described where initial observations were pressed home to the ultimate goal.

3.3 Identification of Virulence Determinants Produced at Successive Stages in the Disease Process and the Host Factors Involved

In the experiments described above the multiple influences of the undefined conditions *in vivo* are accepted as a whole. Also, bacteria grown *in vivo* are collected from the most convenient and prolific source. Differences in the complement of virulence determinants between bacteria harvested at different stages of infection and at different anatomical sites, are largely disregarded. The future goal should be to identify virulence determinants throughout the infection process and the host factors that induce them: a formidable task. Current interest in environmental regulation of virulence genes might, however, provide a spur to such studies.

3.3.1 Studies *In Vitro* on Regulation of Virulence Determinant Production

Bacteria possess regulatory networks which induce or repress diverse and unlinked genes in response to environmental signals (this book, Part III): this applies to genes coding for virulence determinants (Dorman, 1994).

In summary, global regulation of gene expression usually involves a hierarchical network of regulons. These are groups of genes controlled by a common regulator, usually a DNA-binding protein that recognizes control regions of its subservient genes. Ground level regulons are under control of higher regulons and so on, with DNA supercoiling and DNA associated proteins having an influence on some of the top members of the hierarchy. Several ground level regulons dealing with virulence determinants are known (Hormaeche *et al*, 1992; Dorman, 1994; Mekalanos, 1995; Akerley and Miller, 1996; Pettersson *et al.*, 1996; Smith, 1996). The *tox*R system regulates production of both components of cholera toxin, an adhesin (toxin-coregulated pilus) and other virulence determinants. It is modulated by temperature, osmolarity, pH, oxygen status and availability of amino acids. *Bordetella pertussis* forms the *bvg* system which regulates production of adenylate cyclase, filamentous hemagglutinin, toxin and hemolysin and is affected by temperature, $MgSO_4$ and nicotinamide. *Bordetella bronchiseptica* also has the *bvg* regulon which activates and represses gene expression. *Yersinia* spp. have a regulatory system for production of *Yersinia* outer membrane proteins (Yops) which responds to Ca^{2+}. Reducing Ca^{2+} restricts growth at 37°C but Yop production increases. The *vir*R regulon of *Shig. flexneri* responds to temperature and causes production at 37°C of invasion plasmid virulence determinants. The *fur* regulon in *E. coli* and other pathogens responds to iron concentrations and regulates production of siderophores, receptors concerned with iron uptake and some toxins.

Recently, experiments *in vitro* have shown for *Staphylococcus aureus* and diverse Gram negative bacteria including *Ps aeruginosa,* that some genetic regulons which affect production of virulence determinants, are induced only when a significant cell population density has been attained (Guangyong *et al*., 1995; Winson *et al*., 1995). The phenomenon is called quorum sensing. The cell density dependency reflects a cell to cell communication system which relies on the accumulation of signal molecules to critical threshold concentrations. For *Staph. aureus* the signalling molecule is an octapeptide which activates *agr*, a global regulator of the virulence response. For the Gram negative bacteria, the signalling molecules are various N-acyl-L-homoserine lactones. In the case of *Ps aeruginosa*, two transcriptional activators, LasR and RhiR, respond to the signalling molecules to integrate the regulation of virulence determinants, secondary metabolites and adaptation and survival in the stationary phase (Latifi *et al*., 1996).

3.3.2 The Relevance of Regulons Demonstrated *In Vitro* to Behaviour *In Vivo*

The presence *in vivo* of virulence determinants whose production *in vitro* is controlled by a certain regulon, does not necessarily mean that the regulon itself operates identically *in vivo*. Unknown factors may affect it or a different regulatory system may be involved. However, the designated regulon is probably playing some role if deletion of the regulon genes affects virulence and the environmental factors that affect the regulon *in vitro* are present *in vivo*. Let us consider the virulence determinant regulons in this respect. The *tox*R system operates in humans since a deletion mutant produced less colonization and diarrhoea in volunteers than the wild type (Mekalanos, 1995; Smith, 1996). However, the influence of environmental conditions is confusing. In cultures, *V. cholerae* produces more toxin, at low temperature rather than at 37°C, under aerobic rather than anaerobic conditions, and at a low rather than high pH. Yet, it produces toxin and other *tox*R regulated gene products in the human intestine which is at 37°C, anaerobic and at alkaline pH. The *bvg* regulon of *B. pertussis* seems to act in mice since mutants deficient in genes controlled by *bvg* were less virulent (Smith, 1996). It probably also functions in patients but apart from the permissive temperature (37°C), the modulators of its action *in vivo* are unknown. Mutations of the *bvg* gene of *B. bronchiseptica* affect virulence in rats (Akerley *et al*., 1996) but again, the environmental parameters modulating its action *in vivo* are unknown. Ca^{2+} regulation of Yop production by yersiniae almost certainly does not occur *in vivo*, where Yop formation occurs freely at high Ca^{2+} concentrations that are nonpermissive *in vitro* (Hormaeche *et al*., 1992; Pettersson *et al*., 1996). The mechanisms operating *in vivo* are not yet clear but studies with Hela cells suggest that contact with phagocytes may be the key. This contact could lead to export from the bacteria of LorQ, a negative regulon of Yop expression, thereby increasing production of Yops (Pettersson *et al*., 1996). Mutants deficient in the *vir*R regulon of *Shig,. flexneri* failed to invade the conjunctiva of guinea pigs (Smith, 1996). The *vir*R regulon probably functions in patients since it is permissive at 37°C. Regarding the relevance of *fur*, iron is limited *in vivo* and a *fur* deficient mutant of *Ps. aeruginosa* was found less able than the wild type to produce corneal infections of mice (Smith, 1996).

Quorum sensing could affect virulence determinant production *in vivo* by responding to population size of pathogens in different nutritional environments of the host. The mammalian host appears not to produce N-acyl-L-homoserine lactones and other signalling molecules. Hence, quorum sensing is probably not important for production of virulence determinants early in infection when the invading pathogens are few in number. Later, as the population increases it could be significant and, for *Ps aeruginosa* infection at least, it might be responsible for a stationary phase in carrier states. LasR deficient mutants of *Ps aeruginosa* are less virulent for mice than wild types (Tang *et al.*, 1996).

In summary, most but not all virulence determinant regulons investigated *in vitro* appear to function *in vivo* but the environmental modulators of their action at different stages of infection are often not clear.

3.3.3 Gaining More Information on What Happens *In Vivo*

The first step is to recognize the virulence genes that are expressed at progressive stages of infection and in different anatomical sites. The second is to detect differences in environmental parameters which might change expression of virulence genes. The third is to try to relate the two and identify the regulon involved.

Recognition of Virulence Genes Expressed *In Vivo*. The 'classical' way is to examine organisms obtained directly from infected animals for the products of these genes. Recently, new methods have been designed.

The *in vivo* expression technology (IVET) has three variations. The first uses auxotrophic mutants as a selection system (Mekalanos, 1995). DNA segments from a wild type *Salm. typhimurium* are attached to a synthetic operon to provide promoters for a promoterless *pur*A gene and a promoterless *lacZ* operon. The resulting genes are introduced into a *pur*A auxotrophic mutant of *Salm. typhimurium* which, unless complemented, does not grow *in vivo*. In mice, some bacteria grow in the spleen indicating *pur*A expression by the synthetic gene due to promoters provided by the wild type genes, i.e. those expressed *in vivo*. Bacteria with genes not expressed *in vivo* are eliminated. Plating bacteria recovered from the spleen on McConkey lactose agar distinguishes members with genes expressed *in vitro* as well as *in vivo* (red colonies, 95%) and those having genes expressed only *in vivo* (white colonies, 5%). The genes concerned are termed *ivi* (*in vivo* induced) genes. In the second IVET system (Malan *et al.*, 1995), the promoterless *pur*A gene in the first method is replaced by a promoterless chloramphenicol acetyl transferase (*cat*) gene and the mice are dosed with chloramphenicol. Again, plating on McConkey lactose agar distinguishes between organisms containing genes expressed *in vivo* and *in vitro* and those containing genes expressed only *in vivo*. One of the *ivi* of genes of *Salm. typhimurium* recognised by this second method was *fad*B (see later). In the third system, genetic recombination is used to report gene expression (Mekalanos, 1995). A resolvase is produced from a promoterless copy of the *tnp*R gene of the transposable element γδ under the influence of promoters provided by the genes expressed *in vivo* by the pathogen (eg. *V. cholerae*). The resolvase excises a chloramphenicol resistance reporter gene, making the organisms antibiotic sensitive. Tissue homogenates are replica plated on medium alone and medium with chloramphenicol. Colonies sensitive to chloramphenicol contain bacteria whose genes are expressed *in vivo*. This system can be used for relatively few organisms, in different tissues and at diferent stages of infection.

The IVET systems require auxotrophic mutants and gene transfer systems. A method which may have wider application is to use antibodies induced by the products of the genes expressed *in vivo* as the reporting system (Suk *et al.*, 1995; Wallich *et al.*, 1995). The proteins coded by gene expression libraries are separated by SDS-PAGE and immunoblotted with sera obtained from live, infected animals (i.e. antibodies to gene products expressed *in vivo*) and with antisera against heat killed bacteria (i.e. antibodies to gene products expressed *in vitro*). Comparisons of the immunoblots indicate products formed only *in vivo*. For *B. burgdorferi* infections in mice, expression libraries were prepared by ligating DNA fragments to a phage which was then grown in *E.coli*. Six genes expressed only *in vivo*, were demonstrated and one, p21, coded for a new 20 kDa protein (Suk *et al.*, 1995). This method detects only genes whose products are antigenic.

Another method (Collins, 1996), used for *Mycobacterium tuberculosis*, was to confer virulence for mice on an avirulent strain (H37Ra) by complementation with a gene library from a virulent strain (H37Rv) using an integrating cosmid vector. Repeated infection of mice and recovery from the spleens and lungs, selected individual faster growing recombinants from which the H37Rv DNA inserts were retrieved. A 25 kb growth promoting DNA fragment was recovered. The virulence gene in this fragment was not identified. However, in a similar study using infection of guinea pigs with *Mycobacterium bovis*, a virulence gene *rpo*V was identified. It is present in *M.tuberculosis* was well as *M.bovis* and probably has a regulatory role.

Most bacteial pathogens should be amenable to one or other of these new systems for recognizing gene expression *in vivo*. The expressed gene must be shown to contribute to virulence and then its product and biological function identified. Such studies are in progress. Mutations in the *ivi* genes of *Salm. typhimurium*, detected by the auxotrophic IVET system, reduced virulence for mice (Mekalanos, 1995). *Fad* B, the *ivi* gene of *Salm. typhimurium,* demonstrated by the antibiotic selection system encodes an enzyme which could interfer with host defences. It oxidises fatty acids such as arachidonic acid which are either bactericidal themselves or stimulate inflammatory responses (Mekalanos *et al.*, 1995).

Clearly, it is now possible to identify virulence genes at progressive stages of infection.

Environmental Parameters Which Might be Responsible for Changes in Expression of Virulence Genes. Geigy Scientific Tables (Lentner, 1981, 1984) contain much information on temperature, osmolarity, oxygen status, cations, anions, amino acids, sugars and other compounds in blood, saliva, gastric juices, bile, intestinal juices, urine, vaginal secretions, lung secretions, cerebrospinal fluid, synovial fluid, brain, kidney and liver. Factors at potential sites of infection can be measured by standard physiological and pathological procedures (Lentner, 1981, 1984). Elements present in body fluids, cells or subcellular particles can be measured *in situ* by X-ray micro-analysis during electon microscopy (Smith, 1996). Quantitative fluorescence microscopy of carriers of dyes which react to environmental parameters have been used to indicate, e.g., the pH in the phagolysomes of macrophages. Also *lac*Z fusions to genes whose expression responds to such parameters have been used to estimate Ca^{2+} levels in human macrophages containing *Yersinia pestis* and levels of oxygen, pH, Fe, Mg, glucose and mannose in tissue culture cells infected with *Salm. typhimurium*. Both techniques (Smith, 1996) could be used on specimens taken from infected animals. In short, information on the environment *in vivo* is either already available or can be gained, if the will is there.

Relating Environmental Parameters to Virulence Determinant Production. The environmental parameters *in vivo* should be simulated *in vitro* and production of relevant virulence determinants monitored. Possible operation *in vivo* of known regulons should be examined. Studies on the operation of the *bvg* regulon of *B. pertussis* under different concentrations of $MgSO_4$ and nicotinic acid derivatives are along those lines (Smith, 1996). Also, an acid pH *in vitro* and in the phagolysosomes of mouse macrophages has been shown to induce the *pag* gene of *Salm. typhimurium*, essential for intracellular growth (Smith, 1996). The *Yst* regulated toxin of *Y. enterocolitica* is produced at 37°C *in vivo* but not in culture media unless the temperature is below 30°C . If, however, the osmolarity and pH of the medium are adjusted to values normally present in the ileum, toxin production at 37°C occurs (Mikulski *et al.*, 1994). Contact with host cells should be investigated as a trigger for action of regulons *in vivo*. The effect on Yop production by yersiniae (Petterson *et al.*, 1994) has been mentioned before and pilus-host cell adhesion of *E.coli* promotes virulence gene expression (Zhang and Normak, 1996). Studies on nutrition and growth *in vivo* are needed with regard to the possible relevance of quorum sensing.

The IVET system has demonstrated unexpected virulence gene expression *in vivo*. *V. cholerae* expresses many *ivi* genes in the intestines of orally infected mice but none are *tox*R regulated (Mekalanos, 1995). Toxin is, however, produced in the intestine because diarrhoea is apparent. Does this mean that toxin production *in vivo* is controlled by another regulon? As expected, the *irg*A gene of *V. cholerae* was expressed in iron limiting media and in mouse peritoneal cavities but it was not expressed in mouse intestine and rabbit ileal loops (Mekalanos, 1995). Either iron is readily available in the intestine or, in this site, regulation of the *irg*A gene is different from that in the peritoneal cavity.

4. An Example of In-Depth Analysis of Bacterial Behaviour *In Vivo*: Sialylation of Gonococcal Lipopolysaccharide (LPS) by Host Factors

This final section shows the rewards that can accrue from a vigorous follow-up of differences between pathogens grown *in vivo* and *in vitro*. References to original papers can be found in a review (Smith *et al.*, 1995) and a conference book (Zollinger *et al.*, 1996). The sequence of headings is in the context of the main points of the previous section.

4.1 The Seminal Observation on Gonococci Grown *In Vivo*

Gonococci in urethral exudates from patients are resistant to complement-mediated killing by human serum, an important facet of pathogenicity. In most cases, this resistance is lost after one subculture but it is restored by incubating the gonococci with blood cell extracts or urogenital secretions.

4.2 Identification of the Determinant of Serum Resistance

First, the factor in blood cells extracts which induces serum resistance was identified as cytidine 5'-monophospho-N-acetyl neuraminic acid (CMP-NANA). Then, it was shown that, concomitant with CMP-NANA induction of serum resistance in gonococci, sialyl (NANA) groups were transferred to terminal Galβ1-4GlcNAc epitopes on the side chains of high M_r LPS components. And, serum resistance was abolished by removing the sialyl groups with neuraminidase. Finally, the LPS of gonococci in urethral exudates was shown to be sialylated: neuraminidase treatment drastically reduced serum resistance and unmasked the LPS sialylation sites for reaction with a specific monoclonal antibody. Hence, the determinants of serum resistance are sialylated high M_r LPS components. They are present as an irregular surface coat. The gonococcal sialyltransferase which had been indicated for the first time by these studies was demonstrated in detergent extracts of all gonococci examined. Recently, its gene has been cloned. The mechanism of serum resistance is not clear but interference with the action of both natural antibody and complement in the classical pathway is indicated.

4.3 Demonstration that LPS Sialylation Affects Many Facets of Pathogenicity

Comparison of gonococci incubated with and without CMP-NANA showed that sialylation of LPS inhibited: entry to epithelial cells lines; opsonophagocytic killing by human neutrophils; the bactericidal action of antisera against gonococcal proteins; binding of C3; and stimulation of antibodies in animals. Furthermore, LPS sialylation affected the virulence of gonococcal strains for volunteers.

4.4 Confirmation of Results by Use of a Sialyltransferase Deficient Mutant

The mutant (JB1) was obtained by transposon mutagenesis of strain F62 and screening for unlabelled colonies after incubation with CMP-^{14}CNANA.

JB1 and F62 were equally susceptible to fresh human serum but only F62 became resistant on incubation with CMP-NANA or blood cell extracts. JB1 was more susceptible than F62 to opsonophagocytic killing by human neutrophils and not, like F62, made more resistant by incubation with CMP-NANA. Opa protein positive variants of JB1 and F62 entered a conjunctival cell line similarly but, on incubation with CMP-NANA, the entry of the JB1 variant was not diminished in contrast to that of the F62 variant. A rabbit antiserum raised against gonococcal protein I killed both JB1 and F62 but only F62 became resistant on incubation with CMP-NANA. Finally, similar amounts of complement component C3 were absorbed by JB1 and F62 but binding was decreased by incubation with CMP-NANA only for F62. Clearly, these comparisons of the properties of JB1 and F62 confirm previous results and underline the important of LPS sialylation in pathogenicity.

4.5 The Host Factors

We need to know their nature, availability *in vivo* and mode of action.

Clearly, CMP-NANA plays the major role. There is, however, at least one other factor. A low M_r material in blood cell extracts which, on incubation with gonococci and CMP-NANA enhances LPS sialylation and increases serum resistance up to two-fold, has been identified as lactate.

In gonorrhoea, gonococci are seen intracellularly and extracellularly mainly in the urogenital tract but occasionally elsewhere. Both CMP-NANA and lactate are present at the relevant sites. CMP-NANA is in most human cells including those of the blood and cervical epithelium. Normally, however, only minute amounts are found extracellularly. Lactate is present in neutrophils, monocytes, vaginal secretions, ejaculate, blood and many other sites. During urogenital inflammation, both CMP-NANA and lactate would be released to extracellular gonococci from dying epithelial cells and phagocytes. Furthermore, under anaerobic conditions which occur *in vivo*, gonococci form lactate from glucose. Interestingly, anaerobically grown gonococci undergo LPS sialylation and induction to serum resistance by CMP-NANA more efficiently than aerobically grown organisms. The combined effect of CMP-NANA and lactate may be important in the early stages of gonorrhoea when relatively few gonococci must survive the humoral and cellular defences of the inflammatory response. Rapid LPS sialylation will increase resistance to these defences.

As regards mode of action, CMP-NANA is the substrate for the sialylation reaction. The role of lactate is unknown and will be difficult to unravel. However, the task is worthwhile because lactate is ubiquitous *in vivo*. The results for gonococci may have implications, especially if regulatory systems are involved, for other bacterial pathogens growing in lactate-rich situations *in vivo* including anaerobic conditions (eg. abscesses). Lactate does not stimulate the gonococcal sialyltransferase directly and the enhancement process promoted by lactate in live gonococci is separate from the action of CMP-NANA itself. *Pre-incubation* of live gonococci with lactate enhances *subsequent* LPS sialylation and induction of serum resistance by CMP-NANA. Inhibition by chloramphenicol indicates that metabolic events are required. The mechanisms involved are subjects for future research. In this respect, it is interesting that pyruvate as well as lactate enhances LPS sialylation.

4.6 The Wider Implications

The observations on gonococci prompted investigations on meningococci.. Some, but not all, meningococci contain LPS components which are either endogenously sialylated or can be sialylated by exogenous (host) CMP-NANA. Pathogenicity is affected by LPS sialylation but not as profoundly as for gonococci because capsular polysaccharide also contributes. LPS sialylation appears to interfere with adhesion and invasion of epithelial cells by noncapsulated strains; deposition of C3 on and serum killing of most strains and opsonophagocytic killing of some strains. In a group B epidemic, an immunotype capable of LPS sialylation was associated with invasive disease and an immunotype incapable of sialylation with the carrier state. In mouse and rat models of infection, capsulation was the major virulence determinant, but the contribution of LPS sialylation was also significant. Lactate is present in cerebrospinal fluid but its role in sialylation of LPS by CMP-NANA has not yet been investigated.

Finally, some strains of *Haemophilus influenzae, Haemophilus ducreyi* and *Campylobacter jejuni* contain sialylated LPS components but their roles in pathogenicity have not been investigated.

5. Conclusions

We have learned something about the behaviour of bacterial pathogens *in vivo* but there is much more to be found out. The lack of knowledge applies particularly to rates of growth and the underpinning nutrition and metabolism. The subject is neglected and there is little chance of revival in the near future. The position over production of virulence determinants is better. It is now accepted as good practice to confirm that virulence determinants indicated by studies *in vitro* are produced *in vivo* and contribute to virulence. And, in most but not all cases, confirmation is obtained. With regard to the discovery of new determinants, increasingly, pathogens grown *in vivo* are being compared with those grown *in vitro* to recognize hitherto unknown bacterial components induced by the environment *in vivo*. Unfortunately, in most cases these studies are not followed up to prove that the newly recognized components are virulence determinants. Where this has been done, many new facts about pathogenicity have emerged. As regards the influence of changing environmental conditions *in vivo* on virulence determinant production at different stages of infection, current studies *in vitro* on regulation of such determinants have evoked speculation on what might happen. But, as yet, investigations of changes in virulence determinant complement during infection and the environmental factors and regulatory systems that actually control them *in vivo* have not progressed very far. Nevertheless, new methods have been evolved for finding out what happens *in vivo*; and the future looks bright for those who try.

References

Akerley BJ, Miller JF (1996) Understanding signal transduction during bacterial infection. Trends in Microbiology 4:141-145

Clarke AA, Reece KA, Lodge J, Stephen J, Hirst BH, Jepson MA (1996) Invasion of murine intestinal M cells by *Salmonella typhimurium inv* mutants severely deficient for invasion of cultured cells. Infect Immun 64:4363-4368

Collins PM (1996) In search of tuberculosis virulence genes. Trends in Microbiology 4:426-430

Dorman CJ (1994) Genetics of Bacterial Virulence. Blackwell Scientific Publications, Oxford

Guangyong J, Beavis RC, Novick RP (1995) Cell density control of staphylococcal virulence mediated by an octapeptide pheromone. Proc Natl Acad Sci USA 92:12055-12059

Hormaeche CE, Penn CW, Smyth CJ (1992) Molecular Biology of Bacterial Infection. 49th Symposium of the Society for General Microbiology, Cambridge University Press, Cambridge

Latifi A, Foglino M, Tanaka K, Williams P, Lazdunski A (1996) A hierarchical quorum-sensing cascade in *Pseudomonas aeruginosa* links the transcriptional activators LasR and RhiR (VsmR) to the stationary phase sigma factor RpoS. Mol Microbiol 21:1137-1146

Lentner C (1981) Geigy Scientific Tables vol 1 Units of Measurement, Body Fluids, Composition of the Body, Nutrition. Ciba Geigy, Basle

Lentner C (1984) Geigy Scientific Tables vol 3 Physical Chemistry, Composition of Blood, Haematology, Somatometric Data. Ciba Geigy, Basle

Mahan MJ, Tobias JW, Slauch JM, Hanna PC, Collier RJ, Mekalanos JJ (1995) Antibiotic-based selection for bacterial genes that are specifically induced during infection of the host. Proc Natl Acad Sci USA 92:669-673

Mekalanos JJ (1995) Bacterial response to host signals: analysis and applications. The Harvey Lectures (Wiley-Liss Inc) 89:1-13

Mikulskis AV, Delor I, Thi VH, Cornelis GR (1994) Regulation of *Yersinia enterocolitica* enterotoxin Yst gene. Influence of growth phase, temperature, osmolarity, pH and bacterial host factors. Mol Microbiol 14:905-915

Perdomo OJJ, Cavaillon JM, Huerre M, Ohayon H, Gounon P, Sansonetti PJ (1994) Acute inflammation causes epithelial cell invasion and mucosal destruction in experimental shigellosis. J Exp Med 180:1307-1319

Pettersson J, Nordfelth R, Dubinina E, Bergman T, Gustafsson M, Magnusson KE, Wolf-Watz H (1996) Modulation of virulence factor expression by pathogen target cell contact. Science 273:1231-1233

Rott JA, Bolin CA, Brogden KA, Minion FC, Wannemuehler MJ (1995) Virulence Mechanisms of Bacterial Pathogens 2nd Edition. American Society for Microbiology, Washington

Siebers A, Finlay BB (1996) M cells and the pathogenesis of mucosal and systemic infections. Trends in Microbiology 4:22-29

Smith H (1958) The use of bacteria growth *in vivo* for studies on the basis of their pathogenicity. Ann Rev Microbiol 12:77-102

Smith H (1990) Pathogenicity and the microbe *in vivo*. J Gen Microbiol 136:377-393

Smith H (1995) The revival of interest in mechanisms of bacterial pathogenicity. Biol Rev 70:277-316

Smith H (1996) What happens *in vivo* to bacterial pathogens? Annal NY Acad Sci 797:77-92

Smith H, Cole JA, Parsons NJ (1995) Sialylation of neisserial lipopolysaccharide: a major influence on pathogenicity. Microb Pathog 19:365-377

Smith H, Kepple J (1954) Observations on experimental anthrax: demonstration of a specific lethal factor produced *in vivo* by *Bacillus anthracis*. Nature 173:869-870

Suk K, Das S, Sun W, Jwang B, Bathold SW, Flavell RA, Fikrig E (1995) *Borrelia burgdorferi* genes selectively expressed in the infected host. Proc Natl Acad Sci USA 92:4269-4273

Tang HB, DiMango E, Bryan R, Gambello M, Iglewski BH, Goldberg JB, Prince A (1996) Contribution of specific *Pseudomonas aeruginsa* virulence factors to pathogenesis of pneumonia in a neonatal mouse model of infection. Infect Immun 64:37-43

Wallich R, Brenner C, Kramer MD, Simon MM (1995) Molecular cloning and immunological characterization of a novel linear-plasmid-encoded gene, pG, of *Borrelia burgdorferi* expressed only *in vivo*. Infect Immun 63:3327-3335

Weinberg ED (1995) Acquisition of iron and other nutrients *in vivo*. In Virulence Mechanisms of Bacterial Pathogens 2nd Edition JA Rott *et al.*, Eds pp 79-93 American Society for Microbiology, Washington

Winson MK, Camara M, Latifi A, Foglino M, Chhabra SR, Daykin M, Bally M, Chapon V, Salmond GPC, Bycroft BY, Lazdunski A, Stewart GSAB, Williams P (1995) Multiple N-acyl-L-homoserine lactone signal molecules regulate production of virulence determinants and secondary metabolites in *Pseudomonas aeruginosa.* Proc Natl Acad Sci USA 92:9427-9431

Zhang JP, Normark S (1996) Induction of gene-expression in *Escherichia coli* after pilus-mediated adherence. Science 273:1234-1236

Zollinger WD, Frasch CE, Deal CD (1996) Abstracts of the Tenth International Pathogenic Neisseria Conference. National Institutes of Health, Bethesda, USA

NATO ASI Series H

Vol. 89: **Molecular and Cellular Mechanisms of H^+ Transport.**
Edited by B. H. Hirst. 504 pages. 1994.

Vol. 90: **Molecular Aspects of Oxidative Drug Metabolizing Enzymes: Their Significance in Environmental Toxicology, Chemical Carcinogenesis and Health.**
Edited by E. Arınç, J. B. Schenkman, E. Hodgson. 623 pages. 1994.

Vol. 91: **Trafficking of Intracellular Membranes: From Molecular Sorting to Membrane Fusion.**
Edited by M. C. Pedroso de Lima, N. Düzgüneş, D. Hoekstra. 371 pages. 1995.

Vol. 92: **Signalling Mechanisms – from Transcription Factors to Oxidative Stress.**
Edited by L. Packer, K. Wirtz. 466 pages. 1995.

Vol. 93: **Modulation of Cellular Responses in Toxicity.**
Edited by C. L. Galli, A. M. Goldberg, M. Marinovich. 379 pages. 1995.

Vol. 94: **Gene Technology.**
Edited by A. R. Zander, W. Ostertag, B. V. Afanasiev, F. Grosveld. 556 pages. 1996.

Vol. 95: **Flow and Image Cytometry.**
Edited by A. Jacquemin-Sablon. 241 pages. 1996.

Vol. 96: **Molecular Dynamics of Biomembranes.**
Edited by J. A. F. Op den Kamp. 424 pages. 1996.

Vol. 97: **Post-transcriptional Control of Gene Expression.**
Edited by O. Resnekov, A. von Gabain. 283 pages. 1996.

Vol. 98: **Lactic Acid Bacteria: Current Advances in Metabolism, Genetics and Applications.**
Edited by T. F. Bozoglu, B. Ray. 412 pages. 1996.

Vol. 99: **Tumor Biology Regulation of Cell Growth, Differentiation and Genetics in Cancer.**
Edited by A. S. Tsiftsoglou, A. C. Sartorelli, D. E. Housman, T. M. Dexter. 342 pages. 1996.

Vol. 100: **Neurotransmitter Release and Uptake.**
Edited by Ş. Pöğün. 344 pages. 1997.

Vol. 101: **Molecular Mechanisms of Signalling and Membrane Transport.**
Edited by K. W. A. Wirtz. 344 pages. 1997.

Vol. 102: **Interacting Protein Domains**
Their Role in Signal and Energy Transduction
Edited by L. Heilmeyer. 296 pages. 1997.

NATO ASI Series H

MIX
Papier aus verantwortungsvollen Quellen
Paper from responsible sources
FSC® C105338

If you have any concerns about our products, you can contact us on
ProductSafety@springernature.com

In case Publisher is established outside the EU, the EU authorized representative is:
Springer Nature Customer Service Center GmbH
Europaplatz 3, 69115 Heidelberg, Germany

Printed by Libri Plureos GmbH
in Hamburg, Germany